Studies in Big Data

Volume 113

Series Editor

Janusz Kacprzyk, Polish Academy of Sciences, Warsaw, Poland

The series "Studies in Big Data" (SBD) publishes new developments and advances in the various areas of Big Data-quickly and with a high quality. The intent is to cover the theory, research, development, and applications of Big Data, as embedded in the fields of engineering, computer science, physics, economics and life sciences. The books of the series refer to the analysis and understanding of large, complex, and/or distributed data sets generated from recent digital sources coming from sensors or other physical instruments as well as simulations, crowd sourcing, social networks or other internet transactions, such as emails or video click streams and other. The series contains monographs, lecture notes and edited volumes in Big Data spanning the areas of computational intelligence including neural networks, evolutionary computation, soft computing, fuzzy systems, as well as artificial intelligence, data mining, modern statistics and Operations research, as well as self-organizing systems. Of particular value to both the contributors and the readership are the short publication timeframe and the world-wide distribution, which enable both wide and rapid dissemination of research output.

The books of this series are reviewed in a single blind peer review process.

Indexed by SCOPUS, EI Compendex, SCIMAGO and zbMATH.

All books published in the series are submitted for consideration in Web of Science.

Tzung-Pei Hong · Leticia Serrano-Estrada ·
Akrati Saxena · Anupam Biswas
Editors

Deep Learning for Social Media Data Analytics

 Springer

Editors
Tzung-Pei Hong
Department of Computer Science
and Information Engineering
National University of Kaohsiung
Kaohsiung, Taiwan

Akrati Saxena
Department of Mathematics and Computer
Science
Eindhoven University of Technology
Eindhoven, The Netherlands

Leticia Serrano-Estrada
Urban Design and Regional Planning Unit
University of Alicante
Alicante, Spain

Anupam Biswas
Department of Computer Science
and Engineering
National Institute of Technology Silchar
Cachar, Assam, India

ISSN 2197-6503   ISSN 2197-6511   (electronic)
Studies in Big Data
ISBN 978-3-031-10871-6   ISBN 978-3-031-10869-3   (eBook)
https://doi.org/10.1007/978-3-031-10869-3

This Springer imprint is published by the registered company Springer Nature Switzerland AG
The registered company address is: Gewerbestrasse 11, 6330 Cham, Switzerland

Preface

The content of this volume focuses on both theory and practical application of Deep Learning for Social Media Data Analytics. Nowadays, social media platforms are overwhelmed by content constantly being generated by billions of users. These huge volumes of data are mostly unstructured and thus, while offering an enormous potential to influence business, politics, security, and other social aspects, technical challenges emerge that still need to be fully addressed by research. Social media posts, shared in the form of texts, often accompanied by images or videos, have a heterogeneous nature and complex structure due to the wide range of embedded expressions and features, making deep learning and model training rather useful for processing, classifying, analyzing, managing, and storing the large volumes of data involved. This information then becomes resourceful for use on many day-to-day applications. For instance, natural language processing of the text shared in social media allows drawing conclusions often necessary for sentiment analysis or opinion mining, whereas image recognition models are helpful for making sense of posts containing photos with no caption. Likewise, acquiring users' personal interests and identifying social media influencers is crucial for targeted advertisements, whereas group-centric studies, flow of information, and prediction of links are relevant for recommendations. Effective customer services based on chatbots allow businesses to automate customer services. For instance, interactive question-answering systems are often used for which deep learning techniques are incorporated.

In view of its great potential, the research presented in this book emerges from the conviction that there is still much progress to be made towards exploiting deep learning in the context of social media data analytics and its applications to benefit society in a wide range of areas from health to smart cities. The collaborative effort between experts from different fields has enabled the integration of a varied range of approaches to deep learning-based theories, techniques, and methods applied to social media data. It comprises detailed studies that cover some of the most salient areas of research in the field.

The book is organized into four Parts, which include 15 chapters that report on original research in network structure analysis, social media text analysis, user behavior analysis, and social media security analysis. In Part I of the book, Chapter

"Node Classification Using Deep Learning in Social Networks" presents a comprehensive literature review of deep learning research methods designed for node classification, offering a future research agenda in the field, whereas Chapter "NN-LP-CF: Neural Network Based Link Prediction on Social Networks Using Centrality-Based Features" proposes a neural network-based link prediction model using centrality-based features and global topological features to predict missing links. Real-world networks are used for performing the experiments, which are then compared to state-of-the-art methods.

Part II delves into text analytics techniques and applications. Specifically, Chapter "Deep Learning for Code-Mixed Text Mining in Social Media: A Brief Review" presents a thorough review of deep learning techniques for code-mixed text analysis, introducing a range of text-mining tasks that use deep learning to model different solutions. Some of the challenges associated and the deep learning techniques used to overcome them are outlined. Chapter "Convolutional and Recurrent Neural Networks for Opinion Mining on Drug Reviews" focuses on sentiment analysis and opinion mining applied to drug reviews using deep convolutional and recurrent neural networks and hybrid models for predicting categories on health forum data. Following a similar direction, Chapter "Text-Based Sentiment Analysis Using Deep Learning Techniques" proposes a comparative experimental analysis of selected deep learning models for conducting a text-based sentiment analysis, specifically applied to an IMDB movie reviews dataset. Chapter "Social Sentiment Analysis Using Features Based Intelligent Learning Techniques" offers a complementary approach by analyzing and classifying text from Twitter, Amazon mobile reviews and IMDB movie reviews into two classes, namely positive and negative sentiments by adopting various machine learning algorithms.

Part III is concerned with the theoretical background and experimental examples of how deep learning techniques are applied to recommendation systems and user behavior analysis. In this section, Chapter "Modified-PIP with Deep Neural Network (DNN) Architecture: A Coherent Recommendation Framework for Capturing User Behaviour" seeks to develop a high-performance deep learning model for computing similarity behavior from movie recommendation systems using a MovieLens-100K dataset. Chapter "A Survey on Graph Neural Network Based Video Recommendation System" highlights the key benefits and convenience of deep learning-based recommender systems over traditional ones. Relevant applications and issues related to data privacy are explained and future directions of the field are outlined. In a different approach, the research presented in Chapter "Characterisation of Mental Health Conditions in Social Media Using Deep Learning Techniques" covers a comprehensive review of previous work developed in relation to the characterization of social media users' mental health conditions through deep learning techniques and the future directions in this field of knowledge. In line with this topic, Chapter "Predicting Mental Health and Nutritional Status from Social Media Profile Using Deep Learning" focuses on the prediction of mental health and nutritional status of Facebook users through Facebook's Confessions Datasets (Mental Health Dataset category), by designing and applying a deep learning model. In a different approach to user behavior, Chapter "Impact of Artificial Intelligence-Based Chatbots on Customer Engagement and Business Growth" delves into how customer support, service, and business growth, which are strongly linked to the client's perception of trust, are being impacted by the introduction of artificial intelligence-based chatboxes. Several case-study companies and brands are analyzed, and a discussion is presented on how their effort to automate communication with customers has resulted in business growth.

The last part of the book covers the state-of-the-art and deep learning techniques that have recently been developed to address some of the multiple forms of security-related challenges in social media platforms. Specifically, Chapter "Do Not 'Fake It Till you Make It'! Synopsis of Trending Fake News Detection Methodologies Using Deep Learning" focuses on the detection of fake content in social media by developing a comparative study of recent findings, architectures, current insights, and discussions in the field. It also sheds some light on upcoming trends and future directions on the development of more efficient fake news detectors. Chapter "Towards Detecting Fake Spammers Groups in Social Media: An Unsupervised Deep Learning Approach" includes a thorough review of the application of multiple deep learning and machine learning-based unsupervised algorithms to detect false reviews with fake reviewers. Advantages and disadvantages of some of the most popular algorithms are discussed, and the implementation of these to a Google Play Store App dataset is used as a real-time case study. Chapter "A Deep Learning Approach for Anomalous User-Intrusion Detection in Social Media Network System" delves into social media network intrusion detection systems and proposes a novel deep learning approach for detecting and classifying the types of attacks. Chapter "Deep Digging of Anomalous Transactions in Financial Networks with Imbalanced Data" addresses the challenges involved in the detection of anomalies or fraudulent users and transactions. Specifically, the research covers deep learning approaches to fraudulent transaction detection in the financial network domain.

All in all, the editors would like to express their deep appreciation and gratitude to the contributors who have made their research available for this volume, and to the anonymous reviewers who have provided an invaluable service in referring to the chapters. We sincerely hope that this work serves as a reference for researchers, as well as a compilation of innovative ideas and solutions for practitioners interested in recent theories and applications of deep learning for social media data analytics.

Sincerely,

Kaohsiung, Taiwan   Dr. Tzung-Pei Hong
Alicante, Spain   Dr. Leticia Serrano-Estrada
Eindhoven, The Netherlands   Dr. Akrati Saxena
Cachar, India   Dr. Anupam Biswas
April 2022

Contents

Network Structure Analysis

Node Classification Using Deep Learning in Social Networks

Aikta Arya, Pradumn Kumar Pandey, and Akrati Saxena

Abstract Recently, the demand and utility of online social networks are well accepted to share information and connect people from diverse areas. Online social networks have provided a common platform for frequent human interactions, resulting in a significant increase in information about the individual users, their interactions, and relationships. These users can be classified into different classes based on the similarity and differences in users' characteristics and their local and global position in the network. The node classification problem has been recognized due to its real-time applications in recommendation systems, epidemiological diffusion, sociological dynamics of communities, and anomaly detection. Diverse attempts have been made to perform informative node classifications. Furthermore, the deep learning based approaches for node classification in online social networks have provided state-of-the-art results with better insights and high accuracy. In this chapter, we provide a rigorous literature review of deep learning based methods designed for node classification, and conclude the chapter with interesting and futuristic open research directions to fill the gap in the current works and the demand of next-generation online social systems.

Keywords Deep learning · Node classification · Community detection · Role identification · Graph partitioning · Online social networks

A. Arya (✉) · P. K. Pandey
Indian Institute of Technology Roorkee, Roorkee, India
e-mail: a_arya@cs.iitr.ac.in

P. K. Pandey
e-mail: pradumn.pandey@cs.iitr.ac.in

A. Saxena
Department of Mathematics and Computer Science, Eindhoven University of Technology, Eindhoven, The Netherlands
e-mail: a.saxena@tue.nl

1 Introduction

Diverse real-world complex interactive systems are projected as networks, for example, social systems, biological systems, transportation systems, and ecological systems [64]. In the case of social networks, a network is an abstract representation of people as nodes and intricate interactions among them as links [87]. Online social networks (OSNs) are common platforms on which users interact frequently and generate extensive data that can be used to study their behavior. Due to technological advancements, now we are able to store this huge volume of data and have capabilities to process it for retrieving valuable information. In order to comprehend such complex and vast systems, rigorous network analysis methods are required that can help in analyzing the structural architecture, exploring the internal organization, predicting absent links, and understanding the dynamics of the complex system [60]. Among almost all core applications and challenges of network analysis (also known as graph mining), *node classification* has received a huge attention of researchers working on complex networks and deep learning.

Ideally, every social network user is associated with a label having some kind of relevance to the user. However, OSNs do not provide a feature where users can select the relevant labels that can be appropriately applied to them. The label of the users are identified using the information available on users' profile or by conducting surveys. However, many users do not update or share their information frequently, due to privacy concerns; for instance, users do not share their views (on different topics, such as political or religious) publicly online. Hence, a large amount of available data from social networks is not labeled, given its diverse applications (given in detail in Sect. 1.1). In order to classify the nodes (or assign a meaningful label to each node), we perform *node classification* or *node clustering*, that assigns labels to each node based on its similarity with other nodes of the network. These labels are based on various different characteristics of the nodes, such as religious or political views, demographic data, for instance, gender, age or locality, affiliations, and users' interests that capture various aspects of personal preferences and behavior of an individual user.

The process of node classification assigns labels to nodes depending on its similarity with other nodes present in the network. Moreover, defining similarity among nodes is fundamentally essential task that can be done on the basis of 'homophily' among nodes, influence diffusion/maximization and various other properties of nodes. Generally, the nebulous concept of homophily refers to the trend among people of being connected with others that are similar to themselves in multiple ways [85]. Furthermore, the presence of homophily among nodes triggers the formation of communities in the network in which the nodes possessing similar characteristics are united together in the same community. Hence, in such cases we perform community based node classification. Nevertheless, the influence diffusion task can be considered as finding best k people/nodes to focus for maximizing the count of people that will be influenced eventually [83]. In such cases the links in the network are associated with the weights within the range of 0 and 1 that represents the probability by which a node can influence other nodes once activated [16, 49]. Hence, in such scenarios it is more appropriate to perform influence based node classification.

Diverse attempts have been made to provide informative and meaningful node classifications with a high accuracy. Now a days the deep learning (DL) based approaches have became the 'crown' of the field of machine learning and artificial-intelligence [57] for exhibiting significant performance in natural language processing [5], image analysis [55], acoustics [34], graph data mining, and so on. The impressive power of deep learning practices and approaches to extract complex features and patterns out of underlying data is well acknowledged. However, applications of these traditional deep learning frameworks and architectures for the rigorous analysis of graph data consist of many challenges, such as irregular structures, heterogeneity and complexity [111], makes it a non-trivial problem.

After addressing diverse challenges to employ deep learning (DL) practices to graph data, deep learning methods are adopted to solve various problems [10], and significant improvements in accuracy are observed as compared to traditional approaches [54]. In particular, the DL based approaches for node classification can broadly be classified into two classes: node classification based on (I) **community membership** [26, 78], and (II) **role of the nodes** [75]. In community membership based methods, the nodes present inside the network are divided into several groups such that various nodes inscribed in the same group are more densely-associated as compared to various other nodes present in different groups [20]. However, in role detection based approaches, nodes are divided into groups depending on the similarity found in their roles in a given network, where the roles can be defined based on structural similarity or influence propagation behavior of nodes. Capturing actual roles of the nodes is a complex task as two distinct nodes playing the similar roles in the network might be distant and not in close proximity [25]. Most of the role-based works have focused on the structural similarity of the nodes, such as bridge, articulation or hub nodes.

For better illustration of community-based, and role-based node classification, we take the most commonly used data-set, Karate club social network [107], and perform node classification on it as represented in Fig. 1. In this network, every node possesses a label representing the club to which the individual node belongs. In the following illustration in Fig. 1a community based node-classification is performed using node2vec [28] approach and in Fig. 1b Role-based node classification is performed using RolX [32] method. Community-based node classification breaks up this network into some tightly coupled groups, whereas role-based node classification shows that the nodes having similar types of roles based on network structure are clustered together, as shown using the same color in the Figure.

Generally, it is observed that the nodes associated with the identical communities are more plausible to be connected however, the nodes that possess the identical roles in the network might not be connected and are often found to be far from each other. As the definitions of community (clustering) and role based classifications are different and almost orthogonal to each other. A significant effort has been made in both the areas: community and role identifications, and different mechanisms have been introduced to identify the membership of nodes to different communities and

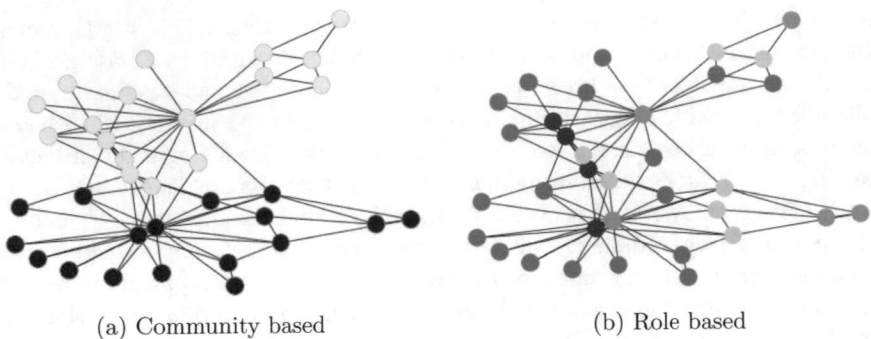

(a) Community based   (b) Role based

Fig. 1 Karate club social network [107] for analyzing the community based and role based node classification. Nodes with same color are categorized under same class; **a** Community based node classification using node2vec [28] method (the color of the node represents its community), **b** Role based node classification using RolX [32] method [Best view in color]

roles, according to the considered classification. Recently, proliferation in the use of deep learning practices in data and network analysis is observed. Several deep-learning based mechanisms have been proposed to identify communities and roles of the nodes in the network. In this chapter, we present a brief review of the classification methods for each category.

1.1 Applications

Node classification has been used in various applications that we discuss next.

- **Link Prediction:** Elementally in networks/graphs, the links/edges can be diversified as (i) intra-community links, when both end nodes of the link associate the same community, and (ii) inter-community links, when both end nodes of the link associate different communities. It has been observed that intra and inter-community links possess different structural properties and the similarity scores, such as Jaccard coefficient, Adamic-Adar index, and RA index, of intra-community connections/links is much higher than the inter-community connections/links [80]. Therefore, the link prediction mechanisms that use community information have provided state-of-the-art accuracy for link prediction [6, 41, 58, 80, 81].
- **Recommendations:** Besides friend recommendation, users categorization is also helpful in recommending content or posts based on the personal preference of users [7, 59]. The user categories are also helpful in suggesting advertisements, musics, videos, and movies based on the interest levels of different types of users [53, 63, 104].
- **Anomaly Detection:** The sub-domain of data-mining associated with discovering infrequent occurrences in the datasets is elementally recalled as anomaly detection.

In anomaly detection, the node classification can be used to classify nodes based on the abnormal structural properties present in the network [3, 21, 73].

- **Identifying social influence:** The rigorous community and role based node classification analysis can help us in identifying local as well as global influence of the nodes, where local influence refers the influence of the node in its community and global influence is the influence of the node in the entire network [24, 72, 82]. The community information has also been used in computing better centrality values of the nodes for identifying influential nodes in the network [1].
- **Network Visualization:** The node classification is also beneficial in network visualization, especially where the entire network is very large and complex to be visualized at once [39, 84]. The network visualization using its low dimensional representation is helpful in getting useful insights about the network [68, 81].
- **Network Reconstruction:** The network embedding methods aim to preserve the structural properties, such as roles, communities, or proximity, of the network. Hence, the generated embedding can be further utilised in reconstruction of networks. The network reconstruction from such embedding can be utilized to validate and evaluate the effectiveness of underlying network embedding [38]. Besides this, the network reconstruction errors can help in identifying anomalous nodes in the network [18].

2 Preliminaries

In this section, we briefly discuss the terminologies used in the chapter.

Definition 1 (*Network*) A network is represented as $\mathcal{G} = (\mathcal{V}, \mathcal{E})$ comprises vertex set $\mathcal{V} = \{v_1, v_2, v_3, ..., v_n\}$ and edge set $\mathcal{E} = \{e_{j,k}\}_{j,k=1}^{n}$. if the node $v_i \in V$ has an attribute as $a_i \in \mathcal{A} \subseteq \mathbb{R}^{nxd}$, then the network $\mathcal{G} = (\mathcal{V}, \mathcal{E}, \mathcal{A})$ is an *attributed network*.

Definition 2 (*Community*) Given a network $\mathcal{G}$ containing m communities $\mathcal{C} = \{\mathcal{C}_1, \mathcal{C}_2, ..., \mathcal{C}_m\}$, where every community $\mathcal{C}_i$ is a sub-graph of $\mathcal{G}$, such that the nodes across two different communities $\mathcal{C}_j$ and $\mathcal{C}_k$ are sparsely connected whereas the nodes within single community are densely connected.

Definition 3 (*Network Embedding*) Given a network $\mathcal{G} = (\mathcal{V}, \mathcal{E})$, the network embedding process maps each node $v \in \mathcal{V}$ in the given network to low-dimensional embedding $X \in \mathbb{R}^{nxr}$ given that $r << n$.

Definition 4 (*Roles*) Roles are described by the set of nodes having a higher 'similarity' to the distinct nodes present inside the set as compared to other nodes present outside. The similarity of the nodes can be defined based on the structure, influence, or any other characteristics of the nodes and the network. In the literature, most of the works have focused on the 'structural-similarity'. The phrase 'structurally-similar' refers to the set of nodes possessing similar-structural properties [75].

In the next section, we discuss state-of-the-art deep learning methods for community detection in the network.

3 Community-Based Node Classification

The community based node- classification has gained attention from many researchers due to its remarkable practical significance. In fact, our world constitutes a vast complex network with various serially connected communities. For instance, in social networks, sponsors of various online shopping platforms suggest products to the users of already detected communities [95, 100]. Application of community detection methods and practices to citation networks [13] helps in identifying recent research trends and determining the evolution, importance as well as inter-connectedness of research topics. Furthermore, incorporation of community detection approaches and frameworks in $\mathcal{P}$rotein-$\mathcal{P}$rotein $\mathcal{I}$nteraction (PPI) [12] networks and metabolic networks [29, 74] discovers proteins having related biological functions and metabolic complexities. Similarly, using community detection approaches in $\mathcal{B}$rain networks [86] reveals anatomical and functional partitioning of different regions in the brain.

Most of the traditional (shallow based) techniques for node classification such as statistical inference [2, 36, 48] and spectral clustering [4, 56] are applicable to small networks or very naive cases. However, real-world networks are much more complex, having abundant nonlinear information, that makes traditional models least applicable to these scenarios as they are computationally expensive. Whereas the less expensive deep learning approaches are more poignant to these scenarios as they provide flexible community detection approaches to represent network embedding of lower dimensions and learn nonlinear network characteristics.

Further, in succeeding part of this section we study diverse community based approaches for node classification as, Generative Adversarial network based approaches, Auto-Encoder-based approaches, Graph convolution network-based approaches, methods combining undirected graphical models & graph convolution network based approaches and Random walk based approaches.

3.1 *Auto-Encoder-Based Approaches*

$\mathcal{A}$uto-$\mathcal{E}$ncoders provides elementary neural-network framework to reconstruct high dimensional data (graphed data) into the form of low dimensional network representations. The autoencoders use the decoder and encoder components to learn, in an unsupervised manner, a new network representation. They possess a symmetrical-architecture having multiple hidden layers in such a way that the outcome of one layer(s) is provided as an input to the succeeding layers. Their main objective is to reduce the in-between error among reconstructed/restructured network data and the original input in order to learn a hidden representation optimally, which is denoted as:

$$Loss(\theta_1, \theta_2) = \sum_{i=1}^{n} l(x_i, g(f(x_i; \theta_1); \theta_2))$$

here $l(*)$ is the loss function and $g(*; \theta_2)$, $f(*; \theta_1)$ are decoder and encoder respectively. As most of the $\mathcal{A}$uto-$\mathcal{E}$ncoders based approaches and practices results in terms of network embedding [11, 101], Various clustering approaches, such as spectral clustering and $\mathcal{K}$-means clustering, are used as an alternative to extract communities. These alternatives directly patch clustering into the existing framework for extracting communities [8, 9].

$\mathcal{A}$uto-$\mathcal{E}$ncoders based approaches can be further classified into four types: denoising, stacked, sparse, and variational autoencoder. The most basic type of autoencoder is a Stacked autoencoder that contains consecutive series of auto-encoders which is frequently used by other encoders. In particular, when the stacked framework targets other properties, such as de-noising or sparsity, we categorize them as de-noising and sparse autoencoder.

Stacked auto-encoders. The semi-DNR [101] sequences auto-encoders into stacks for generating nonlinear reconstruction of the DNR network. This practice requires that every encoder layer must possess fewer neurons than the previous layer for extracting important features of input data and reduce its dimensions. The semi-DNR framework completely utilizes a-priory knowledge of two nodes belonging to the identical communities, in order to introduce pairwise constraints among nodes in the network. The semi-DNR framework uses the following loss function:

$$Loss = l(M, Z) + \lambda Tr(H^T L H)$$

Few other stacked auto-encoders can be enlisted as DeCom [9], CDDTA [96], AAGR [11], DIR [43] and NEC [88].

Sparse auto-encoders. Most specifically, the large-scale networks are very complex to process and store; hence it is important to have sparse representations. Furthermore, finding the optimal network representations by incorporating the sparse constraints to autoencoders has the potential to motivate a new research line. The $\mathcal{G}$raph-$\mathcal{E}$ncoders [91] explicitly controls the size of hidden representations by incorporating regularization with hidden layer. The reconstruction error or loss function after incorporating sparsity constraints can be written as:

$$Loss = \sum_{j=1}^{n} ||z_j - x_j||_2 + \beta K L(\rho | \hat{\rho})$$

here $\hat{\rho}$ and ρ are the parameters of sparsity and β restricts the penalty used in sparsity. Some other stacked auto-encoders based framework can be found in: DFuzzy [8] and CDMEC [98].

De-noising auto-encoders. If we have inputs that possess significant noise, then De-noising auto-encoders can be utilized to provide network representations that can with-stand noise. In [93], the proposed MAGE framework utilizes convolutional networks to incorporate structural and content information and further introduces random noise to the content information iteratively in the entire auto encoder set-up. Hence, by this procedure, the content and structural information can be incorporated into an individual-unified process, and their composite analysis can be done. Furthermore, in [99] GRACE models clusters by incorporating network dynamics and assuming that cluster formation process in the network needs dynamic-embedding to attain a substantially stable state.

Variational auto-encoders. We can easily find various existing approaches that can be categorized as variational auto-encoders [51]. These approaches consider hidden layer representations as an inherent variable (latent) with respect to its preceding distributions. In [66], ARVGA computes the reconstruction error as:

$$\mathcal{L}oss = \mathcal{E}_{q(G|(X,A))}[log p(\hat{A}|G)] - KL[q(G|X,A)||p(G)]$$

Some other variational auto-encoders based framework can be found in: VGAECD [14], improved VGAECD [15], gamma ladder VAE [79], NetVAE [45].

3.2 Generative Adversarial Network-Based Approaches

Motivated by minimax two-player game generative adversarial networks [27] comprise of a generator ($\hat{\mathcal{G}}$) and a discriminator ($\hat{\mathcal{D}}$) module. The generator ($\hat{\mathcal{G}}$) module extracts data distribution whereas, discriminator $\hat{\mathcal{D}}$ module estimates whether a sample data belongs to the original data used for training or the data synthetically generated using $\hat{\mathcal{G}}$ module. The complete training process of GANs can be given as:

$$\min_{\hat{\mathcal{G}}} \max_{\hat{\mathcal{D}}} \mathcal{V}(\hat{\mathcal{G}}, \hat{\mathcal{D}}) = \min_{\hat{\mathcal{G}}} \max_{\hat{\mathcal{D}}} (E_{x \sim P_{data(x)}}[log \hat{\mathcal{D}}(x)] + E_{z \sim P_{z(z)}}[log(1 - \hat{\mathcal{D}}(\hat{\mathcal{G}}(z)))])$$

The fact that GANs are unsupervised (in most cases) motivated for using GANs for community detection. The GANs also possess an important capability for network data analysis as the synthetic data generated using $\hat{\mathcal{G}}$ follows the same distribution as the original data. In [42] a framework referred to as CommunityGAN is proposed, which is inspired by Affiliation Graph Model (AGM) to enhance the performance through the initiation of minimax game among the network discriminator and motif-level network generator. The CommunityGAN framework can discover overlapping communities as well as it can also learn distinct graph representation.

Another framework SEAL proposed in [110] comprises a discriminator module that estimates whether a community (provided as an input) is real or not as well as a generator module for generating various communities. In [90], authors proposed a framework referred to as AGAE that integrates clustering to the (deep) graph-embedding process to self-supervise the network training procedure with the help of an adversarial regularizer. Various other approaches discussed in [23, 31, 37, 89] uses GANs for deriving representations of nodes that can be further employed for community detection. In [102] the authors noticed that most GANs based frameworks differentiate among the output of node embedding and samples of Gaussian-distribution without making any changes in real-data, which makes it inappropriate for adversarial learning. The authors referred to their framework as (JANE) Joint adversarial network embedding.

3.3 Graph Convolution Network-Based Approaches

The Graph Convolution networks [52] (GCNs), the widely used stream of graphical neural network approaches [94] for learning node representations have motivated many researchers because of its influencing results in unsupervised and supervised node classification approaches. Various approaches have been devised for leveraging its power for effectively extracting high-dimensional data of complex networks for community detection.

In [44] authors pointed out that the node representations output of GCNS is not community-oriented in any case, and detecting communities is an inherently unsupervised task. To address this particular issue, they gave a model referred to as JGECD for detecting communities using joint GCN embedding. The JGECD framework comprises three elementary modules: a topology reconstruction module, a community detection module and a dual-encoder module. Authors formally defined the probability that the pth node is associated with qth community as:

$$\mathcal{U}_{pq} = \frac{exp(\theta_q^T h_p)}{\sum_{q=1}^{k} exp(\theta_q^T h_p)}$$

Furthermore, the extended version of JGECD proposed in [108] employs MRFas-GCN [46] as an encoder in order to have unsupervised community detection. The authors employed a dual encoder having community-centric properties to remodel network attributes and network topology. The decoder employed to remodel network topology can be given as follows:

$$\hat{A} = sigmoid(\mathcal{D}\mathcal{U}\mathcal{W}\mathcal{U}^T\mathcal{D}^T)$$

Here $\mathcal{W}$ represents weight matrix, $\mathcal{D}$ represents degree matrix, and $\mathcal{U}$ is an output from encoder representing the likely-hood matrix of nodes associated with different communities. In [112] a temporal-heterogeneous model based on GCN is proposed, which is referred to as HTGCN. The proposed framework extracts communities out of heterogeneous-temporal graphs or networks. In a digest, the framework firstly employs heterogeneous-GCN to extract the heterogeneous-network feature representations, and then compressed aggregation process is applied in order to manifest dynamic as well as static properties of the resultant communities.

3.4 Integrating Un-directed Graphical Models and Graph Convolution Networks

In recent years, a significant number of works has been introduced in order to combine undirected graphical models with GCN for detecting communities (overlapping / non-overlapping). The basic shortcoming with GCN is that it employs local smoothing of features (ignoring community properties) to get node embedding which is not community-oriented. However, undirected-graphical models possess various global objectives to define community. Hence, the shortcomings of GCN can be resolved using undirected graphical models. Hence both of these approaches can be combined to perform community detection.

That major effort made in this stream is proposed in [46] which is referred to as MRFasGCN. The proposed framework merges MRF and GCN in order to perform detect communities in a semi-supervised manner. After MRFasGCN several other approaches combining CRF to GCN were also proposed. In [71] another approach referred to as GMNN is proposed that integrates statistical rational-learning with graph neural networks. The GMNN framework learns node representation in order to model label dependency among nodes for accomplishing the sole purpose of semi-supervised community-based node classification. Furthermore, in [22] the authors pointed out that already existing GCNs do not preserve similarities (among nodes) in the network. To solve this problem, they proposed the addition of one more layer of CRF to GCN in order to enforce the characteristic that similar nodes should have similar features.

3.5 Random Walk Based Methods

The elementary requirement of any network embedding based approach is to retain and preserve the underlying network structure. Moreover, preserving the structural information of neighbourhood, constituting the local characteristic of the node, is another important characteristic of network embedding. Even though, the adjacency vector of the node preserves the high dimensional first order neighbourhood information, it is not sufficient as it is found to be discrete, sparse and high dimensional (due to sparseness of large graphs/networks). Such network representations are not adequate for subsequent applications for instance, the word representations in natural language processing also face such problems. The introduction of Word2Vector [62] effectively represented word in contiguous, dense and low-dimensional vectors. The elementary idea behind Word2vector is the reconstruction of neighbourhood word vector derived from the co-occurrence rate of the neighbourhood words. The analogy of Word2vector was further borrowed by several random walk based embedding mechanisms for various networks in which the random paths over a network are generated by considering an individual node of the network as a word and the respective designated random path as a sentence. The neighbourhood vector is defined similar to the neighbourhood word vector in Word2Vector using co-occurrence rate. The analogy of Word2Vector was further employed by diverse network embedding mechanisms, for instance DeepWalk [68], Node2Vec [28], metapath2vec [17], TriDNR[67], and so on.

DeepWalk [68] employs an effective neural language model named as SkipGram [61] for network embedding. The SkipGram focuses on maximizing the probability of co-occurrence between the words that are present in the window w. The DeepWalk initialises with sampling multiple paths of the given network utilizing truncated random walk. Every individual path that gets sampled from the given network is related with the respective sentence from the corpus in which an individual node corresponds with individual words. Afterwards, the application of SkipGram on paths maximizes the probability of inclusion of node's neighbourhood conditioned on the network embedding. Hence the nodes having similar neighbourhood vector shares the similar embedding. DeepWalk uses the following objective function:

$$\min_{x} - \log P(\{v_{j-w}, ..., v_{j-1}, v_{j+1}, ..., v_{j+w}\}|x_j) \tag{1}$$

where w is given size of the window that indirectly helms the size of random walk. The incorporation of SkipGram further removes the ordering constraint in Eq. 1 as:

$$\min_{x} - \log \sum_{-w \leq k \leq w} P(v_{j+k}|x_j) \tag{2}$$

where $P(v_{j+k}|x_j)$ is defined with the help of softmax function as:

$$P(v_{j+k}|x_j) = \frac{exp(x_{j+k}^T x_j)}{\sum_{l=1}^{|V|} exp(y_l^T x_j)} \tag{3}$$

However it is very expensive to compute the calculations given in Eq. 3 due to normalization factor. Hence, the solution to this problem is to approximate the computation involved in full softmax in an inexpensive way using (a) negative sampling function [62] (b) hierarchical softmax function [62].

The negative sampling function [62] reduces the time complexity of calculations given in Eq. 3 by utilizing logistic regression in order to distinguish the nodes from noises. The computation of each log $P(v_{j+k}|x_j)$ given in Eq. 2 is calculated as,

$$\log \sigma(x_{j+k}^T x_j) + \sum_{t=1}^{K} E_{v_t \sim P_n}[\log \sigma(-x_{v_t}^T x_j)] \tag{4}$$

where $P_n(v_j)$ is the noise distribution of node v_j in the negative samples and K is the total number of nodes that are actually sampled.

The hierarchical softmax function [62] resolves the problem of expensive computation of log $P(v_{j+k}|x_j)$ given in Eq. 3 by generating a binary tree where nodes are designed to leaf nodes. Now, instead of calculating it for all the nodes we just need to evaluate for the paths from root to designated leaves. Hence the optimization problem transforms into: given a specific path maximize its probability. Furthermore, the Eq. 3 transforms as:

$$P(v_{j+k}|x_j) = \prod_{t=1}^{\log(|V|)} P(a_t|X_j) \qquad (5)$$

where the path to designated leaf v_j is a sequence of nodes $(a_0, a_1, a_2, ..., a_{\log(|V|)})$ having a_0 as root node and $a_{\log(|V|)} = v_j$.

The proliferation in use of DeepWalk [68] has inspired many to come up with diverse graph embeddings based on random walks using distinct deep learning models such as SkipGram [61] and Long-Short Term Memory (LSTM) [35]. Most of the motivated works later follow up with the concept of DeepWalk with minor changes in the settings that either preserve proximity [17, 19, 67] or employ sampling method using random walk [28, 103].

node2vec [28] provides a framework for contiguous (non-discrete) feature representations of individual nodes present in the network in the form of network embeddings. The novel work learns the mapping of nodes and converts and maps them to the low dimensional space along with preserving the neighbourhood information of nodes in the network with maximum likelihood. node2vec utilizes a flexible approach for defining the neighbourhood network of a node and propose a procedure for biased random walks in order to effectively probe through diverse neighbourhood networks.

The proposed node2vec algorithm pertains on the notion of extending SkipGram architecture [61, 68]. Moreover, the objective function is transformed as:

$$\max_f \sum_{u \in V} \log Pr(N_S(u)|f(u)) \qquad (6)$$

where $N_S(u)$ represents the neighbourhood network of node $u \in V$ constructed using given sampling strategy S f is the feature matrix of $|V|$ x d parameters. Furthermore they employ various assumptions in order to further simplify the objective function given in Eq. 6 as:

$$\max_f \sum_{u \in V} \left[-\log Z_u + \sum_{n_j \in N_S(u)} f(n_j).f(u) \right] \qquad (7)$$

here, Z_u is comparatively expensive function (in terms of computations) hence needs to be further approximated (most specifically using negative sampling function [62]). Furthermore, the objective function given in Eq. 7 is further optimized using stochastic gradient ascent over model parameters.

metapath2vec [17] formalises a framework for generating network embedding in heterogeneous network environment. The proposed model utilizes meta-paths based random walks in order to generate heterogeneous neighborhood network of a given node. Furthermore, metapath2vec constructs node embedding using heterogeneous SkipGram model.

The metapath2vec model describes the objective function that maximizes the probability in terms of local structure (similar to the objective functions of [28, 62, 68]) as:

$$arg \max_{\theta} \prod_{v \in V} \prod_{c \in N(v)} P(c|v; \theta) \tag{8}$$

To incorporate the heterogeneous neighbourhood information in the network embedding metapath2vec uses heterogeneous SkipGram model that constitutes metapath-based random walks. Hence in heterogeneous environment the objective function transforms as follows.

$$arg \max_{\theta} \sum_{v \in V} \sum_{t \in T_V} \sum_{c_t \in N_t(v)} P(c_t|v; \theta) \tag{9}$$

where $N_t(v)$ is the neighbourhood of v having t types of nodes and $P(c_t|v; \theta)$ is the softmax function as discussed in Eq. 3. Furthermore, to solve the problem of computational complexity of softmax function metapath2vec employs negative sampling as given in Eq. 4.

TriDNR [67] provides a framework for network representation that utilizes node label (if available), node content and node structural information in order to learn an effective and optimal node representation. TriDNR enforced learning from deep neural language model specifically at three levels such as: (a) at the level of the node label, the TriDNR approach maps labels to words and the objective function maximizes the probability of observed word sequence on providing a given class label, (b) at the level of the node content, the TriDNR approach maps nodes with the words and the objective function maximizes the co-occurrence of observed word sequence, given an individual node. (c) at the level of the node structure, the TriDNR approach maps node with another node and the objective function maximizes the probability of encountering neighbouring nodes, given a initial node in a random walk. All these information are compositely given to the neural network model to learn an optimal and effective representation of a given node. The objective function of TriDNR is given as:

$$\mathcal{L} = (1 - \alpha) \sum_{j=1}^{N} \sum_{s \in S} \sum_{-a \leq k \leq a, k \neq 0} \log P(v_{j+k}|V_j) \tag{10}$$

where α balances the weight among node label, content and structure information, a is the window size of the sequence.

Roles of different nodes in a network is another important aspect to be considered in network analysis for classifying the nodes. Next section is dedicated for the discussion of different role based node classification methods.

4   Role-Based Node Classification

Most of the approaches available for node classification are designed based on the
elementary proximity of the nodes, i.e., the underlying network embedding vectors
are based on the community structure. Therefore, they fail to apprehend the role-based
structural information [76]. There are various challenges in studying role-oriented
node-classification, and next, we highlight some of these challenges.

1. Precise and accurate role definitions are intricate to design for real-world complex
   social networks.
2. The interaction patterns of nodes with different roles and the distribution of nodes
   possessing identical roles in a large-scale social network is a very complex and
   cumbersome task to apprehend.
3. Based on roles, two distinguished nodes with identical structures might be very
   distant in the network, and it makes the task of defining effective loss function
   difficult.

The challenges discussed above were addressed by many researchers in the form
of various role-based node classification techniques, which can be split into two
classes of shallow and DL-based approaches. The shallow approaches comprise
matrix factorization based simple methods. The distinct matrix factorization based
techniques perform node classification by learning network embedding vector using
low-rank approximation and higher-order similarity matrices of the network, such as
non-negative matrix factorization, singular value decomposition, and NetSMF [70].
     Next, we discuss some of the recently published methods for classifying nodes
based on their roles.
     DRNE [92] captures regular equivalence in the network. The regular equivalence
can be taken as an alternative for role equivalence which is defined as a notion
by which two regularly equivalent nodes have a neighborhood that is also regularly
equivalent. Furthermore, it also proposes layer-normalized LSTM (Long Short-Term
Memory) [35] for representing each node in the network using an aggregate representation of its neighborhood in a recursive manner. After learning individual node
representation, node classification can be performed based on their roles. It learns
its network embedding vectors using the following loss function:

$$\mathcal{L}_{eq} = \sum_{v \in V} ||H_v - Agg(\{H_u | u \in N(v)\})||_F^2$$

where Agg is the aggregation operation performed on neighborhood embed-dings
using layer normalized LSTM. In an individual recursive step, the already learned
embedding of a single node only preserves the local structure of its neighborhood.
With iteratively updated learned parameters, the learned individual node embedding
incorporates their structural information globally. Hence it follows the consistent
notion of regular equivalence. The Authors have also considered the case for which
multiple solutions are available as long as they satisfy the recursive process. To avoid

such a trivial solution (where all embed-dings are zero) they have designed a slightly different regularizer as:

$$\mathcal{L}_{reg} = \sum_{v \in V} ||log(d_v + 1) - MLP(Agg(\{H_u | u \in N(v)\}))||_F^2$$

The regularizer is weighted with a weighted parameter γ_{deg} and entire model is trained using an elementary loss function:

$$\mathcal{L} = \mathcal{L}_{eq} + \gamma_{deg}\mathcal{L}_{reg}$$

GAS [30] comprise of graph auto-encoders explicitly guided with elementary structural information in order to learn individual node representations, which are also role oriented. The proposed idea leverage on extracted structural features of lower dimensions to train Graph Neural Networks (GNNs). The GNNs are closely connected with Weisfeiler-Lehman (WL) test and possess the power to capture the network structure with remarkable accuracy [97]. GAS implements a L-layer Graph-Convolution encoder having each layer encoded as:

$$H^{(l)} = \sigma(\tilde{A}H^{(l-1)}\theta^{(l-1)})$$

Here $\tilde{A} = A + I$ and $\theta^{(l-1)}$ is the $(l-1)$th layer parameter matrix. The input $H^{(0)}$ can be consider as an embedding-lookup table or $\tilde{A}$. Instead of using original GCN [52], GAS applies the sum-pooling propagation rule for distinguishing local structures in a better way. Therefore, it is possible that more power-full GNNs such as Graph Isomorphic Network [97] might further enhance the resultant performance. Using a few crucial structural features as an input to train/supervise the model is the key to GAS. The feature extraction process is very much similar to ReFeX [33] except that the aggregation operation is performed only once, and normalization is performed instead of discarding or binning. The MLP model is used as a decoder for elementary approximation of the required features to some extent, i.e., $\hat{F} = MLP_{dec}(H)$ and the loss function is given as:

$$\mathcal{L} = ||F - \hat{F}||_F^2$$

RESD [109] extracts the required higher-order structural feature representations of individual nodes present in the network. Furthermore, as the feature extraction process is very sensitive to various noise/redundancy present in real-world networks, most of the feature extraction approaches fail to represent the roles of the individual nodes with better accuracy. Hence, RESD uses Variation Auto Encoder (VAE) [50] to minimize noise, enhance the robustness of network embedding for node classification, and model the relationships of extracted features. Due to the application of VAE framework, the important structural information of individual nodes (such as degree) might be lost; hence the authors have used a degree-regularized constraint to preserve important structural information. Basically RESD also uses ReFeX [33] to extract required features (F_{ReFeX}). The robust network representation having reduced-noise levels are learned using VAE as:

$$z_i = MLP_{enc}(F_i)$$
$$\mu_i = W_\mu Z_i + b_\mu$$
$$log(\sigma_i) = W_\sigma Z_i + b_\sigma$$
$$H_i = \mu_i + \sigma_i \odot \epsilon, \epsilon \sim Gaussian(0, I)$$

The VAE framework discussed in [50] is generally trained using feature-reconstruction process. A degree-oriented regularizer similar to the one present in DRNE [92] is used in RESD that helps in preserving important structural information. The aggregated objective function is given as:

$$\mathcal{L} = ||F - \hat{F}||_F^2 + \gamma_{deg}\mathcal{L}_{deg}$$

GraLSP [47]: In spite of various achievements, GNNs fail to identify some important structural patterns, which play a crucial role in various network applications. In [47], a GNN based framework is proposed that explicitly integrates local structural patterns with neighborhood aggregation using various anonymous random walks. Basically, GraLSP portrays local network structures using different anonymous random walks, which is a flexible and powerful tool for estimating the structural patterns. These random walks are then processed for feature aggregation in which different mechanisms to infer the influence of structural architecture features, such as amplification, receptive radius, are designed. In addition, they have also devised an objective function that captures structural similarities, which is further jointly optimized with individual node proximity objectives.

Empirically, For any random node v_i, it generates w random walks $W_i = \{w_{i1}, w_{i2}, ..., w_{iw}\}$ starting from node v_i having length l_w for capturing structural patterns, after this the walks W_i are anonymized [40] as $aw(w)$. The individual anonymous walk $aw(w)$ is used as elementary network embedding lookup-table as $u_{aw(w)}$. The process of aggregation of neighbourhood representation is performed as follows:

$$(H_{neighbour})_i^{(l)} = MEAN_{w \in W_i, j \in [1, \lfloor \frac{2l_w}{|w|} \rfloor] 1}(\alpha_{i,w}^{(l)}(a_{i,w}^{(l)} \odot H_{w_j}^{(l-1)}))$$
$$H_i^{(l)} = ReLU(W_{self}^{(l)} H_i^{(l-1)} + W_{neighbour}^{(l)}(H_{neighbour})_i^{(l)})$$

here $W_{neighbour}$ and W_{self} are trained parameter matrices. Furthermore, the learned attention parameter based on the local structure of individual node is given as:

$$\alpha_{i,w}^{(l)} = \frac{exp(SLP_{att}(u_{aw(w)}))}{\sum_{w' \in W_i} exp(SLP_{att}(u_{aw(w')}))}$$

where SLP(*) is the single-layer perceptron and $a_{i,w}^{(l)}$ represents amplification parameter which is computed as:

$$a_{i,w}^{(l)} = SLP_{amp}(u_a w(w))$$

From the view-point of preserving the important information of proximity among nodes, the authors have leveraged the loss function used in DeepWalk [68] as:

$$\mathcal{L}_{proximity} = -\sum_{v_i \in V} \sum_{v_j \in N_i} (log\sigma(H_i H_i^T) - \gamma_{neg} E_{v_k \sim P_n(v)}[log\sigma(H_i H_i^T)])$$

After L aggregation iterations, the network embed-dings are found in the state $H = H^L$. In a similar process to preserve structural resemblance (similarity) among nodes, GraLSP uses the loss function as given below:

$$\mathcal{L}_{structural} = -\sum_{v_i \in V, w_j, w_k, w_s \in W_i} log\sigma(u_j^T u_k - u_k^T u_s),$$

such that. $\hat{p}(w_j|v_i) > \hat{p}(w_j|G)$, $\hat{p}(w_k|v_i) > \hat{p}(w_k|G)$, $\hat{p}(w_n|v_i) < \hat{p}(w_n|G)$.

Here $\hat{p}(.)$ is elementary empirical distribution of various $aw(w)$ (anonymous walks):

$$\hat{p}(w_j|v_i) = \frac{\sum_{w \in W_i} \Pi(aw(w) = w_j)}{w}$$

$$\hat{p}(w_j|G) = \frac{\sum_{i=1}^n \hat{p}(w_j|v_i)}{n}$$

Finally, the combined loss function having trade-off parameter $\gamma_{structural}$ is given as:

$$\mathcal{L} = \mathcal{L}_{proximity} + \gamma_{structural} \mathcal{L}_{structural}$$

GCC [69] is a self-supervised pre-training framework based on GNNs that captures the topological structure and properties of universal network ranging across multiple networks. During its pre-training process, GCC discriminates sub-graph instances inside the network as well as across the networks leveraging on contrastive learning for empowering GNNs to learn transferable (among multiple networks) and intrinsic structural representations of the network. A pre-trained GNNs framework encodes the node-centric sub-graph of an individual node as H_i to discriminate similar sub-graphs from the ones which are dissimilar. The GCC framework achieves the power to capture local structural patterns when the scale of the pre-training data-set gets significantly large, and the resultant representations obtained thereafter can measure structural similarities with a higher accuracy. The other GNNs based proposed frameworks either train in supervise way or concentrate on proximity and community, such as presented in [105, 106]. Hence the authors in GCC have proposed a role-based network embedding/ classification approach.

Empirically, GCC captures its k-hop reachable neighborhood as $G_i{}^k$ and uses Graph Isomorphic Network (GIN) [97] encoder having convolution layer as:

$$H^{(l)} = MLP_{GIN}((A + (1 + \epsilon).\,I)H^{(l-1)})$$

Here initial attributes of GIN are set to eigenvectors of $G_i{}^k$ and ϵ is a fixed parameter. A slightly similar sub-graph instance of node v_i is induced according to the sequence of nodes found in the random walk initiated from node v_i. K dissimilar sub-graph instances are impelled in a slightly different manner initiated from other nodes that might be present in some other networks. The network representations of these $K + 1$ instances are derived as $\{x_0, x_1, ...x_K\}$ using GIN encoder. GCC uses a contrastive learning framework called InfoNCE [65] and the loss function adapted by them can be given as:

$$\mathcal{L} = \sum_{v_i \in V} -log \frac{exp(H_i x^+/\iota)}{\sum_{j=0}^{K} exp(H_i x_j/\iota)}$$

where ι is a given hyper-parameter.

DeepGL [77] learns edge and node features that can be further generalized over the entire network. Initially, a set of base features, for instance, graph-let features, are captured from the input graph, and a hierarchical graph representation having a multi-layered structure is learned automatically. Each successive layer of the multi-layer structure relies on the output of its previous layer to learn higher-order features. The DeepGL framework is designed in such a way that the learned relational functions, where each function represents a feature, are naturally generalized over the entire network. Therefore, it can be also be utilized in the form of transfer-learning tasks in graphs. Moreover, this approach also supports attributed/labeled graphs to learn inductive graph representations (interpretable) in a space-efficient manner.

The recent deep learning approaches for role-based node classification discussed above comprise modeling role-based methods used in traditional shallow approaches with DL techniques for mapping crucial structural architecture information to the non-linear network representations for node classification. The DRNE discussed earlier reconstructs node degrees after learning regular equivalence. GAS supervises the training of GCN using structural features of the network. RESD aggregates variants of ReFeX and DRNE using VAE framework. GraLSP uses the concept of anonymous walks for extracting structural similarities. GCC pre-trains the model with contrastive learning that encodes node-centric sub-graphs for node classification.

Finally, the DeepGL framework is designed in such a way that the learned relational functions, where each function represents a feature, are naturally generalized over the entire network.

5 Conclusion and Future Direction

This chapter brings forth the overview of the state-of-the-art deep learning based approaches for solving the challenging and interesting problem of *node-classification*. In recent years, deep learning based approaches in any domain have shown significant improvement in their performance. This trend also continued in the domain of the node-classification problem, which ended up in a significant increase in the count of publications in high-impact journals and conferences. The application of deep learning practices in the domain of node-classification resulted in significant improvement in efficiency, applicability, robustness, and effectiveness. The deep learning based new approaches are more scalable and flexible to use as compared to traditionally used practices for node classification. We have discussed the contribution of deep learning based approaches for solving the node classification problem. Furthermore, we have also summarized the wide range of applications of node classification in online social networks.

The techniques for solving node-classification problem needs further investigation. Hence, various node embed dings must employ structural community related information such that the resultant feature representation helps in the role as well as community based node classification. Until now, node classification approaches have only been proposed for single-layer networks; these approaches must be extended for multi-layer networks also. Moreover, the distinct approaches discussed in this chapter are primarily developed for static networks. However, to capture the actual behavior of the real-world systems, we must introduce new approaches that can grasp the natural dynamicity and continuous streaming behavior of time-evolving real-world networks. Furthermore, The currently existing methods for node classification are specifically designed for *un-signed networks*. However, the aspects of relationship among two nodes can be positive as well as negative, and the negative ties among distinct nodes require distinguished community learning techniques. Hence, new approaches must be introduced for solving the problem of node classification in *signed networks*.

Acknowledgements This work is supported by The Department of Science and Technology, Government of India, sponsored project having Grant no. 'DST-1401-CSE'.

References

1. Ahsan, M., Singh, T., Kumari, M.: Influential node detection in social network during community detection. In: 2015 International Conference on Cognitive Computing and Information Processing (CCIP), pp. 1–6. IEEE (2015)
2. Airoldi, E.M., Blei, D.M., Fienberg, S.E., Xing, E.P.: Mixed membership stochastic blockmodels. J. Mach. Learn. Res. (2008)
3. Akoglu, L., Tong, H., Koutra, D.: Graph based anomaly detection and description: a survey. Data Min. Knowl. Disc. 29(3), 626–688 (2015)
4. Amini, A.A., Chen, A., Bickel, P.J., Levina, E.: Pseudo-likelihood methods for community detection in large sparse networks. Ann. Stat. 41(4), 2097–2122 (2013)
5. Bahdanau, D., Cho, K., Bengio, Y.: Neural machine translation by jointly learning to align and translate (2014). arXiv:1409.0473
6. Bai, S., Fang, S., Li, L., Liu, R., Chen, X.: Enhancing link prediction by exploring community membership of nodes. Int. J. Mod. Phys. B 33(31), 1950382 (2019)
7. Balabanović, M., Shoham, Y.: Fab: content-based, collaborative recommendation. Commun. ACM 40(3), 66–72 (1997)
8. Bhatia, V., Rani, R.: Dfuzzy: a deep learning-based fuzzy clustering model for large graphs. Knowl. Inf. Syst. 57(1), 159–181 (2018)
9. Bhatia, V., Rani, R.: A distributed overlapping community detection model for large graphs using autoencoder. Futur. Gener. Comput. Syst. 94, 16–26 (2019)
10. Cai, H., Zheng, V.W., Chang, K.C.C.: A comprehensive survey of graph embedding: Problems, techniques, and applications. IEEE Trans. Knowl. Data Eng. 30(9), 1616–1637 (2018)
11. Cao, J., Jin, D., Dang, J.: Autoencoder based community detection with adaptive integration of network topology and node contents. In: International conference on knowledge science, engineering and management, pp. 184–196. Springer (2018)
12. Chen, J., Yuan, B.: Detecting functional modules in the yeast protein-protein interaction network. Bioinformatics 22(18), 2283–2290 (2006)
13. Chen, P., Redner, S.: Community structure of the physical review citation network. J. Informet. 4(3), 278–290 (2010)
14. Choong, J.J., Liu, X., Murata, T.: Learning community structure with variational autoencoder. In: 2018 IEEE International Conference on Data Mining (ICDM), pp. 69–78. IEEE (2018)
15. Choong, J.J., Liu, X., Murata, T.: Optimizing variational graph autoencoder for community detection. In: 2019 IEEE International Conference on Big Data (Big Data), pp. 5353–5358. IEEE (2019)
16. Domingos, P., Richardson, M.: Mining the network value of customers. In: Proceedings of the seventh ACM SIGKDD international conference on Knowledge discovery and data mining, pp. 57–66 (2001)
17. Dong, Y., Chawla, N.V., Swami, A.: metapath2vec: Scalable representation learning for heterogeneous networks. In: Proceedings of the 23rd ACM SIGKDD international conference on knowledge discovery and data mining, pp. 135–144 (2017)
18. Fan, H., Zhang, F., Li, Z.: Anomalydae: Dual autoencoder for anomaly detection on attributed networks. In: ICASSP 2020-2020 IEEE International Conference on Acoustics, Speech and Signal Processing (ICASSP), pp. 5685–5689. IEEE (2020)
19. Fang, H., Wu, F., Zhao, Z., Duan, X., Zhuang, Y., Ester, M.: Community-based question answering via heterogeneous social network learning. In: Proceedings of the AAAI Conference on Artificial Intelligence, vol. 30 (2016)
20. Fortunato, S.: Community detection in graphs. Phys. Rep. 486(3–5), 75–174 (2010)
21. Fu, W., Song, L., Xing, E.P.: Dynamic mixed membership blockmodel for evolving networks. In: Proceedings of the 26th Annual International Conference on Machine Learning, pp. 329–336 (2009)
22. Gao, H., Pei, J., Huang, H.: Conditional random field enhanced graph convolutional neural networks. In: Proceedings of the 25th ACM SIGKDD International Conference on Knowledge Discovery & Data Mining, pp. 276–284 (2019)

23. Gao, H., Pei, J., Huang, H.: Progan: Network embedding via proximity generative adversarial network. In: Proceedings of the 25th ACM SIGKDD International Conference on Knowledge Discovery & Data Mining, pp. 1308–1316 (2019)
24. Ghalmane, Z., El Hassouni, M., Cherifi, C., Cherifi, H.: Centrality in modular networks. EPJ Data Sci. **8**(1), 15 (2019)
25. Gilpin, S., Eliassi-Rad, T., Davidson, I.: Guided learning for role discovery (glrd) framework, algorithms, and applications. In: Proceedings of the 19th ACM SIGKDD international conference on Knowledge discovery and data mining, pp. 113–121 (2013)
26. Girvan, M., Newman, M.E.: Community structure in social and biological networks. Proc. Natl. Acad. Sci. **99**(12), 7821–7826 (2002)
27. Goodfellow, I., Pouget-Abadie, J., Mirza, M., Xu, B., Warde-Farley, D., Ozair, S., Courville, A., Bengio, Y.: Generative adversarial nets. Advances in Neural Information Processing Systems vol. (2014)
28. Grover, A., Leskovec, J.: node2vec: Scalable feature learning for networks. In: Proceedings of the 22nd ACM SIGKDD international conference on Knowledge discovery and data mining, pp. 855–864 (2016)
29. Guimera, R., Amaral, L.A.N.: Functional cartography of complex metabolic networks. Nature **433**(7028), 895–900 (2005)
30. Guo, X., Zhang, W., Wang, W., Yu, Y., Wang, Y., Jiao, P.: Role-oriented graph auto-encoder guided by structural information. In: International Conference on Database Systems for Advanced Applications, pp. 466–481. Springer (2020)
31. He, D., Zhai, L., Li, Z., Jin, D., Yang, L., Huang, Y., Philip, S.Y.: Adversarial mutual information learning for network embedding. In: IJCAI, pp. 3321–3327 (2020)
32. Henderson, K., Gallagher, B., Eliassi-Rad, T., Tong, H., Basu, S., Akoglu, L., Koutra, D., Faloutsos, C., Li, L.: Rolx: structural role extraction & mining in large graphs. In: Proceedings of the 18th ACM SIGKDD International Conference on Knowledge Discovery and Data Mining, pp. 1231–1239 (2012)
33. Henderson, K., Gallagher, B., Li, L., Akoglu, L., Eliassi-Rad, T., Tong, H., Faloutsos, C.: It's who you know: graph mining using recursive structural features. In: Proceedings of the 17th ACM SIGKDD International Conference on Knowledge Discovery and Data Mining, pp. 663–671 (2011)
34. Hinton, G., Deng, L., Yu, D., Dahl, G.E., Mohamed, A.r., Jaitly, N., Senior, A., Vanhoucke, V., Nguyen, P., Sainath, T.N., et al.: Deep neural networks for acoustic modeling in speech recognition: The shared views of four research groups. IEEE Signal Process. Mag. **29**(6), 82–97 (2012)
35. Hochreiter, S., Schmidhuber, J.: Long short-term memory. Neural Comput. **9**(8), 1735–1780 (1997)
36. Holland, P.W., Laskey, K.B., Leinhardt, S.: Stochastic blockmodels: first steps. Soc. Netw. **5**(2), 109–137 (1983)
37. Hong, H., Li, X., Wang, M.: Gane: a generative adversarial network embedding. IEEE Trans. Neural Netw. Learn. Syst. **31**(7), 2325–2335 (2019)
38. Hou, M., Ren, J., Zhang, D., Kong, X., Zhang, D., Xia, F.: Network embedding: taxonomies, frameworks and applications. Comput. Sci. Rev. **38**, 100296 (2020)
39. Hu, Y., Shi, L.: Visualizing large graphs. Wiley Interdiscip. Rev.: Comput. Stat. **7**(2), 115–136 (2015)
40. Ivanov, S., Burnaev, E.: Anonymous walk embeddings. In: International Conference on Machine Learning, pp. 2186–2195. PMLR (2018)
41. Jeon, H., Kim, T.: Community-adaptive link prediction. In: Proceedings of the 2017 International Conference on Data Mining, Communications and Information Technology, pp. 1–5 (2017)
42. Jia, Y., Zhang, Q., Zhang, W., Wang, X.: Communitygan: Community detection with generative adversarial nets. In: The World Wide Web Conference, pp. 784–794 (2019)
43. Jin, D., Ge, M., Li, Z., Lu, W., He, D., Fogelman-Soulie, F.: Using deep learning for community discovery in social networks. In: 2017 IEEE 29th International Conference on Tools with Artificial Intelligence (ICTAI), pp. 160–167. IEEE (2017)

44. Jin, D., Li, B., Jiao, P., He, D., Shan, H.: Community detection via joint graph convolutional network embedding in attribute network. In: International Conference on Artificial Neural Networks, pp. 594–606. Springer (2019)
45. Jin, D., Li, B., Jiao, P., He, D., Zhang, W.: Network-specific variational auto-encoder for embedding in attribute networks. In: IJCAI, pp. 2663–2669 (2019)
46. Jin, D., Liu, Z., Li, W., He, D., Zhang, W.: Graph convolutional networks meet markov random fields: Semi-supervised community detection in attribute networks. In: Proceedings of the AAAI Conference on Artificial Intelligence, vol. 33, pp. 152–159 (2019)
47. Jin, Y., Song, G., Shi, C.: Gralsp: Graph neural networks with local structural patterns. In: Proceedings of the AAAI Conference on Artificial Intelligence, vol. 34, pp. 4361–4368 (2020)
48. Karrer, B., Newman, M.E.: Stochastic blockmodels and community structure in networks. Phys. Rev. E **83**(1), 016107 (2011)
49. Kempe, D., Kleinberg, J., Tardos, É.: Maximizing the spread of influence through a social network. In: Proceedings of the ninth ACM SIGKDD International Conference on Knowledge Discovery and Data Mining, pp. 137–146 (2003)
50. Kingma, D.P., Welling, M.: Auto-encoding variational bayes (2013). arXiv:1312.6114
51. Kingma, D.P., Welling, M.: Auto-encoding variational bayes in 2nd international conference on learning representations. In: ICLR 2014-Conference Track Proceedings (2014)
52. Kipf, T.N., Welling, M.: Semi-supervised classification with graph convolutional networks (2016). arXiv:1609.02907
53. Kipf, T., Welling, M.: Graph convolutional matrix completion. In: ACM SIGKDD International Conference on Knowledge Discovery and Data Mining (2018)
54. Kowsari, K., Brown, D.E., Heidarysafa, M., Meimandi, K.J., Gerber, M.S., Barnes, L.E.: Hdltex: Hierarchical deep learning for text classification. In: 2017 16th IEEE International Conference on Machine Learning and Applications (ICMLA), pp. 364–371. IEEE (2017)
55. Krizhevsky, A., Sutskever, I., Hinton, G.E.: Imagenet classification with deep convolutional neural networks. Adv. Neural. Inf. Process. Syst. **25**, 1097–1105 (2012)
56. de Lange, S., de Reus, M., Van Den Heuvel, M.: The laplacian spectrum of neural networks. Front. Comput. Neurosci. **7**, 189 (2014)
57. LeCun, Y., Bengio, Y., Hinton, G.: Deep learning. Nature **521**(7553), 436–444 (2015)
58. Li, L., Fang, S., Bai, S., Xu, S., Cheng, J., Chen, X.: Effective link prediction based on community relationship strength. IEEE Access **7**, 43233–43248 (2019)
59. Li, X., Chen, H.: Recommendation as link prediction in bipartite graphs: a graph kernel-based machine learning approach. Decis. Support Syst. **54**(2), 880–890 (2013)
60. Martínez, V., Berzal, F., Cubero, J.C.: A survey of link prediction in complex networks. ACM Comput. Surv. (CSUR) **49**(4), 1–33 (2016)
61. Mikolov, T., Chen, K., Corrado, G., Dean, J.: Efficient estimation of word representations in vector space (2013). arXiv:1301.3781
62. Mikolov, T., Sutskever, I., Chen, K., Corrado, G.S., Dean, J.: Distributed representations of words and phrases and their compositionality. Advances in Neural Information Processing Systems, vol. 26 (2013)
63. Monti, F., Bronstein, M.M., Bresson, X.: Geometric matrix completion with recurrent multi-graph neural networks (2017). arXiv:1704.06803
64. Newman, M.: Networks. Oxford university Press (2018)
65. Oord, A.v.d., Li, Y., Vinyals, O.: Representation learning with contrastive predictive coding (2018). arXiv:1807.03748
66. Pan, S., Hu, R., Long, G., Jiang, J., Yao, L., Zhang, C.: Adversarially regularized graph autoencoder for graph embedding (2018). arXiv:1802.04407
67. Pan, S., Wu, J., Zhu, X., Zhang, C., Wang, Y.: Tri-party deep network representation. Network **11**(9), 12 (2016)
68. Perozzi, B., Al-Rfou, R., Skiena, S.: Deepwalk: Online learning of social representations. In: Proceedings of the 20th ACM SIGKDD International Conference on Knowledge Discovery and Data Mining, pp. 701–710 (2014)

69. Qiu, J., Chen, Q., Dong, Y., Zhang, J., Yang, H., Ding, M., Wang, K., Tang, J.: Gcc: Graph contrastive coding for graph neural network pre-training. In: Proceedings of the 26th ACM SIGKDD International Conference on Knowledge Discovery & Data Mining, pp. 1150–1160 (2020)
70. Qiu, J., Dong, Y., Ma, H., Li, J., Wang, C., Wang, K., Tang, J.: Netsmf: Large-scale network embedding as sparse matrix factorization. In: The World Wide Web Conference, pp. 1509–1520 (2019)
71. Qu, M., Bengio, Y., Tang, J.: Gmnn: Graph markov neural networks. In: International Conference on Machine Learning, pp. 5241–5250. PMLR (2019)
72. Rajeh, S., Savonnet, M., Leclercq, E., Cherifi, H.: Characterizing the interactions between classical and community-aware centrality measures in complex networks. Sci. Rep. **11**(1), 1–15 (2021)
73. Ranshous, S., Shen, S., Koutra, D., Harenberg, S., Faloutsos, C., Samatova, N.F.: Anomaly detection in dynamic networks: a survey. Wiley Interdiscip. Rev.: Comput. Stat. **7**(3), 223–247 (2015)
74. Ravasz, E., Somera, A.L., Mongru, D.A., Oltvai, Z.N., Barabási, A.L.: Hierarchical organization of modularity in metabolic networks. Science **297**(5586), 1551–1555 (2002)
75. Rossi, R.A., Ahmed, N.K.: Role discovery in networks. IEEE Trans. Knowl. Data Eng. **27**(4), 1112–1131 (2014)
76. Rossi, R.A., Jin, D., Kim, S., Ahmed, N.K., Koutra, D., Lee, J.B.: On proximity and structural role-based embeddings in networks: misconceptions, techniques, and applications. ACM Trans. Knowl. Discov. from Data (TKDD) **14**(5), 1–37 (2020)
77. Rossi, R.A., Zhou, R., Ahmed, N.K.: Deep inductive graph representation learning. IEEE Trans. Knowl. Data Eng. **32**(3), 438–452 (2018)
78. Rosvall, M., Bergstrom, C.T.: An information-theoretic framework for resolving community structure in complex networks. Proc. Natl. Acad. Sci. **104**(18), 7327–7331 (2007)
79. Sarkar, A., Mehta, N., Rai, P.: Graph representation learning via ladder gamma variational autoencoders. In: Proceedings of the AAAI Conference on Artificial Intelligence, vol. 34, pp. 5604–5611 (2020)
80. Saxena, A., Fletcher, G., Pechenizkiy, M.: Hm-eiict: Fairness-aware link prediction in complex networks using community information. J. Comb. Optim. 1–18 (2021)
81. Saxena, A., Fletcher, G., Pechenizkiy, M.: Nodesim: node similarity based network embedding for diverse link prediction. EPJ Data Sci. **11**(1), 24 (2022)
82. Saxena, A., Iyengar, S.: Centrality measures in complex networks: A survey (2020). arXiv:2011.07190
83. Scripps, J., Tan, P.N., Esfahanian, A.H.: Node roles and community structure in networks. In: Proceedings of the 9th WebKDD and 1st SNA-KDD 2007 Workshop on Web mining and Social Network Analysis, pp. 26–35 (2007)
84. Shah, N., Koutra, D., Jin, L., Zou, T., Gallagher, B., Faloutsos, C.: On summarizing large-scale dynamic graphs. IEEE Data Eng. Bull. **40**(3), 75–88 (2017)
85. Solomon, R.S., Srinivas, P., Das, A., Gamback, B., Chakraborty, T.: Understanding the psycho-sociological facets of homophily in social network communities. IEEE Comput. Intell. Mag. **14**(2), 28–40 (2019)
86. Sporns, O., Betzel, R.F.: Modular brain networks. Annu. Rev. Psychol. **67**, 613–640 (2016)
87. Strogatz, S.H.: Exploring complex networks. Nature **410**(6825), 268–276 (2001)
88. Sun, H., He, F., Huang, J., Sun, Y., Li, Y., Wang, C., He, L., Sun, Z., Jia, X.: Network embedding for community detection in attributed networks. ACM Trans. Knowl. Discov. from Data (TKDD) **14**(3), 1–25 (2020)
89. Sun, Y., Wang, S., Hsieh, T.Y., Tang, X., Honavar, V.: Megan: A generative adversarial network for multi-view network embedding (2019). arXiv:1909.01084
90. Tao, Z., Liu, H., Li, J., Wang, Z., Fu, Y.: Adversarial graph embedding for ensemble clustering. In: International Joint Conferences on Artificial Intelligence Organization (2019)
91. Tian, F., Gao, B., Cui, Q., Chen, E., Liu, T.Y.: Learning deep representations for graph clustering. In: Proceedings of the AAAI Conference on Artificial Intelligence, vol. 28 (2014)

92. Tu, K., Cui, P., Wang, X., Yu, P.S., Zhu, W.: Deep recursive network embedding with regular equivalence. In: Proceedings of the 24th ACM SIGKDD International Conference on Knowledge Discovery & Data Mining, pp. 2357–2366 (2018)
93. Wang, C., Pan, S., Long, G., Zhu, X., Jiang, J.: Mgae: Marginalized graph autoencoder for graph clustering. In: Proceedings of the 2017 ACM on Conference on Information and Knowledge Management, pp. 889–898 (2017)
94. Wu, Z., Pan, S., Chen, F., Long, G., Zhang, C., Philip, S.Y.: A comprehensive survey on graph neural networks. IEEE Trans. Neural Netw. Learn. Syst. **32**(1), 4–24 (2020)
95. Xie, J., Szymanski, B.K.: Towards linear time overlapping community detection in social networks. In: Pacific-Asia Conference on Knowledge Discovery and Data Mining, pp. 25–36. Springer (2012)
96. Xie, Y., Wang, X., Jiang, D., Xu, R.: High-performance community detection in social networks using a deep transitive autoencoder. Inf. Sci. **493**, 75–90 (2019)
97. Xu, K., Hu, W., Leskovec, J., Jegelka, S.: How powerful are graph neural networks? (2018) arXiv:1810.00826
98. Xu, R., Che, Y., Wang, X., Hu, J., Xie, Y.: Stacked autoencoder-based community detection method via an ensemble clustering framework. Inf. Sci. **526**, 151–165 (2020)
99. Yang, C., Liu, M., Wang, Z., Liu, L., Han, J.: Graph clustering with dynamic embedding (2017). arXiv:1712.08249
100. Yang, J., McAuley, J., Leskovec, J.: Community detection in networks with node attributes. In: 2013 IEEE 13th International Conference on Data Mining, pp. 1151–1156. IEEE (2013)
101. Yang, L., Cao, X., He, D., Wang, C., Wang, X., Zhang, W.: Modularity based community detection with deep learning. In: IJCAI, vol. 16, pp. 2252–2258 (2016)
102. Yang, L., Wang, Y., Gu, J., Wang, C., Cao, X., Guo, Y.: Jane: Jointly adversarial network embedding. In: IJCAI, pp. 1381–1387 (2020)
103. Yang, Z., Cohen, W., Salakhudinov, R.: Revisiting semi-supervised learning with graph embeddings. In: International Conference on Machine Learning, pp. 40–48. PMLR (2016)
104. Ying, R., He, R., Chen, K., Eksombatchai, P., Hamilton, W.L., Leskovec, J.: Graph convolutional neural networks for web-scale recommender systems. In: Proceedings of the 24th ACM SIGKDD International Conference on Knowledge Discovery & Data Mining, pp. 974–983 (2018)
105. You, J., Gomes-Selman, J., Ying, R., Leskovec, J.: Identity-aware graph neural networks (2021). arXiv:2101.10320
106. You, J., Ying, R., Leskovec, J.: Position-aware graph neural networks. In: International Conference on Machine Learning, pp. 7134–7143. PMLR (2019)
107. Zachary, W.W.: An information flow model for conflict and fission in small groups. J. Anthropol. Res. **33**(4), 452–473 (1977)
108. Zhang, B., Yu, Z., Zhang, W.: Community-centric graph convolutional network for unsupervised community detection. In: IJCAI, pp. 551–556 (2020)
109. Zhang, W., Guo, X., Wang, W., Tian, Q., Pan, L., Jiao, P.: Role-based network embedding via structural features reconstruction with degree-regularized constraint. Knowl.-Based Syst. **218**, 106872 (2021)
110. Zhang, Y., Xiong, Y., Ye, Y., Liu, T., Wang, W., Zhu, Y., Yu, P.S.: Seal: Learning heuristics for community detection with generative adversarial networks. In: Proceedings of the 26th ACM SIGKDD International Conference on Knowledge Discovery & Data Mining, pp. 1103–1113 (2020)
111. Zhang, Z., Cui, P., Zhu, W.: Deep learning on graphs: a survey. IEEE Trans. Knowl. Data Eng. (2020)
112. Zheng, Y., Chen, S., Zhang, X., Wang, D.: Heterogeneous graph convolutional networks for temporal community detection (2019). arXiv: 1909.10248

NN-LP-CF: Neural Network Based Link Prediction on Social Networks Using Centrality-Based Features

Shashank Sheshar Singh, Divya Srivastva, Ajay Kumar, and Vishal Srivastava

Abstract Link prediction has drawn significant attention from researchers in recent years due to the rapid growth of social networks. The link prediction problem identifies the missing and future links in online social networks. Most traditional methods focus on the node and edge centrality measures to predict missing links using local and global features. These methods do not consider the advantage of both local and global features. However, some quasi-local metrics have been proposed to overcome the disadvantages of local and global centrality measures. This paper uses local and global topological features to generate likelihood feature scores for the proposed neural network model NN-LP-CF. The proposed model generates a likelihood score to predict missing links. The experimental results demonstrate the superiority of NN-LP-CF over traditional centrality measures over different performance metrics.

Keywords Link prediction · Neural networks · Centrality measures · Social networks

1 Introduction

Online social networks have immense application potential in the field of sociology and computer science. These social media platform not only connects people but is used for information sharing and user interaction. The link prediction problem has drawn a lot of attention from researchers in recent years. Link prediction is

S. S. Singh (✉)
Department of Computer Science and Engineering, Thapar Institute of Engineering & Technology, Patiala, India
e-mail: shashank.sheshar@gmail.com

D. Srivastva · A. Kumar · V. Srivastava
Department of Computer Science and Engineering, Bennett University, Greater Noida, India
e-mail: divya.srivastava@bennett.edu.in

A. Kumar
e-mail: ajay.kumar@bennett.edu.in

V. Srivastava
e-mail: vishal.srivastava@bennett.edu.in

© The Author(s), under exclusive license to Springer Nature Switzerland AG 2022
T.-P. Hong et al. (eds.), *Deep Learning for Social Media Data Analytics*,
Studies in Big Data 113, https://doi.org/10.1007/978-3-031-10869-3_2

the problem of identifying missing links and future links from the network [1]. In addition to predicting connections, link prediction can be useful to predict the growth of a social network. Link prediction problems have application potential for other social network analysis problems (SNA), such as community detection [2], influence maximization [3], network evolution [4], and opinion formation [5].

Social networks are social constructs that are used to capture several relationships (like friendship, collaboration, common interest, etc.) among different social entities. Examples of social networks include Facebook, Twitter, Instagram, etc. With the surge of Internet application, these networks are growing continuously at a very high pace i.e. everyday, millions of users are joining these networks and hence adding even more links among others in the networks. Evolution of these networks show addition or deletion of nodes and edges in the networks. Link prediction talks about links evolution in these networks. The social networks are represented by graphs during implementation where a node (or vertex) represents a social entity and an edge (or link) shows the relationships between two nodes (social entities). A graph (or network), at any instant, can have two types of links: Existing links and Non-existing links. Link prediction answers the question that what is the probability that a non-existing link becomes existing link at a given instant of time. In other words, Link prediction finds missing/future links in a given graph. Formally, Novell and Klienberg [6] defined link prediction as follows. Given snapshots of graphs at different time instants $G(t_0), G(t_1),...G(t_{n-1}), G(t_n)$, LP finds the links that may appear in the graph at next instant of time $G(t_{n+1})$. Link prediction has several application, like friend recommendations in Facebook network, product recommendation in e-commerce websites, protein-protein interaction in biological networks, connectivity prediction in transport networks, identifying hidden relationship in terrorist networks and a lot more.

This paper presents a neural network-based link prediction model using centrality-based features as shown in Fig. 1. The NN-LP-CF incorporates both local and global topological information to predict missing links. The main contribution of the paper is given as follows:

- The centrality measures have been utilized to produce a feature set for the neural network model.
- The NN-LP-CF model takes these features and generates a likelihood score for the prediction model to predict missing links.
- The experiments are performed on real-world networks and compared with the state-of-the-art methods to analyze the performance over different prediction matrices. The experimental results show the superiority of NN-LP-CF model.

The remainder of the paper is organised as follows. Section 2 discusses the background such as social network, social graph, and centrality measures. Section 3 is devoted for discussion link prediction algorithms and proposed neural network model. Section 4 presents the experimental analysis and comparison with state-of-the-art algorithms. Finally, Sect. 5 concludes the finding of proposed work with future directions.

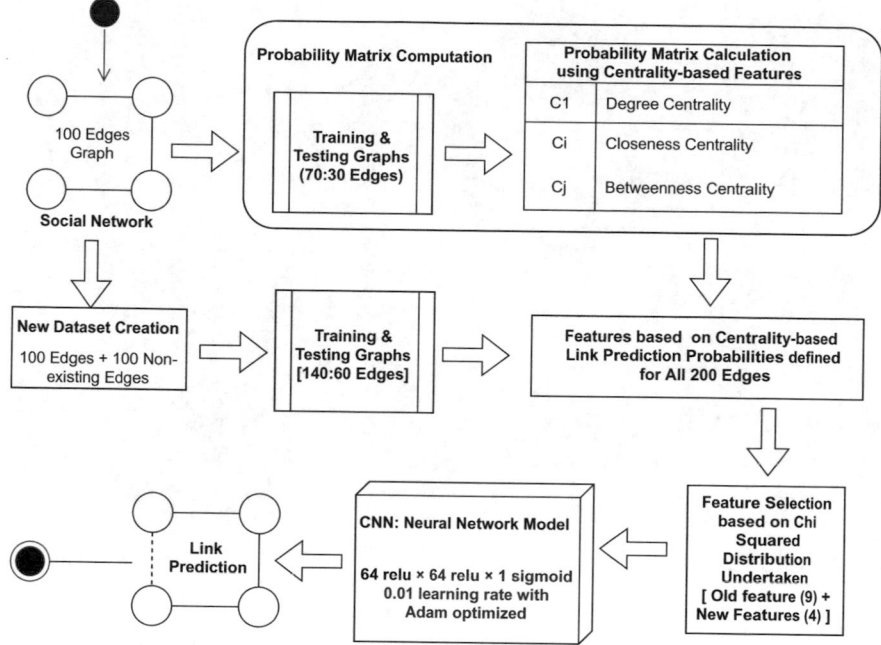

Fig. 1 The NN-LP-CF link prediction model

2 Preliminaries

2.1 Social Network

Aristotle once said "man is by nature a social animal". Society generally has a great impact on individual's life rather it precedes the individual. Being social is an important aspect of an individual, it may be in-person or over a network. Being connected in-person is simple but what if it is over a network? or say a "social network?". A social network is a social structure that consists of a group of social members, a set of relationships, and social interaction among those members. People's interactions with one another are shown as networks of individuals and groups. Friendships and enmities are developed at the individual level, students play together, studies together, coworkers form study groups, research groups, common interest groups and organisations collaborate on initiatives together. Even the nation's states wage wars against other nations. These all represent a social network. Some of the common social network platform are shown in Fig. 2. It shows the famous social networking platform like; Facebook, Skype, Twitter, Instagram, YouTube, etc. where different people access various platform.

A social network [7] is made up of a group of nodes (also known as vertices or actors) that are linked together by particular relationships (also known as arcs,

Fig. 2 Social network

connections, ties, or edges). The nodes represent actors, such as; individuals, organisations, groups, teams, or communities, and the edges among them reflect social interactions such as receiving/sending text messages, multimedia messages and etc.. The most prevalent points are people who frequent social networking sites like Facebook. The links in this case reflect likes or friends. Social networks allow for the analysis of social network theories, as well as the examination of network dynamics, global and local patterns, and the identification of influential entities.

2.2 Social Graph

Social Graph is a structure reflects social relationship among different entities and provides a worldwide mapping of people's interrelationships. Almost every internet users use this word to represent the relationship among them. Facebook had the largest social network dataset in the world as of 2010. It continues to be the most popular and has the most users of all websites. The other popular websites that are using social graphs are, Instagram, YouTube, Blogs sites, Linked-in, Commercial websites and etc. have a significant edge over other service providers and own their own whole social graph. An undirected graph $G(V, E)$ can be used to describe a social network, where V denotes the collection of vertices, which in a social network means people, enterprises, organisations, groups, and so on, and E defines the set of edges. The edges show the connections, relationships, and flows that exist between the nodes.

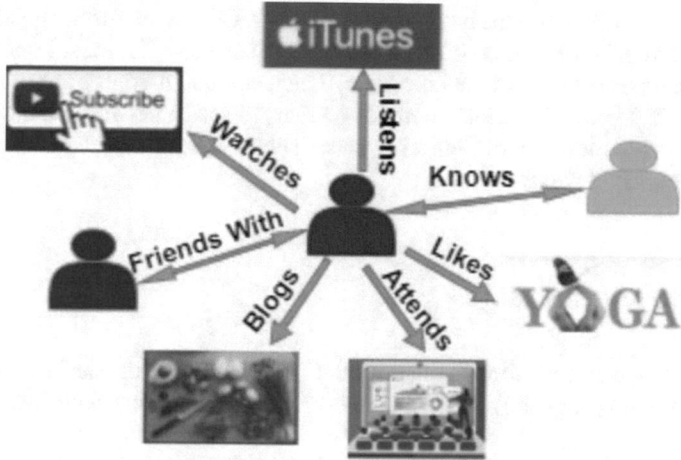

Fig. 3 Social graph

An instance of social graph is shown in Fig. 3. It shows the different types of relationship (one way, two way) between an individual and different online platforms. The figure shows that, the individual in center has two way relationship for different people. The relationship between one is that he/she knows person in green while in the other, he/she is a friend with person in blue. For one way relationship, the person in center likes to listens music on iTunes, writes/reads blogs, subscriber of YouTube, attends conferences and likes yoga. Thus, the links shows that how an individual is related with other services over the web. All the activities over the network is analyzed for better understanding of network and is termed as "social network analysis" (SNA).

2.3 Social Network Analysis

Social Network Analysis [8], is the process of analyzing the various activities over the social network. It is a social structure made up of a limited number of social actors, such as people or organisations, who are linked by interpersonal interactions. These ties, also known as bonds, can be personal or professional in character, and can range from casual friendships, acquaintances, or coworkers to intimate familial ties. Aside from relationships, social networks frequently indicate the flow of information, interactions, and commonalities among a group of people.

Kinship structure [9], relationships among members of deviant groups, corporate dominance, social mobility [10], class structure [11], science citations [12], international trade exploitation [13], and many other topics have all been studied using social network analysis. This analysis helps in analysing the clients better. It helps in

categorizing the clients on the basis of their interests. Clients are further given sugges-
tions/recommendations based on their respective interests. The most important part
of social analysis is to detect the fake news, illegal content, inappropriate content etc.
over the web. There are various methods to analyze social network based on nodes
and links [14]. Among them Centrality is one such measure used for understanding
the behaviour of network.

2.4 Centrality Measures

It is a set of metrics that aims to quantify the "importance", "impact" or "influence"
of a certain node (or group) inside a network. There are various centrality measures
as follows:

2.4.1 Degree Centrality

Degree centrality appoints a score to every hub dependent on the measure of connections it holds. It educates us about the number regarding immediate, one-jump
interfaces every hub needs to other organization hubs. It's utilized to find profoundly
associated individuals, notable individuals, individuals who are probably going to
have the most information, or individuals who can undoubtedly connect with the
remainder of the organization. The most fundamental measurement of hub con-
nectedness is degree centrality. When taking a gander at transnational information
or record action, it's occasionally advantageous to check out in-degree (number of
approaching connections) and out-degree (number of outward connections) as discrete estimations. Mathematically, the degree centrality of a node u can be defined
as the degree of that node.

$$C_{deg}(u) = D(u) \tag{1}$$

where, $C_{deg}(u)$ is the degree centrality of the node u and $D(u)$ is the degree of the
node. For comparison, it can be normalized by the maximum degree of a node as

$$C_{deg}(u) = \frac{D(u)}{|V| - 1} \tag{2}$$

where, $|V|$ is the total number of nodes of the graph.

2.4.2 Betweenness Centrality

The measure of times a hub is on the briefest way between different hubs is estimated by betweenness centrality. This measurement distinguishes which hubs in an
organization go about as "spans" associating different hubs. This is refined by first

recognizing the most brief ways as a whole and afterward counting how often every hub falls on one of them. It's utilized to find the individuals who have something to do with how a framework functions. Betweenness can be useful in assessing correspondence elements, yet it ought to be utilized with alert. A high betweenness count could propose that somebody is accountable for two distinct groups in an organization, or that they are on the edges of the two bunches. The betweenness centrality of a node u can be computed mathematically as

$$C_{between}(u) = \sum_{s!=u!=t} \frac{GD_u(s, t)}{GD(s, t)} \tag{3}$$

where, $GD_u(s, t)$ is the total number of geodesic paths between the nodes s and t containing the node u and $GD(s, t)$ is the total possible geodesic paths between s and t.

2.4.3 Closeness Centrality

Closeness centrality allocates a score to every hub dependent on the fact that they are so near the remainder of the organization's hubs. This measurement tracks down the most brief pathways associating all hubs and afterward grants a score to every hub dependent on the complete of those ways. It's used to find individuals who are most appropriate to quick influence the whole organization. Closeness centrality can help with the revelation of good 'telecasters,' but in a thickly associated network, all hubs will ordinarily have equivalent scores. Closeness could be more helpful for discovering powerhouses in a solitary bunch. The closeness centrality of a given node u can be computed by the following equation

$$C_{close}(u) = \sum_{v \in V} \frac{1}{GD(u, v)} \tag{4}$$

where, $C_{close}(u)$ is the closeness centrality of node u and $GD(u, v)$ is geodesic distance between u and v. The normalized closeness centrality is

$$C_{close(u)} = \sum_{v \in V} \frac{|V| - 1}{GD(u, v)} \tag{5}$$

2.4.4 EigenCentrality

EigenCentrality surveys a hub's impact dependent on the quantity of linkages it needs to different hubs in the organization, like degree centrality. EigenCentrality then, at that point, makes it a stride further by thinking about how profoundly associated a hub is, just as the number of connections their associations have, etc. EigenCentrality can find hubs with sway over the whole organization, in addition to those quickly associated with it, by figuring a hub's lengthy associations. EigenCentrality is a decent 'all-around' SNA score, helpful for understanding human informal organizations just as malware spread organizations. For a given graph, the relative centrality measure x_u of the node u can be calculated as

$$x_u = \frac{1}{\lambda} \sum_{v \in N(u)} x_v = \frac{1}{\lambda} \sum_{v \in G} a_{u,v} x_v \qquad (6)$$

where, $N(u)$ is the neighbors of the node u and λ is the largest eigen value of the adjacency matrix $A = a_{u,v}$ with $a_{u,v} = 1$ when an edge exists between the nodes u and v, and 0, otherwise.

2.4.5 PageRank

PageRank is a variety of EigenCentrality that allocates a score to hubs dependent on their associations and the associations of their associations. PageRank contrasts in that it furthermore considers connect bearing and weight, suggesting that connections can just pass on impact in one manner and in changing sums. This measurement distinguishes hubs with an organization sway that stretches out past their immediate associations. PageRank is valuable for getting references and authority since it considers bearing and connection weight. More data: PageRank is a notable positioning calculation that supports the first Google web crawler. The PageRank centrality of node u is equal to the leading eigenvector x_u of matrix A^{col} (The leading eigenvalue is 1):

$$x = A^{col} x \qquad (7)$$

where, $A^{col} = A D^{-1}$ with D as diagonal matrix and A as the adjacency matrix.

3 Link Prediction

Lots of approaches to link prediction in social networks are available in the literature, in which some of seminal review works are given in [15–17]. Earlier Newman [18] posed the problem of link prediction where collaboration networks in Physics and Biology are considered. In the article, nodes represent the authors in the network and a link between two authors represents the co-authorship relation i.e. whenever, any two authors work on a paper, there exists a link between them. Further, Liben-Nowell and Kleinberg [6] simulate the link prediction in social networks. They simulate it by mapping a social entity or a person to a node and interaction between two entities to a link between two nodes. This was the first paper that gave the formal definition of link prediction. They computed similarity score between each pair of nodes by using topological structure of network. Hasan et al. [19] explored more about link prediction and applied supervised learning on it. For this, they simulated link prediction as binary classification problem where scores calculated by several structural similarity methods employed as features to the machine and different supervised learning algorithms applied to make predictions. Experimental results showed better accuracy compared to the unsupervised methods or heuristics. Later on, they explored the link prediction problem comprehensively and came up with a seminal survey paper [15]. Social network evolution is somehow related to link prediction because the evolution process of networks consists of node as well as link evolution's. The research on social network evolution nearly takes after the task of link prediction where several seminal studies have been proposed like Barabasi and Albert work [20] on random network published in Science, Kleinberg work [21] in Nature Communication, [22] in KDD, and [6] in EPL. In recent years, several different types of approaches to link prediction have been proposed that include deep learning methods [23–25], embedding-based methods [26–28], probabilistic and maximum likelihood approaches [29–31], and so on.

Recently, graph neural networks (GNNs) have attracted researchers attention, involved in social network analysis, especially link prediction. Recent works [32] have proved GNNs as a powerful tool for learning graph structural and topological features. Employing this learning in GNN frameworks have shown superior performance in link prediction task [28, 33]. Zhang et al. [34] classified GNN-based link prediction methods in two main groups. First, Graph Autoencoder and its variational version where GNN is applied on complete network to learn each node represen tation. Then the link prediction task is performed by aggregating the node level representation to edge level representation. Second, Zhang et al. [35] introduced SEAL framework for link prediction which systematically transforms link prediction problem into subgraph classification problem. This framework extracts h-hops enclosing subgraphs around a given target link and construct an attribute matrix consisting of structural node labels, latent embeddings, and explicit attributes of nodes for the subgraph. Then the subgraph information with its attribute matrix collectively feed into GNN framework to perform link classification.

In this work, we have present a neural network-based model NN-LP-CF for predicting missing links on social networks as shown in Fig. 1. First, the proposed method estimates the probability matrix for all link prediction indexes based on different centrality's calculated using a training graph. Then, a new graph is produced by added an equal number of non-existing edges to existing edges to compute features based on centrality-based link prediction probabilities for all edges of the new graph. Next, features have been selected based on the Chi-Squared distribution undertaken by considering old and new features. Then training data fed into a 64 Relu × 64 Relu × 1 Sigmoid neural network model for training of the model. Finally, testing data have been used to predict the probabilities of missing links.

4 Experimental Discussion

We have performed experiments on six real-world datasets Karate, Dolphin, Adjnoun, Football, Polbooks, and Macaque, to analyze the performance of the NN-LP-CF against the compared methods. We have used nine centrality-based link prediction methods to compare with the performance of the proposed model. Four metrics, AUPR, AUC, Average Precision, and F1-Score [4] have been utilized to analyze the performance. All these experiments were performed on a 64-bit Linux Mint 19.3 PC with Intel(R) Core(TM) i7-3632QM CPU @ 2.20 GHz processor and 16 GB memory. The experimental results are shown in Tables 1, 2, 3, and 4.

- **AUPR.** The comparison of NN-LP-CF against the centrality-based link prediction methods regarding AUPR metric is shown in Table 1. The proposed method outperforms each centrality method except edge betweenness on the Karate dataset. NN-LP-CF is the best performing method under AUPR on all the remaining datasets Dohphin, Adjnoun, Football, Polbooks, and Macaque datasets.
- **Average Precision.** The comparison of NN-LP-CF against the centrality-based link prediction methods regarding average precision metric is shown in Table 2. The proposed method is the second-best performing method edge betweenness on the Karate dataset. NN-LP-CF outperforms all the compared methods under average precision on all the remaining datasets Dohphin, Adjnoun, Football, Polbooks, and Macaque datasets.
- **AUC.** The comparison of NN-LP-CF against the centrality-based link prediction methods regarding AUC metric is shown in Table 3. The proposed method is the least performing method on the Karate dataset because of less clustering coefficient. NN-LP-CF outperforms all the compared methods except harmonic centrality measure on the Dolphin dataset. The proposed method outperforms compared methods except for harmonic and closeness centrality measures on the Adjnoun dataset. Similarly, NN-LP-CF is the best performing method for the remaining datasets Football, Polbooks, and Macaque.
- **F1-Score.** The comparison of NN-LP-CF against the centrality-based link prediction methods regarding F1-score metric is shown in Table 4. The proposed method is the second-best performing method on the Dolphin dataset after degree centrality. NN-LP-CF outperforms all the compared methods on the Karate and Adjnoun datasets. The proposed method is comparable with the state-of-the-art methods on the remaining datasets Football, Polbooks, and Macaque.

Table 1 The AUPR Comparison against the State-of-the-art Methods

Dataset	Ratio	Degree	Eigen	Katz	Closeness	Subgraph	Harmonic	PageRank	Betweenness	Load	NN-LP-CF
Karate	0.1	0.0539	0.0626	0.0642	0.0447	0.0484	0.0471	0.0579	0.0722	0.0518	0.0702
	0.2	0.0821	0.0911	0.0893	0.0781	0.0751	0.0682	0.0844	0.0889	0.0712	0.0936
	0.3	0.1037	0.1097	0.1163	0.1107	0.1012	0.0846	0.1395	0.1283	0.0881	0.1936
	0.4	0.1193	0.1227	0.1298	0.1100	0.1148	0.1076	0.1245	0.1322	0.1120	0.2492
	0.5	0.1320	0.1262	0.1192	0.1076	0.1209	0.1222	0.1318	0.1265	0.1221	0.3381
Dolphin	0.1	0.0446	0.0412	0.0422	0.0379	0.0369	0.0371	0.0351	0.0270	0.0304	0.0623
	0.2	0.0634	0.0649	0.0727	0.0629	0.0683	0.0616	0.0631	0.0533	0.0579	0.0777
	0.3	0.0857	0.0843	0.0944	0.0841	0.0816	0.0775	0.0825	0.0839	0.0789	0.1004
	0.4	0.1008	0.0990	0.0987	0.0980	0.1028	0.0962	0.0945	0.1030	0.0934	0.1876
	0.5	0.1164	0.1026	0.1179	0.1145	0.1125	0.1046	0.1160	0.1119	0.1070	0.1997
Adjnoun	0.1	0.0248	0.0239	0.0275	0.0253	0.0260	0.0221	0.0177	0.0196	0.0239	0.0298
	0.2	0.0420	0.0415	0.0411	0.0403	0.0426	0.0385	0.0338	0.0358	0.0419	0.0450
	0.3	0.0557	0.0546	0.0544	0.0536	0.0596	0.0489	0.0487	0.0498	0.0563	0.0711
	0.4	0.0647	0.0669	0.0633	0.0625	0.0672	0.0639	0.0609	0.0646	0.0663	0.0697
	0.5	0.0713	0.0731	0.0762	0.0728	0.0775	0.0689	0.0742	0.0709	0.0729	0.1735
Football	0.1	0.0574	0.0630	0.0595	0.0605	0.0886	0.0921	0.0455	0.0580	0.0647	0.1739
	0.2	0.1099	0.1046	0.1085	0.1102	0.1355	0.1440	0.0942	0.1043	0.1072	0.2398
	0.3	0.1495	0.1526	0.1480	0.1546	0.1651	0.1694	0.1380	0.1413	0.1264	0.2687
	0.4	0.1807	0.1864	0.1817	0.1775	0.1752	0.1810	0.1800	0.1744	0.1481	0.2660
	0.5	0.1914	0.2028	0.2028	0.2075	0.1819	0.1892	0.1932	0.1976	0.1628	0.2598
Polbooks	0.1	0.0519	0.0523	0.0517	0.0587	0.0743	0.0771	0.0421	0.0477	0.0695	0.1321
	0.2	0.0989	0.0933	0.0974	0.1079	0.1235	0.1251	0.0852	0.0896	0.1248	0.2088
	0.3	0.1333	0.1264	0.1362	0.1459	0.1559	0.1639	0.1247	0.1211	0.1552	0.2271
	0.4	0.1565	0.1587	0.1655	0.1674	0.1700	0.1711	0.1476	0.1549	0.1697	0.2501
	0.5	0.1763	0.1737	0.1730	0.1830	0.1798	0.1841	0.1685	0.1629	0.1764	0.2326
Macaque	0.1	0.2299	0.1265	0.1855	0.1533	0.2767	0.2819	0.0861	0.0739	0.2487	0.3709
	0.2	0.3560	0.2265	0.3037	0.2526	0.4029	0.4041	0.1550	0.1353	0.3812	0.5047
	0.3	0.4282	0.3000	0.3897	0.3333	0.4754	0.4762	0.2153	0.1938	0.4706	0.5457
	0.4	0.4798	0.3744	0.4367	0.4180	0.5227	0.5192	0.2588	0.2491	0.5189	0.5760
	0.5	0.5058	0.4400	0.4828	0.4605	0.5456	0.5289	0.3050	0.2845	0.5516	0.5756

Table 2 The Avg Precision Comparison against the State-of-the-art Methods

Dataset	Ratio	Degree	Eigen	Katz	Closeness	Subgraph	Harmonic	PageRank	Betweenness	Load	NN-LP-CF
Karate	0.1	0.0700	0.0805	0.0837	0.0593	0.0629	0.0612	0.0745	0.0892	0.0688	0.0844
	0.2	0.0930	0.1023	0.1014	0.0883	0.0844	0.0767	0.0965	0.1022	0.0814	0.0759
	0.3	0.1098	0.1164	0.1234	0.1164	0.1059	0.0883	0.1451	0.1338	0.0927	0.0822
	0.4	0.1187	0.1213	0.1284	0.1087	0.1109	0.1049	0.1241	0.1319	0.1099	0.0872
	0.5	0.1241	0.1177	0.1121	0.1039	0.1129	0.1110	0.1253	0.1211	0.1149	0.0854
Dolphin	0.1	0.0488	0.0439	0.0468	0.0430	0.0391	0.0391	0.0363	0.0286	0.0317	0.0684
	0.2	0.0614	0.0633	0.0711	0.0617	0.0656	0.0588	0.0613	0.0510	0.0549	0.0656
	0.3	0.0764	0.0749	0.0857	0.0739	0.0708	0.0671	0.0732	0.0748	0.0679	0.0689
	0.4	0.0814	0.0791	0.0801	0.0781	0.0816	0.0759	0.0761	0.0826	0.0725	0.0521
	0.5	0.0843	0.0725	0.0855	0.0821	0.0791	0.0731	0.0839	0.0789	0.0752	0.0561
Adjnoun	0.1	0.0270	0.0267	0.0308	0.0279	0.0285	0.0235	0.0190	0.0216	0.0258	0.0327
	0.2	0.0429	0.0418	0.0420	0.0405	0.0433	0.0382	0.0340	0.0360	0.0422	0.0459
	0.3	0.0536	0.0527	0.0527	0.0511	0.0573	0.0463	0.0467	0.0476	0.0539	0.0549
	0.4	0.0586	0.0612	0.0576	0.0561	0.0612	0.0573	0.0553	0.0585	0.0601	0.0619
	0.5	0.0608	0.0622	0.0644	0.0615	0.0656	0.0582	0.0628	0.0598	0.0611	0.0465
Football	0.1	0.0597	0.0653	0.0620	0.0628	0.0897	0.0933	0.0468	0.0596	0.0654	0.1787
	0.2	0.1096	0.1037	0.1082	0.1094	0.1339	0.1423	0.0934	0.1037	0.1054	0.2409
	0.3	0.1445	0.1475	0.1429	0.1494	0.1592	0.1633	0.1327	0.1362	0.1201	0.2637
	0.4	0.1672	0.1729	0.1680	0.1642	0.1611	0.1664	0.1665	0.1609	0.1338	0.2516
	0.5	0.1640	0.1746	0.1747	0.1794	0.1535	0.1610	0.1649	0.1698	0.1342	0.2291
Polbooks	0.1	0.0545	0.0550	0.0541	0.0616	0.0782	0.0807	0.0445	0.0506	0.0730	0.1388
	0.2	0.0992	0.0939	0.0984	0.1088	0.1246	0.1259	0.0856	0.0901	0.1258	0.2116
	0.3	0.1298	0.1229	0.1326	0.1427	0.1522	0.1604	0.1212	0.1179	0.1516	0.2237
	0.4	0.1451	0.1471	0.1544	0.1557	0.1578	0.1593	0.1362	0.1435	0.1577	0.2381
	0.5	0.1517	0.1488	0.1477	0.1575	0.1544	0.1592	0.1437	0.1378	0.1514	0.2064
Macaque	0.1	0.2344	0.1292	0.1885	0.1566	0.2809	0.2860	0.0881	0.0759	0.2531	0.3735
	0.2	0.3586	0.2284	0.3058	0.2547	0.4055	0.4066	0.1567	0.1369	0.3838	0.5058
	0.3	0.4303	0.3016	0.3911	0.3347	0.4772	0.4779	0.2167	0.1951	0.4724	0.5465
	0.4	0.4815	0.3755	0.4380	0.4191	0.5240	0.5205	0.2600	0.2502	0.5203	0.5766
	0.5	0.5069	0.4407	0.4840	0.4615	0.5466	0.5298	0.3060	0.2854	0.5525	0.5761

Table 3 The AUC Comparison against the State-of-the-art Methods

Dataset	Ratio	Degree	Eigen	Katz	Closeness	Subgraph	Harmonic	PageRank	Betweenness	Load	NN-LP-CF
Karate	0.1	0.6922	0.6951	0.7084	0.6572	0.7128	0.6560	0.7312	0.6958	0.6830	0.5446
	0.2	0.6530	0.6919	0.6702	0.6429	0.6744	0.6608	0.6647	0.6382	0.6580	0.4794
	0.3	0.6291	0.6445	0.6705	0.6482	0.6351	0.6148	0.6631	0.6249	0.6141	0.4498
	0.4	0.6026	0.6020	0.6101	0.5930	0.6228	0.5956	0.6102	0.6046	0.6038	0.4378
	0.5	0.5734	0.5805	0.5639	0.5486	0.5642	0.5703	0.5631	0.5618	0.5692	0.4259
Dolphin	0.1	0.7624	0.7622	0.7498	0.7601	0.7448	0.7669	0.7475	0.7251	0.7527	0.7648
	0.2	0.7250	0.7266	0.7215	0.7299	0.7438	0.7363	0.7232	0.7168	0.7364	0.6369
	0.3	0.7028	0.6929	0.7025	0.7027	0.6890	0.6836	0.6837	0.6884	0.6962	0.5525
	0.4	0.6556	0.6531	0.6494	0.6550	0.6635	0.6507	0.6431	0.6540	0.6498	0.4611
	0.5	0.6105	0.6012	0.6159	0.6208	0.6130	0.6017	0.6117	0.6112	0.6091	0.4520
Adjnoun	0.1	0.6879	0.6716	0.6907	0.6871	0.6717	0.6820	0.6632	0.6482	0.6708	0.6754
	0.2	0.6639	0.6650	0.6515	0.6627	0.6590	0.6648	0.6639	0.6494	0.6589	0.6571
	0.3	0.6427	0.6414	0.6330	0.6447	0.6479	0.6321	0.6350	0.6398	0.6399	0.6068
	0.4	0.6158	0.6204	0.6121	0.6203	0.6176	0.6243	0.6148	0.6209	0.6197	0.5867
	0.5	0.5874	0.5923	0.5950	0.5918	0.5996	0.5873	0.5904	0.5903	0.5927	0.4820
Football	0.1	0.8147	0.8191	0.8199	0.8065	0.8264	0.8325	0.7965	0.7991	0.8079	0.8450
	0.2	0.8060	0.7984	0.8051	0.8067	0.8189	0.8154	0.7993	0.8017	0.8027	0.8361
	0.3	0.7943	0.7981	0.7968	0.7935	0.8009	0.8001	0.7898	0.7948	0.7849	0.8080
	0.4	0.7689	0.7689	0.7672	0.7702	0.7708	0.7685	0.7681	0.7629	0.7636	0.7781
	0.5	0.7212	0.7301	0.7275	0.7297	0.7228	0.7280	0.7248	0.7255	0.7196	0.7332
Polbooks	0.1	0.8653	0.8651	0.8712	0.8725	0.8779	0.8766	0.8542	0.8510	0.8776	0.8905
	0.2	0.8471	0.8432	0.8520	0.8581	0.8503	0.8548	0.8377	0.8415	0.8595	0.8745
	0.3	0.8235	0.8136	0.8228	0.8280	0.8208	0.8338	0.8193	0.8175	0.8268	0.8306
	0.4	0.7815	0.7793	0.7865	0.7804	0.7747	0.7756	0.7788	0.7810	0.7805	0.7905
	0.5	0.7305	0.7223	0.7282	0.7347	0.7230	0.7228	0.7260	0.7191	0.7272	0.7303
Macaque	0.1	0.8500	0.7388	0.8043	0.7870	0.8789	0.8825	0.6239	0.5648	0.8455	0.8912
	0.2	0.8436	0.7515	0.8125	0.7829	0.8647	0.8740	0.6271	0.5798	0.8466	0.8873
	0.3	0.8352	0.7535	0.8106	0.7733	0.8519	0.8521	0.6312	0.5910	0.8476	0.8683
	0.4	0.8226	0.7507	0.7960	0.7781	0.8358	0.8376	0.6331	0.6053	0.8305	0.8462
	0.5	0.7944	0.7496	0.7798	0.7675	0.8149	0.8048	0.6375	0.6112	0.8183	0.8140

Table 4 The F1-Score Comparison against the State-of-the-art Methods

Dataset	Ratio	Degree	Eigen	Katz	Closeness	Subgraph	Harmonic	PageRank	Betweenness	Load	NN-LP-CF
Karate	0.1	0.0758	0.0731	0.1165	0.0594	0.0558	0.0550	0.1061	0.1116	0.0578	0.0937
	0.2	0.1091	0.1185	0.1285	0.1062	0.1170	0.1020	0.1407	0.1663	0.1059	0.1235
	0.3	0.1473	0.1514	0.1669	0.1391	0.1366	0.1278	0.2032	0.1705	0.1438	0.0928
	0.4	0.1643	0.1644	0.1707	0.1573	0.1850	0.1670	0.1743	0.1799	0.1741	0.1033
	0.5	0.1703	0.1833	0.1647	0.1482	0.1694	0.1807	0.1518	0.1682	0.1738	0.1318
Dolphin	0.1	0.0657	0.0724	0.0640	0.0587	0.0564	0.0564	0.0472	0.0445	0.0587	0.0642
	0.2	0.1111	0.0988	0.1092	0.1057	0.1137	0.1061	0.0909	0.0620	0.1142	0.1045
	0.3	0.1659	0.1196	0.1654	0.1575	0.1478	0.1439	0.1063	0.1108	0.1565	0.1192
	0.4	0.1889	0.1363	0.1799	0.1841	0.1906	0.1804	0.1125	0.1242	0.1835	0.1003
	0.5	0.1955	0.1164	0.2009	0.2067	0.1992	0.1860	0.1203	0.1504	0.1947	0.1128
Adjnoun	0.1	0.0398	0.0358	0.0361	0.0336	0.0348	0.0325	0.0269	0.0240	0.0329	0.0422
	0.2	0.0680	0.0644	0.0622	0.0593	0.0656	0.0598	0.0460	0.0299	0.0612	0.0717
	0.3	0.0954	0.0890	0.0877	0.0834	0.0932	0.0833	0.0562	0.0435	0.0877	0.0985
	0.4	0.1125	0.1090	0.1072	0.1059	0.1096	0.1100	0.0772	0.0428	0.1081	0.1142
	0.5	0.1229	0.1239	0.1279	0.1213	0.1264	0.1191	0.0877	0.0533	0.1254	0.0840
Football	0.1	0.1072	0.1100	0.1127	0.1090	0.0554	0.0540	0.0391	0.0631	0.0549	0.0864
	0.2	0.1861	0.1811	0.1897	0.1938	0.1103	0.1069	0.0648	0.0914	0.1138	0.1523
	0.3	0.2348	0.2510	0.2404	0.2409	0.1689	0.1654	0.0816	0.0775	0.1720	0.2120
	0.4	0.2650	0.2829	0.2885	0.2675	0.2297	0.2282	0.0908	0.0891	0.2319	0.2674
	0.5	0.2888	0.2836	0.2982	0.2880	0.2761	0.2787	0.0844	0.0960	0.2765	0.3026
Polbooks	0.1	0.1045	0.0711	0.0910	0.1069	0.0737	0.0680	0.0284	0.0330	0.0746	0.0944
	0.2	0.1778	0.1072	0.1113	0.1759	0.1370	0.1324	0.0425	0.0458	0.1402	0.1719
	0.3	0.2325	0.1272	0.2224	0.2002	0.1901	0.1889	0.0624	0.0452	0.1957	0.2299
	0.4	0.2611	0.1402	0.2637	0.2345	0.2264	0.2228	0.0923	0.0734	0.2361	0.2717
	0.5	0.2828	0.1126	0.2723	0.2648	0.2608	0.2555	0.1041	0.0902	0.2660	0.2792
Macaque	0.1	0.2719	0.1881	0.2431	0.2323	0.2769	0.2802	0.0415	0.0188	0.2413	0.2530
	0.2	0.4090	0.2849	0.3809	0.3388	0.4301	0.4397	0.0381	0.0175	0.4092	0.4181
	0.3	0.4830	0.3517	0.4477	0.3944	0.5137	0.5113	0.0446	0.0173	0.5087	0.5151
	0.4	0.5227	0.4106	0.4833	0.4545	0.5487	0.5477	0.0546	0.0171	0.5452	0.5616
	0.5	0.5290	0.4599	0.5099	0.4910	0.5581	0.5411	0.0521	0.0187	0.5658	0.5539

5 Conclusion

This paper presents a neural network based link prediction algorithm using centrality-based features. The proposed neural network model has utilized the centrality measures as a feature set to estimate the likelihood score for predicting missing links. These centrality measures use both local and global topological features, which will give more stable and accurate predictions. The experimental results on real social networks have been performed to validate the performance of the proposed model against centrality-based methods. The experimental results show the superiority of NN-LP-CF against compared methods over prediction matrices. A significant amount of future contribution can be made to this work, such as adopting users' opinions, information diffusion, and contextual features in addition to topological features.

References

1. Kumar, A., Singh, S.S., Singh, K., Biswas, B.: Link prediction techniques, applications, and performance: a survey. Phys. A: Stat. Mech. Appl. 124289 (2020)
2. Mishra, S., Singh, S.S., Mishra, S., Biswas, B.: Tcd2: tree-based community detection in dynamic social networks. Expert Syst. Appl. **169**, 114493 (2021)
3. Singh, S.S., Srivastva, D., Verma, M., Singh, J.: Influence maximization frameworks, performance, challenges and directions on social network: a theoretical study. J. King Saud Univ. - Comput. Inf. Sci. (2021)
4. Singh, S.S., Kumar, A., Mishra, S., Biswas, B.: Community-based link prediction using information diffusion: Clp-id. Inf. Sci. **514**, 402–433 (2020)
5. Singh, S.S., Kumar, A., Singh, K., Biswas, B.: Community based context-aware influence maximization in social networks: C2im. Phys. A **514**, 796–818 (2019)
6. Liben-Nowell, D., Kleinberg, J.: The link prediction problem for social networks. In: Proceedings of the Twelfth International Conference on Information and Knowledge Management, CIKM '03, pp. 556–559. ACM, New York, NY, USA (2003)
7. Kim, J., Hastak, M.: Social network analysis: characteristics of online social networks after a disaster. Int. J. Inf. Manag. **38**(1), 86–96 (2018)
8. Saqr, M., Alamro, A.: The role of social network analysis as a learning analytics tool in online problem based learning. BMC Med. Educ. **19**(1), 1–11 (2019)
9. Duncan, O.D.: Methodological Issues in the Analysis of Social Mobility. Routledge (2018)
10. Ikram, M., Sroufe, R., Rehman, E., Zulfiqar Ali Shah, S., Mahmoudi, A.: Do quality, environmental, and social (qes) certifications improve international trade? a comparative grey relation analysis of developing vs. developed countries. Phys. A: Stat. Mech. App. **545**, 123486 (2020)
11. Bridge, G.: Gentrification, class and community: a social network approach. In: The Urban Context, pp. 259–286. Routledge (2020)
12. Wang, Z., Glänzel, W., Chen, Y.: The impact of preprints in library and information science: an analysis of citations, usage and social attention indicators. Scientometrics **125**(2), 1403–1423 (2020)
13. Holden, C., Lee, K.: Corporate power and social policy: the political economy of the transnational tobacco companies. Global Soc. Policy **9**(3), 328–354 (2009)
14. Cordeiro, M., Sarmento, R.P., Brazdil, P., Gama, J.: Evolving networks and social network analysis methods and techniques. Social Media and Journalism-Trends, Connections, Implications, pp. 101–134 (2018)
15. Al Hasan, M., Zaki, M.J.: A Survey of Link Prediction in Social Networks, pp. 243–275. Springer US, Boston, MA (2011)

16. Martínez, V., Berzal, F., Cubero, J-C.: A survey of link prediction in complex networks. ACM Comput. Surv. **49**(4), 69:1–69:33 (2016)
17. Kumar, A., Singh, S.S., Singh, K., Biswas, B.: Link prediction techniques, applications, and performance: a survey. Phys. A: Stat. Mech. Appl. **553**, 124289 (2020)
18. Newman, M.E.J.: Clustering and preferential attachment in growing networks. Phys. Rev. E **64**, 025102 (2001)
19. Al Hasan, M., Chaoji, V., Salem, S., Zaki, M.: Link prediction using supervised learning. In: Proceedings of SDM 06 workshop on Link Analysis, Counterterrorism and Security (2006)
20. Barabasi, A.-L., Albert, R.: Emergence of scaling in random networks. Science **286**(5439), 509–512 (1999)
21. Kleinberg, J.M.: Navigation in a small world. Nature **406**(6798), 845 (2000)
22. Leskovec, J., Kleinberg, J., Faloutsos, C.: Graphs over time: Densification laws, shrinking diameters and possible explanations. In: Proceedings of the Eleventh ACM SIGKDD International Conference on Knowledge Discovery in Data Mining, KDD '05, pp. 177–187. ACM, New York, NY, USA (2005)
23. Li, X., Du, N., Li, H., Li, K., Gao, J., Zhang, A.: A Deep Learning Approach to Link Prediction in Dynamic Networks, pp. 289–297
24. Zhang, M., Chen, Y.: Weisfeiler-lehman neural machine for link prediction. In: Proceedings of the 23rd ACM SIGKDD International Conference on Knowledge Discovery and Data Mining, KDD '17, pp. 575–583. Association for Computing Machinery, New York, NY, USA (2017)
25. Wang, H., Wang, J., Wang, J., Zhao, M., Zhang, W., Zhang, F., Xie, X., Guo, M.: Graph-gan: Graph representation learning with generative adversarial nets (2017). CoRR, abs/arXiv:1711.08267
26. Grover, A., Leskovec, J.: Node2vec: Scalable feature learning for networks. In: Proceedings of the 22Nd ACM SIGKDD International Conference on Knowledge Discovery and Data Mining, KDD '16, pp. 855–864. ACM, New York, NY, USA (2016)
27. Wang, D., Cui, P., Zhu, W.: Structural deep network embedding. In: Proceedings of the 22nd ACM SIGKDD International Conference on Knowledge Discovery and Data Mining, KDD '16, pp. 1225-1234. Association for Computing Machinery, New York, NY, USA (2016)
28. Kipf, T.N., Welling, M.: Variational graph auto-encoders (2016). CoRR, abs/ arXiv:1611.07308
29. Clauset, A., Moore, C., Newman, M.E.J.: Hierarchical structure and the prediction of missing links in networks. Nature **453**(7191), 98–101 (2008)
30. Guimerà, R., Sales-Pardo, M.: Missing and spurious interactions and the reconstruction of complex networks. Proc. Natl. Acad. Sci. **106**(52), 22073–22078 (2009)
31. Getoor, L., Friedman, N., Koller, D., Taskar, B.: Learning probabilistic models of link structure. J. Mach. Learn. Res. **3**(null), 679–707 (2003)
32. Schlichtkrull, M., Kipf, T.N., Bloem, P., van den Berg, R., Titov, I., Welling, M.: Modeling relational data with graph convolutional networks. In: Gangemi, A., Navigli, R., Vidal, M-E., Hitzler, P., Troncy, R., Hollink, L., Tordai, A., Alam, M. (Eds.), The Semantic Web, pp. 593–607. Springer International Publishing, Cham (2018)
33. Kipf, T.N., Welling, M.: Semi-supervised classification with graph convolutional networks (2016). CoRR, abs/ arXiv:1609.02907
34. Zhang, M., Chen, Y.: Weisfeiler-lehman neural machine for link prediction. In: Proceedings of the 23rd ACM SIGKDD International Conference on Knowledge Discovery and Data Mining, Halifax, NS, Canada, August 13 - 17, 2017, pp. 575–583 (2017)
35. Zhang, M., Chen, Y.: Link prediction based on graph neural networks (2018). CoRR, abs/arxiv:1802.09691

Social Media Text Analysis

Deep Learning for Code-Mixed Text Mining in Social Media: A Brief Review

Rrubaa Panchendrarajan and Akrati Saxena

Abstract The advent of social media in day-to-day life has made communications between people more often and easier than ever before. Analyzing the content in social media has opened up a massive amount of research and commercial opportunities. However, the content in social media is noisy and multi-lingual, which postures computational challenges ahead. Especially, the non-native English speakers and writers tend to mix their native language with English while generating social media content. Thus it requires a comprehensive prepossessing of text, including the identification of language for many language processing applications. In the area of language processing, deep learning has shown to be very successful, and the latest research works have witnessed the adoption of deep learning solutions to cater to the challenges in analyzing code-mixed text. Here, we highlight a comprehensive study of deep learning techniques used for analyzing the code-mix text of social media to understand the state-of-the-art and existing research challenges. We will discuss several applications of code-mixed text analysis and future directions.

Keywords Code-mix · Text mining · Deep learning · Natural language processing · Social media

1 Introduction

Code-mixing or code-switching denotes the practice of using two or more languages within a single utterance or discussion. Often this can be observed in inter-person communication and social media interaction of multi-lingual communities. People who are generally fluent in more than one language tend to switch languages for infor-

R. Panchendrarajan (✉)
Faculty of Computing, Sri Lanka Institute of Information Technology, Malabe, Sri Lanka
e-mail: rrubaa.p@sliit.lk

A. Saxena
Department of Mathematics and Computer Science, Eindhoven University of Technology,
Eindhoven, The Netherlands
e-mail: a.saxena@tue.nl

mal communications. The Indian sub-continent residents who are fluent in English commonly have a transition in language between English and their native languages, such as Hindi, Tamil, and Telugu. For instance, the utterance *Can you see the sadgi of Modi Ji.*[1] is written in Hinglish (Hindi + English), and the direct translation of it in English would be *Can you see the simplicity of Mr. Modi.* Similar behavior can be seen in European countries, such as Spain as well, where people mix Spanish and English languages and form a new code-mixed language. This shows a new trend of communication and attracts the attention of the research community to focus on the next paradigm in text mining.

Mining text in social media has opened up various research and business opportunities, including marketing, product recommendations, and analyzing social trends. While the research community has witnessed an enormous amount of research work towards processing social media monolingual text, code-mix data too requires equal attention, and appropriate solutions have to be developed. Unlike the monolingual language, code-mixed text poses various challenges, including borrowing words and phrases from multiple languages and mixing grammars and morphological characteristics of multiple languages. For example, in the same code-mixed instance *Can you see the sadgi of Modi Ji.*, the word *ji* does not have any corresponding English term which is usually used among the Hindi speaking community to show respect. Even the term *sadgi* can be written using multiple spellings, such as *saadgi, saadagi* etc. This shows the imperative to develop effective solutions to overcome the challenges in processing code-mixed data.

Deep learning has played a crucial role in developing state-of-the-art solutions in many research fields, including text mining, over the past two decades. Various text mining tasks use deep learning solutions due to their powerful nature of understanding the linguistic features of the language and handling long sequential data. These solutions are adopted by the code-mixed text mining research community, and promising results can be seen across many code-mixed text mining tasks. This book chapter presents a comprehensive review of different deep learning techniques used for analyzing the code-mixed data. We introduce a range of text mining tasks that use deep learning to model state-of-the-art solutions. We further discuss the challenges associated with each task, different deep learning techniques used to overcome these challenges, open research problems, and future directions in each problem domain.

2   Code-Mixed Text Mining

This section presents a diverse set of deep learning techniques used for code-mixed text mining tasks. Figure 1 summarizes the deep learning techniques discussed in this chapter.

---

[1] The example is inspired from the tweet https://twitter.com/Anonymouss_92/status/1470375298201882624.

Fig. 1 Summary of discussed deep learning techniques. Font size of the technique is corresponding to its use

2.1 Word Embedding

Transforming text into a meaningful representation that machines can understand is a pivotal step in any text mining problem. The past two decades have witnessed that word embedding vector representation plays a dominant role in the success of many text mining applications. Word embedding projects the word into a vector representation that is lower in dimension such that the words conveying similar meaning are projected nearer in the vector-space. These vector representations are jointly learned using a neural network on text mining tasks. Language modeling is broadly used for learning word embedding vector representation, where the model attempts to predict a word given its surrounding or context words. The first layer of the neural network trained for a specific task projects the input words into a vector representation and is adjusted during the learning to minimize the loss on the task. Word2Vec [1], Glove [2], FastText [3] and BERT [4] are some quite popular pre-trained word embedding vector representations available for English.

The first challenge associated with code-mixed text is to efficiently transform the text into a meaningful vector representation. The following three techniques are used as a solution to generate vector representation for code-mixed text.

1. *Monolingual word embedding*—Using the existing monolingual word embedding combined with language detection. Once the language of the word is known, its corresponding monolingual representation can be used to convert the words into vector representation.
2. *Bilingual word embedding*—Learning word embedding vector representation for two languages in a joint embedding vector space [5]. This can be done by either learning the vector representation directly from bi-lingual data or by using a pro-

jection mechanism to project the monolingual vector representations to a common vector space.

3. *Cross-lingual or Multi-lingual word embedding*—Learning word embedding vector representation for multiple languages in a joint embedding vector space [5]. Similar to bilingual word embedding, this can be done by either learning the vector representation directly from multi-lingual data or by using a projection mechanism to project the monolingual vector representations to a common vector space.

2.2   Language Detection

Identifying the language used in a code-mixing text is a skeletal task for downstream text mining applications. Language detection can be performed at word level as well as at utterance level. While Support Vector Machines (SVM) [6] were widely used during the emergence of this problem, Chang et al. [7] showed that Recurrent Neural Networks (RNN) with cross-lingual embedding outperform SVM in determining the languages of the code-mixed data. References [8, 9] used the similar approach of using bi-LSTM (bidirectional Long short-term memory) for language detection. Jaech et al. [8] further modified the solution with a convolutional neural network (CNN) dedicated at the bottom of the network to learn the word embedding representation. Words are input to the CNN at the character level, which enables them to build a vector representation of each word using its character sequence. Samih et al. [9] added a Conditional Random Field (CRF) layer preceding the LSTM to capture the relationship between language tags of words. CRF is a sequence labeling algorithm that models the labels of the entire sequence instead of its elements independently. They used pre-trained cross-lingual embedding to convert the words to a vector representation.

Choudhury et al. [10] studied different approaches to utilize the knowledge of language detection models trained on monolingual data to enhance the performance of language detection in code-mixed text. They trained a neural network with an embedding layer and a dense layer with monolingual data alternating between the languages and then retrained it on code-mixed data if available. This approach significantly enhanced the performance of the model in determining the language used in code-mixed data, and they observed a reasonably high accuracy for the model solely trained on monolingual data as well. Mandal et al. [11] modified the bi-LSTM with CRF architecture with multichannel added at the bottom of the network. Inspired by the performance of multichannel networks in image classification, they used four channels, three with CNN and one with an LSTM. The input word vector was provided to all four channels, and the outputs of the channels were concatenated and given to the bi-LSTM with CRF to predict the language.

Transformer-based neural architectures have caused a stir in the text mining community with obtaining state-of-the-art performance for many classification tasks. Especially, BERT (Bidirectional Encoder Representations from Transformers) [4], and its extensions are widely used for their architectural benefits, such as encoder-decoder model and self attentions which enables focusing on parts of the input sequence. These pre-trained transformer models are then fine-tuned with a new layer on top to generate state-of-the-art classification models in the text mining domain. Thara et al. [12] exploited a similar idea of adapting BERT and its variation for language detection in Malayalam-English data. They observed performance enhancement even with few epochs in fine-tuning the pre-trained models. This shows the evolution in the adaption of deep learning models for the language detection tasks, starting from LSTM with cross-lingual word embedding to transformer-based architectures.

2.3 Sentiment Analysis

Sentiment analysis is the task of determining the polarity associated with a statement as negative, positive, or neutral [14]. This leverages extensive applications related to business, politics, and other social trends [15]. With the proliferation in the practice of using code-mixed data for social media communications, it is essential to devise solutions to infer the sentiment from code-mixed text and effectively handle the challenges that arose due to the mix of multiple languages. Table 1 shows examples of social media posts from [13] dataset with Hindi-English code-mixed text that convey three different sentiments. While many machine learning models were explored as a solution for this problem, Ghosh et al. [16] attempted to apply a multilayer perceptron for sentiment analysis in Facebook posts. They highlighted the necessity of extensive pre-processing of text and feature extraction before input to the multilayer perceptron for better performance.

Choudhary et al. [17] adopted siamese networks for sentiment analysis in code-mixed data. Siamese networks differ from other deep learning models due to their property of parameter sharing. The network is consisting of two or more identical subnetworks with shared parameters. The authors modeled the siamese network with two subnetworks, each consisting of a combination of bi-LSTM and feed-forward network. Two sentences with the same sentiment polarity are input to each sub-

Table 1 Examples from hinglish code-mixed sentiment analysis dataset [13]. For each post, the original tweet is shown in italic and its English translation is mentioned in the next line

Social Media Post	Sentiment
World best PM Narendar Modi Ji mira bhart mahan Mr. Narendar Modi is the best PM in the world. My India is great	Positive
apka koi v yojna-Plan sahi Trah se upyog nahi ho paya h..... Any of your plans has not been used properly	Negative
modi saab ye konsi country hai??? Mr. Modi, which country is it?	Neutral

network, and differences between yielded vector representation of both sentences are minimized while training the network. This enables the network to project sentences with the same polarity to a common vector space. Lal et al. [18] proposed a dual encoder architecture, where the first encoder is responsible for determining the overall sentiment expressed in a sentence and the second encoder with attention mechanism focuses on individual sub-words. CNN is used to generate the sub-word vector representation that will be given as input to the dual encoder, and the output of the dual encoder is passed through several fully connected layers for making the prediction.

With the evident performance of attention-based architectures, they were also used to analyze sentiment in code-mixed data [19, 20]. Jamatia et al. [19] showed that a bi-LSTM with an attention layer outperforms many traditional machine learning algorithms. Javdan et al. [20] proposed an architecture with multiple attention layers. The proposed model comprises two parallel sub-networks, each responsible for learning character and word-level features, respectively. Each sub-network is made of an embedding layer with LSTM and attention layer that are responsible for learning important word-level and character-level features, respectively. The outputs of subnetworks are combined and input to another attention layer with multiple dense layers on top of it. This last attention layer is intended to learn dominant features among the word and character level features.

While word embedding and character embedding are widely used to convert words into vector representation, some works [21, 22] highlighted the shortcomings of both representations and showed that sub-word level embedding representations are more powerful compared to them in representing code-mixed data. Mandalam et al. [21] used sub-word embedding with LSTM for sentiment analysis in *Dravidian* code-mixed data [23]. Chundi et al. [22] used the combination of CNN and bi-LSTM on top of the sub-word embedding layer. Here, CNN learns important features from these sub-word vector representations. Similar to the exploration in language detection task [10], Yadav et al. [24] attempted to effectively transfer the knowledge obtained from monolingual data to code-mixed data. Monolingual word embeddings are trained from large monolingual data, and they learned a projection matrix that can transfer one language's word embedding vector space to another language's vector space. This projection matrix is learned either in supervised or unsupervised settings, and it enables to generate the cross-lingual word embedding effectively. Finally, they applied a bi-LSTM on the embedding layer to make the prediction. It can be observed that attention mechanism and effective word embedding are widely used in state-of-the-art solutions for sentiment analysis in the code-mixed text.

2.4 Hate Speech Detection

Hateful and abusive content on social networks poses a major threat to the online community. This has motivated the research community to automatically detect hate speech in social media content to minimize the spread and the damage to society. With the adoption of heavy use of code-mixed text, it is essential to tackle the challenges associated with it.

Various CNN architectures were explored as the initial deep learning solution to determine hate speech expressed in code-mixed text. This includes a standalone CNN [25, 26], two parallel CNNs dedicated to process word level and character level information [27], CNN combined with LSTM [28] and CNN combined with bi-LSTM [29]. References [26, 28] demonstrated that the CNN-based architectures trained on English text followed by the multilingual text perform better compared to training the same model directly on the multilingual code-mixed data. Srivastava et al. [30] showed that applying transferring learning only to the embedding layer would be sufficient to handle the challenges associated with hate speech inference in code-mixed data. They retrained the pre-trained embedding models, such as *ELMO* [31] and *BERT* [4], on code-mixed data and used it as the embedding layer for a standalone CNN to make the prediction.

LSTM models are more powerful in handling sequential data. A similar transfer learning approach has been used by Kapoor et al. [32] with an LSTM based architecture. They trained an LSTM with multiple dense layers on top of it using an English dataset. The trained dense layers are reused on top of another LSTM model to retrain it on multilingual data. This enabled them to utilize the learned features that are transferred in the final stage of the model. Santosh et al. [33] applied attention mechanism in the LSTM model for hate speech inference in code-mixed text. Here, the input to the LSTM from sentences is generated through a 1-D convolution layer. Chopra et al. [34] modified this model further by incorporating the social network information into the graph embedding. Follower and retweet graphs obtained from the social network are transformed into embedding by obtaining an embedding representation for the users in the network. Deepwalk [35] and node2vec [36] techniques are used to transform the user to an embedding representation. Finally, the graph embedding layer following a CNN, LSTM, and attention layer are combined via a softmax layer for the prediction. This shows that the adoption of deep learning techniques for hate speech inference for code-mixed data has been evolved from simple CNN-based architectures to incorporating transfer learning and social network data.

2.5 Stance Detection

Stance detection is a salient branch of opinion mining that determines whether a piece of opinion is against, favorable, or neutral about a target entity [38]. The target entity can be a product, a person, or an organization. Table 2 shows examples of Hindi-English code-mixed social media posts that express different stances about the target *Demonetisation* that happened in India in 2016 [37]. While the solutions of other classification tasks, such as language detection and sentiment analysis in code-mixed data, can be adopted to solve this problem, a very little effort has been made towards using deep learning techniques for stance detection. Skanda et al. [38] used deep learning only at the feature extraction level where the words are

Table 2 Examples from Hindi-English Code-mixed posts expressing different stances about the target *Notebandi (Demonetisation)* [37]

Social Media Post	Stance
sir aap to against corruption ke biradri wale ho to modi ka birodh kyo kar rahe ho. Notebandi wale kam ka to samarthan karo Sir you are a member of against corruption community, then why are you protesting against Modi. You should support demonetisation	Favor
@Narendramodi_PM Notebandi ka aap ka nirnay galat hai ! @Narendramodi_PM Your decision of demonetisation is wrong	Against
#jeetupatvari tum confuse ho , Kabhi kahte tarika galat h , Kabhi kahate ho notebandi galat h #jeetupatvari You are confused, Sometime you say that the implementation is wrong and sometime you say that the demonetisation is wrong	Neutral

converted to a vector representation using word2vec [1] model, and the Logistic regression method is used for training a stance detection model. Shalini et al. [39] replaced the Logistic regression model with a standalone CNN and bi-LSTM, and showed that CNN improves the performance in detecting stance in code-mixed data. This shows the necessity of further enhancing the performance of stance detection task in code-mixed data with deep learning techniques in the future. Especially, the techniques used in other classification tasks, such as attention-based models and transfer learning, can be further experimented.

2.6 Question Answering

With the abundance of content available in online social networks, the need for answering user queries to retrieve precise information has become an essential task. Question answering is a branch of natural language processing (NLP) that attempts to find or generate an answer for the given question from a data repository. While this problem has been evolved over the past two decades to accommodate various challenges, handling code-mixed data remains an open problem. A limited effort has been made to handle code-mixed data while automatically answering questions. Gupta et al. [40] modeled the problem as a two-step approach, where they first generate the candidate answers using a score-based approach and then rank the candidate answers using a Siamese CNN architecture. The architecture comprises two parallel CNNs with shared weights, where one CNN is dedicated to processing a question and the correct answer, and the other CNN is dedicated to processing the same question and a wrong answer. Outputs vectors of the CNNs are passed to a fully connected layer to generate similarity scores, and differences between the scores are minimized during the training. This helps the model in identifying the difference between correct and wrong answers.

Gupta et al. [41] experimented a different two-step approach. First, they classify the question into a predefined set of categories and then extract the relevant answer. They apply deep learning solutions to classify the questions into predefined categories. A neural network comprises an embedding layer following a CNN and Recurrent neural network (RNN) is used to classify the question. Finally, a score-based approach is used to rank the most appropriate answer for the given question. They further modified the model with layers of bidirectional gated recurrent units (bi-GRU) [42]. They extracted sentences from the data source that may contain the answers using a score-based approach. This collection of sentences related to the question is referred to as a snippet. There are dedicated layers in the proposed solution to project the question and snippet into a vector form. Both vector representations are passed through a final *answer extraction layer* made of pointer networks to determine the answer. Pointer networks use an attention mechanism to choose a member of the input as output. Gupta et al. [41] used two pointer networks to determine the index of the starting and ending word of the answer in the snippet, respectively.

As we discussed earlier, transformer-based architectures, such as BERT, has given state-of-the-art result in many text mining applications. Gupta et al. [43] used a multilingual variant of BERT [4] called mBERT[2] trained on a text corpus of 104 languages to perform question answering using code-mixed data. They experimented with two approaches to adopt the mBERT model for question answering in code-mixed data. The first approach is known as *zero-shot* learning, and the pre-trained mBERT model is directly used to answer questions in the English-Hindi dataset without any further training. The second approach fine tunes the pre-trained model in the English-Hindi dataset. While zero-shot learning does not improve the performance over baseline models, fine-tuning gives a compelling performance. This showed the powerful nature of the mBERT model in transferring between languages, and it opens up research opportunities to further explore transfer learning techniques to improve performance.

2.7 Machine Translation

Translating user-generated content to high resource languages like English is an essential task for many compelling applications, such as search engines, recommendation systems, and product marketing. Adoption of this problem to code-mixed text enables the usage of solutions available for high-recourse languages upon translation. Meanwhile, it adds another level of complexity to the problem, as it requires the translation of the text from more than one language. Nakayama et al. [44] used a simpler version of the BERT [4] model and used the same hyperparameters of the pre-trained BERT model available publicly to perform translation of code-mixed data. Given a sentence with two languages, they mask the words from the 2nd lan-

---

[2] Refer https://github.com/google-research/bert/blob/master/multilingual.md for further details.

guage and train the neural network model to infer the word in the 1st language. This helps the model to complete the sentence in the monolingual text of the 1st language.

Encoder-decoder architectures have demonstrated state-of-the-art results for the monolingual text translation task. The encoder generates a context vector of fixed-length representation for the input word sequence of one language. The decoder receives the context vector and regenerates the sentence in another language. These encoder and decoder are composed of bi-LSTM to handle sequential data. Once encoded, the decoder can be replaced to train the model to translate the sentence into any language. The decoder can accommodate an attention layer for learning the important places of the input sequence to be considered while decoding. This encoder-decoder architecture with attention mechanism is used as the core component in Google's translation system [45]. Mahata et al. [46] adopted this solution for code-mixed text translation by adding a new component to determine the language in the text first. An LSTM model is used to detect languages in the code-mixed text. This enables to use the monolingual translation system to translate text sequences of one language to the sentence in another language. Kugathasan et al. [47] used a similar encoder-decoder architecture with LSTM. The decorder component is modified with an algorithm called *Teacher Forcing Algorithm* which uses the ground truth from a prior time step as part of the input. This enables the decoder LSTM to quickly and efficiently adjust the parameters according to the correct output instead of relying on the loss incurred due to the wrong prediction in the previous time step.

The encoder-decoder models were also explored with other architectures in the literature. Dowlagar et al. [48] replaced the LSTM with CNN to develop an encoder-decoder model with the attention mechanism. They showed that the replacement of LSTM with CNN improves the performance in machine translation tasks for code-mixed data. [49, 50] used pre-trained multilingual transformer-based models to utilize the powerful nature of these models in handling any language with fine-tuning them on a little amount of task-specific data. Gupta et al. [50] used the cross-lingual variation of BERT referred as mBERT. Elmadany et al. [49] experimented two approaches, (i). a transformer model trained from scratch in multilingual data and (ii). pre-trained multilingual models mT5 [51] and mBART [52]. mT5 and mBART were obtained by training their original monolingual models T5 and BART on a large amount of multilingual data. Both models comprise a transformer-based encoder-decoder architecture. It is quite evident that the encoder-decoder architecture performs a vital role in state-of-the-art performance in the task of machine translation for code-mixed data.

2.8 Named Entity Recognition (NER)

Extracting named entities from text is a fundamental task of text mining, and it seeks to identify pronouns in text and categorize them into entity classes, such as a person, organization, location, time, and so on [53]. Figure 2 presents an example of tweet from a code-mixed NER dataset [54]. Extracted named entities are used as features for

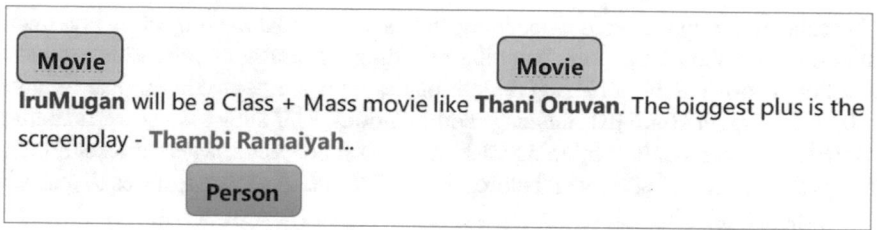

Fig. 2 An example tweet from a code-mixed NER dataset [54]. The tweet is written in Tanglish (Tamil + English). Coloured words represent the entities and their corresponding entity types are indicated in boxes

many text mining tasks, including machine translation, question answering, search engines, and automatic summarization. These benefits make the NER task for code-mixed data a necessary problem to be solved. The nature of the NER task is also referred to as *sequence labeling* or *sequence tagging*, where a label or tag is given to each word in a sequence. In NER, each word is given a named entity category, and words that do not belong to any of the predefined categories are assigned with a specific label.

Recurrent neural networks are shown to be performing very well in sequence labeling tasks as they are capable of efficiently handling long sequential data. Singh et al. [55] analyzed the performance of three different models, (i) a decision tree, (ii) Conditional Random Field (CRF), and (iii) an LSTM to perform NER in code-mixed data. They observed a slightly high performance when using the LSTM model compared to the other two. Gupta et al. [53] performed a similar analysis with a bidirectional gated recurrent unit (bi-GRU) and showed that it outperforms the other machine learning models, such as Conditional Random Field (CRF) and Support Vector Machine (SVM). Glove [2] word embedding is used to transform the words into vector representation before inputting to bi-GRU. Narayanan et al. [56] showed that using character-level embedding combined with a bi-LSTM performs better compared to using word-level embedding for NER task. This could be due to the utilization of character-level morphological features that enhances the performance of the model.

Many state-of-the-art models in sequence labeling task have combined CRF and a neural network model. Sabty et al. [57, 58] combined bi-LSTM and CRF, and experimented with various pre-trained embedding, including Word2Vec [1], Glove [2], BERT [4], ELMO, [31] and FastText [3]. They observed the model performs better when they use FastText word embedding, an extension of Word2Vec proposed by the Facebook. Gaddamidi et al. [59] showed that the performance of bi-LSTM-CRF with Fasttext embedding can be further improved by providing character-level embedding as well as an additional input to the model. References [60, 61] combined CRF with a transformer-based encoder to generate a multilingual embedding representation for the words using their corresponding monolingual word embedding, sub-word embedding, and character embedding. They used Fasttext [3], and Glove

[2] as the monolingual word embedding and used a standalone projection layer with attention to generate the corresponding multilingual representation. This proposed solution is shown to be achieving the stat-of-the-art performance in English-Spanish [60], and English-Hindi [61] language combinations. This shows the powerful nature of these hybrid models combining CRF and neural networks, where the neural networks support learning critical features while CRF utilizes these features to perform the sequence labeling task.

2.9 Part of Speech (POS) Tagging

Similar to NER, Part of Speech (POS) tagging is a fundamental task in text mining, and many downstream tasks require POS tagged word sequence as input [63]. The task aims to assign a relevant part of speech to each word in a sequence, such as *noun, verb, adjective, adverb, preposition, article*, etc. Figure 3 shows an example of Hinglish sentence and its corresponding POS tags. While CRF-based models [63, 64] were widely used for POS tagging code-mixed text, Singh et al. [65] compared the performance of CRF and LSTM in Hindi-English code-mixed data and showed that CRF model still retains the better performance compared to a standalone LSTM. Patel et al. [66] analyzed various simple neural network architectures to solve the same problem. They compared an RNN, LSTM, two layers of LSTM, and GRU models in the same language combination and showed that GRU outperforms the other models.

Ball et al. [67] further experimented the significance of embedding representation in enhancing the performance of POS tagging tasks in Hindi-English code-mixed text. They used a bi-LSTM for predicting the POS tags of code-mixed word sequences. Input to this bi-LSTM comprises four embedding components: Hindi and English word embeddings, and Hindi and English character embeddings obtained through another bi-LSTM. This architecture is trained using the monolingual data, where the components dedicated for only one language are activated during the training. During the inference, all the components are activated to make the prediction. This enables the model to infer language-specific features while training and to avoid explicit language identification during the inference. Bhattu et al. [62] modified this architecture with a CRF added on top of the second bi-LSTM dedicated to

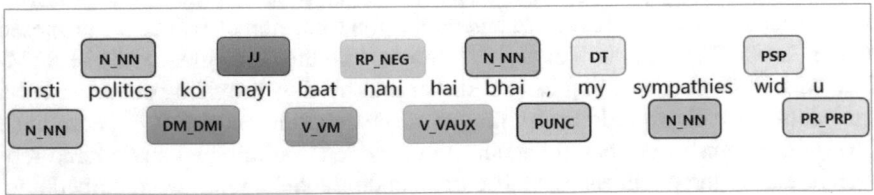

Fig. 3 An example of Hinglish (Hindi + English) sentence and its corresponding POS tags [62]

make the prediction. They directly trained the model in multilingual code-mixed data and experimented the performance with three different word embedding models, Word2Vec [1], Glove [2], and FastText [3]. Further, the first bi-LSTM, which generates the character embedding, is also replaced with CNN for performance comparison. The results showed that using bi-LSTM for character embedding extraction and FastText for word embedding performs well compared to other combinations.

Adoption of the transformer-based architectures to achieve state-of-the-art performance can be seen in the POS tagging task for code-mixed data as well. Dowlagar et al. [68] used a BERT-based model and explored multitasking to jointly train the model to perform both POS tagging and language detection. Input word sequence is given to the BERT model with a CNN on top of it to encode the entire sequence. Two multi-layer perceptrons receive the encoded vector representation and infer the POS tags and language of each word in the sequence independently. This multi-tasking enables the model to effectively learn shared representation without overfitting and to learn fast by leveraging auxiliary information. This state-of-the-art solution would encourage many researchers to experiment transformer-based solutions to further enhance the performance.

2.10 Parsing

Language parsing is the task of analyzing and interpreting the semantic structure of the text according to the rules of formal grammar. Extracting the semantic structure of word sequences helps in finding the real meaning of the sequence. The structure is often represented as a parse tree. For instance, Figure 4 presents an example of a code-mixed sentence in Hinglish and its corresponding parse tree generated using a code-mixed parser [69]. Parse structure is being used in various applications, including named entity recognition, machine translation, and grammar correction [70]. Code-mixed text also poses grammar or semantic rules, and existing works have shown that linguistic rules can be used to develop code-mix word sequences [71]. This opens up opportunities to explore the problem of developing a language parser for code-mix data.

One of the earlier works on developing a language parser for code-mixed data using deep learning technique includes the solution proposed by Bhat et al. [73]. The authors utilized the monolingual training data available to experiment transfer learning approach. They trained two neural models from the respective monolingual text data of the languages that are used in the code-mixed text. The model comprises a feed-forward neural network combined with a single hidden layer. Once the monolingual models are trained, they pass the code-mixed sentence to each model independently and use *linear interpolation* technique to obtain a final decision from monolingual models independent decisions. These interpolation weights are learned using a separate development set, and authors obtained a reasonable performance in parsing code-mixed data. The same authors later showed that jointly learning POS tagging and parsing improves the performance in both tasks [69]. They used a shared

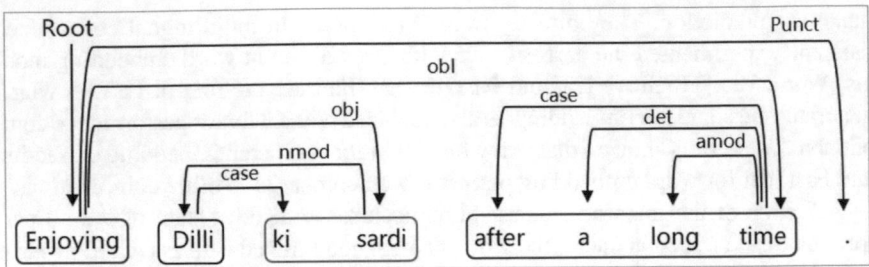

Fig. 4 An example of Hinglish (Hindi + English) sentence and its corresponding parse tree with relationships [72] generated using a code-mixed language parser [69]

bi-LSTM with two parallel feed-forward neural networks dedicated to performing POS tagging and parsing, respectively. The input to the shared bi-LSTM is composed of word embedding and character embedding obtained from another bi-LSTM. The proposed architecture provided a state-of-the-art solution for parsing code-mixed data with Hindi-English languages.

Ghosh et al. [74] enhanced the above state-of-the-art solution by using cross-lingual embedding. Cross-lingual embeddings are obtained by projecting the mono-lingual word embedding of languages used in the dataset into a common vector space, and the projection matrix is trained using a supervised approach proposed by Artetxe et al. [75]. Partanen et al. [76] used the same approach to develop the cross-lingual embedding representation and validate its performance using a bi-LSTM with a multi-layer perception on top for building the candidate parse trees. Zhang et al. [77] used a similar solution of cross-lingual embedding combined with a bi-LSTM. They modified this solution by adding a biaffine scoring layer proposed by Dozat et al. [78] on top of the bi-LSTM. Biaffine attention mechanism was initially proposed for monolingual language parsing and shown to be producing the best per-formance in monolingual as well in the code-mixed text. Similarly, the multi-lingual language model mBERT was used by Srinivasan et al. [71] with a deep neural net-work. The proposed network comprises 8 identical layers, each layer consisting of two sub-layers, an attention layer, and a feed-forward layer. This shows the vital role of embedding in developing a powerful language parser for code-mixed text.

3 Conclusion and Future Direction

This chapter discusses a diverse set of deep learning techniques used to process the code-mixed text in social media. We presented a comprehensive study of deep learn-ing techniques used for various text mining tasks while highlighting the challenges associated with the tasks and corresponding deep learning solutions proposed to overcome the challenges. We also discussed the open research problems and possi-ble future directions to incorporate effective deep learning solutions.

Our study highlights that *embedding* plays a vital role in obtaining state-of-the-art performance in many text mining tasks, especially in sentiment analysis and parsing code-mixed text. Among the three different *embeddings*, *multi-lingual* embedding that projects the words from multiple languages into a common vector space is being widely used due to its powerful nature of generalization and handling any language with fine-tuning it in a small amount of data. Similar to word embedding, *transformer-based* solutions are broadly used for many text mining tasks in code-mixed data, including language detection, machine translation, and question answering. *mBERT* is an example of transformer-based architecture trained on multi-lingual data, and we could see that this model is being extensively used across many text mining tasks.

Different other deep learning techniques are used to overcome the challenges that arise while processing the code-mixed data. Our study shows that *transfer learning* technique which attempts to train the model using monolingual data and later fine-tuning it on multilingual data enables us to build state-of-the-art solutions in text mining tasks, such as hate speech inference in code-mixed data. Moreover, the *attention* mechanism and *multi-tasking*, especially training models to perform POS tagging and language parser together, can be seen as alternatives to overcome the challenges in processing code-mixed data. However, we observed that a small amount of work had been done in adopting deep learning solutions for tasks, such as stance detection for code-mixed data. Moreover, in many downstream applications of social network mining, such as chatbot/interaction systems, influence analysis, and user behavior analysis, the code-mixed data analysis is yet to be explored. Besides, it opens up research opportunities to explore the techniques used in other tasks to develop effective solutions for processing code-mixed text.

References

1. Mikolov, T., Chen, K., Corrado, G., Dean, J.: Efficient estimation of word representations in vector space (2013). arXiv:1301.3781
2. Pennington, J., Socher, R., Manning, C.D.: Glove: Global vectors for word representation. In: Proceedings of the 2014 Conference on Empirical Methods in Natural Language Processing (EMNLP), pp. 1532–1543 (2014)
3. Bojanowski, P., Grave, E., Joulin, A., Mikolov, T.: Enriching word vectors with subword information. Trans. Assoc. Comput. Linguist. **5**, 135–146 (2017)
4. Devlin, J., Chang, M-W., Lee, K., Toutanova, K.: Bert: Pre-training of deep bidirectional transformers for language understanding (2018). arXiv:1810.04805
5. Ruder, S., Vulić, I., Søgaard, A.: A survey of cross-lingual word embedding models. J. Artif. Intell. Res. **65**, 569–631 (2019)
6. Noble, W.S.: What is a support vector machine? Nat. Biotechnol. **24**(12), 1565–1567 (2006)
7. Chang, J.C., Lin, C-C.: Recurrent-neural-network for language detection on twitter code-switching corpus (2014). arXiv:1412.4314
8. Jaech, A., Mulcaire, G., Ostendorf, M., Smith, N.A.: A neural model for language identification in code-switched tweets. In: Proceedings of The Second Workshop on Computational Approaches to Code Switching, pp. 60–64 (2016)
9. Samih, Y., Maharjan, S., Attia, M., Kallmeyer, L., Solorio, T.: Multilingual code-switching identification via lstm recurrent neural networks. In: Proceedings of the Second Workshop on Computational Approaches to Code Switching, pp. 50–59 (2016)

10. Choudhury, M., Bali, K., Sitaram, S., Baheti., A.: Curriculum design for code-switching: experiments with language identification and language modeling with deep neural networks. In: Proceedings of the 14th International Conference on Natural Language Processing (ICON-2017), pp. 65–74 (2017)

11. Mandal, S., Singh, A.K.: Language identification in code-mixed data using multichannel neural networks and context capture (2018). arXiv:1808.07118

12. Thara, S., Poornachandran, P.: Transformer based language identification for malayalam-english code-mixed text. IEEE Access **9**, 118837–118850 (2021)

13. Joshi, A., Prabhu, A., Shrivastava, M., Varma, V.: Towards sub-word level compositions for sentiment analysis of hindi-english code mixed text. In: Proceedings of COLING 2016, the 26th International Conference on Computational Linguistics: Technical Papers, pp. 2482–2491 (2016)

14. Saxena, A., Reddy, H., Saxena, P.: Recent developments in sentiment analysis on social networks: Techniques, datasets, and open issues. In: Principles of Social Networking, pp. 279–306. Springer (2022)

15. Saxena, A., Reddy, H., Saxena., P.: Introduction to sentiment analysis covering basics, tools, evaluation metrics, challenges, and applications. In: Principles of Social Networking, pp. 249–277. Springer (2022)

16. Ghosh, S., Ghosh, S., Das, D.: Sentiment identification in code-mixed social media text (2017). arXiv:1707.01184

17. Choudhary, N., Singh, R., Bindlish, I., Shrivastava, M.: Sentiment analysis of code-mixed languages leveraging resource rich languages (2018). arXiv:1804.00806

18. Lal, Y.K., Kumar, V., Dhar, M., Shrivastava, M., Koehn, P.: De-mixing sentiment from code-mixed text. In: Proceedings of the 57th Annual Meeting of the Association for Computational Linguistics: Student Research Workshop, pp. 371–377 (2019)

19. Jamatia, A., Swamy, S.D., Gambäck, B., Das, A., Debbarma, S.: Deep learning based sentiment analysis in a code-mixed english-hindi and english-bengali social media corpus. Int. J. Artif. Intell. Tools **29**(05), 2050014 (2020)

20. Mukherjee, S., Prasan, V., Nediyanchath, A., Shah, M., Kumar, N.: Robust deep learning based sentiment classification of code-mixed text. In: Proceedings of the 16th International Conference on Natural Language Processing, pp. 124–129 (2019)

21. Venkata Mandalam, A., Sharma, Y.: Sentiment analysis of dravidian code mixed data. In: Proceedings of the First Workshop on Speech and Language Technologies for Dravidian Languages, pp. 46–54 (2021)

22. Chundi, R., Hulipalled, V.R., Simha, J.B.: Saekcs: Sentiment analysis for english–kannada code switchtext using deep learning techniques. In: 2020 International Conference on Smart Technologies in Computing, Electrical and Electronics (ICSTCEE), pp. 327–331. IEEE (2020)

23. Chakravarthi, B.R., Priyadharshini, R., Muralidaran, V., Suryawanshi, S., Jose, N., Sherly, E., McCrae, J.P.:. Overview of the track on sentiment analysis for dravidian languages in code-mixed text. In: Forum for Information Retrieval Evaluation, pp. 21–24 (2020)

24. Yadav, S., Chakraborty, T.: Unsupervised sentiment analysis for code-mixed data (2020). arXiv:2001.11384

25. Badjatiya, P., Gupta, S., Gupta, M., Varma, V.: Deep learning for hate speech detection in tweets. In: Proceedings of the 26th international conference on World Wide Web companion, pp. 759–760 (2017)

26. Mathur, P., Shah, R., Sawhney, R., Mahata, D.: Detecting offensive tweets in hindi-english code-switched language. In: Proceedings of the Sixth International Workshop on Natural Language Processing for Social Media, pp. 18–26 (2018)

27. Ho Park, J., Fung, P.: One-step and two-step classification for abusive language detection on twitter (2017). arXiv:1706.01206

28. Mathur, P., Sawhney, R., Ayyar, M., Shah, R.: Did you offend me? classification of offensive tweets in hinglish language. In: Proceedings of the 2nd Workshop on Abusive Language Online (ALW2), pp. 138–148 (2018)

29. Vashistha, N., Zubiaga, A.: Online multilingual hate speech detection: experimenting with hindi and english social media. Information **12**(1), 5 (2021)
30. Srivastava, A., Hasan, M., Yagnik, B., Walambe, R., Kotecha, K.: Role of artificial intelligence in detection of hateful speech for hinglish data on social media (2021). arXiv:2105.04913
31. Peters, M.E., Neumann, M., Iyyer, M., Gardner, M., Clark, C., Lee, K., Zettlemoyer, L.: Deep contextualized word representations. In: Proceedings of the 2018 Conference of the North American Chapter of the Association for Computational Linguistics: Human Language Technologies, Volume 1 (Long Papers), New Orleans, Louisiana. Association for Computational Linguistics (2018)
32. Kapoor, R., Kumar, Y., Rajput, K., Shah, R.R., Kumaraguru, P., Zimmermann, R.: Abuse and offense detection for code-switched languages: Mind your language. In: Proceedings of the AAAI Conference on Artificial Intelligence, vol. 33, pp. 9951–9952 (2019)
33. Santosh, T.Y.S.S., Aravind, K.V.S.: Hate speech detection in hindi-english code-mixed social media text. In: Proceedings of the ACM India Joint International Conference on Data Science and Management of Data, pp. 310–313 (2019)
34. Chopra, S., Sawhney, R., Mathur, P., Shah, R.R.: Hindi-english hate speech detection: Author profiling, debiasing, and practical perspectives. In: Proceedings of the AAAI Conference on Artificial Intelligence, vol. 34, pp. 386–393 (2020)
35. Perozzi, B., Al-Rfou, R., Skiena, S.: Deepwalk: online learning of social representations. In: Proceedings of the 20th ACM SIGKDD International Conference on Knowledge Discovery and Data Mining, pp. 701–710 (2014)
36. Grover, A., Leskovec, J.: node2vec: Scalable feature learning for networks. In: Proceedings of the 22nd ACM SIGKDD international conference on Knowledge discovery and data mining, pp. 855–864 (2016)
37. Swami, S., Khandelwal, A., Singh, V., Akhtar, S.S., Shrivastava, M.: An english-hindi code-mixed corpus: Stance annotation and baseline system (2018). arXiv:1805.11868
38. Srinidhi Skanda, V., Anand Kumar, M., Soman, K.P.: Detecting stance in kannada social media code-mixed text using sentence embedding. In: 2017 International Conference on Advances in Computing, Communications and Informatics (ICACCI), pp. 964–969. IEEE (2017)
39. Shalini, K., Anand Kumar, M., Soman, K.: Deep-learning-based stance detection for Indian social media text. In: Emerging Research in Electronics, Computer Science and Technology, pp. 57–67. Springer (2019)
40. Gupta, V., Chinnakotla, M., Shrivastava, M.: Transliteration better than translation? answering code-mixed questions over a knowledge base. In: Proceedings of the Third Workshop on Computational Approaches to Linguistic Code-Switching, pp. 39–50 (2018)
41. Gupta, D., Kumari, S., Ekbal, A., Bhattacharyya, P.: Mmqa: A multi-domain multi-lingual question-answering framework for english and hindi. In: Proceedings of the Eleventh International Conference on Language Resources and Evaluation (LREC 2018) (2018)
42. Gupta, D., Ekbal, A., Bhattacharyya, P.: A deep neural network framework for english hindi question answering. ACM Trans. Asian and Low-Res. Lang. Inf. Process. (TALLIP) **19**(2), 1–22 (2019)
43. Gupta, S., Khade, N.: Bert based multilingual machine comprehension in english and hindi (2020). arXiv:2006.01432
44. Nakayama, S., Kano, T., Tjandra, A., Sakti, S., Nakamura, S.: Recognition and translation of code-switching speech utterances. In: 2019 22nd Conference of the Oriental COCOSDA International Committee for the Co-ordination and Standardisation of Speech Databases and Assessment Techniques (O-COCOSDA), pp. 1–6. IEEE (2019)
45. Wu, Y., Schuster, M., Chen, Z., Le, Q.V., Norouzi, M., Macherey, W., Krikun, M., Cao, Y., Gao, Q., Macherey, K. et al.: Google's neural machine translation system: Bridging the gap between human and machine translation (2016). arXiv:1609.08144
46. Mahata, S.K., Mandal, S., Das, D., Bandyopadhyay, S.: Code-mixed to monolingual translation framework. In: Proceedings of the 11th Forum for Information Retrieval Evaluation, pp. 30–35 (2019)

47. Kugathasan, A., Sumathipala, S.: Neural machine translation for sinhala-english code-mixed text. In: Proceedings of the International Conference on Recent Advances in Natural Language Processing (RANLP 2021), pp. 718–726 (2021)
48. Dowlagar, S., Mamidi, R.: Gated convolutional sequence to sequence based learning for english-hinigilsh code-switched machine translation. In: Proceedings of the Fifth Workshop on Computational Approaches to Linguistic Code-Switching, pp. 26–30 (2021)
49. Elmadany, A.R., Abdul-Mageed, M. et al.: Investigating code-mixed modern standard arabic-egyptian to english machine translation. In: Proceedings of the Fifth Workshop on Computational Approaches to Linguistic Code-Switching, pp. 56–64 (2021)
50. Gupta, A., Vavre, A., Sarawagi, S.: Training data augmentation for code-mixed translation. In: Proceedings of the 2021 Conference of the North American Chapter of the Association for Computational Linguistics: Human Language Technologies, pp. 5760–5766 (2021)
51. Xue, L., Constant, N., Roberts, A., Kale, M., Al-Rfou, R., Siddhant, A., Barua, A., Raffel, C.: mt5: A massively multilingual pre-trained text-to-text transformer (2020). arXiv:2010.11934
52. Liu, Y., Jiatao, G., Goyal, N., Li, X., Edunov, S., Ghazvininejad, M., Lewis, M., Zettlemoyer, L.: Multilingual denoising pre-training for neural machine translation. Trans. Assoc. Comput. Linguist. **8**, 726–742 (2020)
53. Gupta, D., Ekbal, A., Bhattacharyya, P.: A deep neural network based approach for entity extraction in code-mixed Indian social media text. In: Proceedings of the Eleventh International Conference on Language Resources and Evaluation (LREC 2018) (2018)
54. Rao, P.R.K., Devi, S.L.: Cmee-il: Code mix entity extraction in Indian languages from social media text@ fire 2016-an overview. FIRE (Working Notes), 289 (2016)
55. Singh, V., Vijay, D., Akhtar, S.S., Shrivastava, M.: Named entity recognition for hindi-english code-mixed social media text. In: Proceedings of the Seventh Named Entities Workshop, pp. 27–35 (2018)
56. Narayanan, A., Rao, A., Prasad, A., Das, B.: Character level neural architectures for boosting named entity recognition in code mixed tweets. In: 2020 International Conference on Emerging Trends in Information Technology and Engineering (ic-ETITE), pp. 1–6. IEEE (2020)
57. Sabty, C., Sherif, A., Elmahdy, M., Abdennadher, S.: Techniques for named entity recognition on arabic-english code-mixed data. Int. J. Transdiscip. AI **1**(1), 44–63 (2019)
58. Sabty, C., Elmahdy, M., Abdennadher, S.: Named entity recognition on arabic-english code-mixed data. In: 2019 IEEE 13th International Conference on Semantic Computing (ICSC), pp. 93–97. IEEE (2019)
59. Gaddamidi, S., Prasath, R.: Performance analysis of named entity recognition approaches on code-mixed data. In: International Conference on Information, Communication and Computing Technology, pp. 153–167. Springer (2021)
60. Indra Winata, G., Lin, Z., Shin, J., Liu, Z., Fung, P.: Hierarchical meta-embeddings for code-switching named entity recognition (2019). arXiv:1909.08504
61. Priyadharshini, R., Chakravarthi, B.R., Vegupatti, M., McCrae, J.P.: Named entity recognition for code-mixed indian corpus using meta embedding. In: 2020 6th International Conference on Advanced Computing and Communication Systems (ICACCS), pp. 68–72. IEEE (2020)
62. Bhattu, S.N., Nunna, S.K., Somayajulu, D.V.L.N., Pradhan, B.: Improving code-mixed pos tagging using code-mixed embeddings. ACM Trans. Asian and Low-Res. Lang. Inf. Process. (TALLIP) **19**(4), 1–31 (2020)
63. Gupta, D., Tripathi, S., Ekbal, A., Bhattacharyya, P.: Smpost: parts of speech tagger for code-mixed indic social media text (2017). arXiv:1702.00167
64. Ghosh, S., Ghosh, S., Das, D.: Part-of-speech tagging of code-mixed social media text. In: Proceedings of the Second Workshop on Computational Approaches to Code Switching, pp. 90–97 (2016)
65. Singh, K., Sen, I., Kumaraguru, P.: A twitter corpus for hindi-english code mixed pos tagging. In: Proceedings of the Sixth International Workshop on Natural Language Processing for Social Media, pp. 12–17 (2018)
66. Patel, R.N., Pimpale, P.B., Sasikumar, M.: Recurrent neural network based part-of-speech tagger for code-mixed social media text (2016). arXiv:1611.04989

67. Ball, K., Garrette, D.: Part-of-speech tagging for code-switched, transliterated texts without explicit language identification. In: Proceedings of the 2018 Conference on Empirical Methods in Natural Language Processing, pp. 3084–3089 (2018)
68. Dowlagar, S., Mamidi, R.: A pre-trained transformer and cnn model with joint language id and part-of-speech tagging for code-mixed social-media text. In: Proceedings of the International Conference on Recent Advances in Natural Language Processing (RANLP 2021), pp. 367–374 (2021)
69. Bhat, I.A., Bhat, R.A., Shrivastava, M., Sharma, D.M.: Universal dependency parsing for hindi-english code-switching (2018). arXiv:1804.05868
70. Thara, S., Poornachandran, P.: Code-mixing: A brief survey. In: 2018 International conference on advances in computing, communications and informatics (ICACCI), pp. 2382–2388. IEEE (2018)
71. Srinivasan, A., Dandapat, S., Choudhury, M.: Code-mixed parse trees and how to find them. In: Proceedings of the The 4th Workshop on Computational Approaches to Code Switching, pp. 57–64 (2020)
72. de Marneffe, M.-C., Manning, C.D., Nivre, J., Zeman, D.: Universal dependencies. Comput. Linguist. **47**(2), 255–308 (2021)
73. Bhat, I.A., Bhat, R.A., Shrivastava, M., Sharma., D.M.: Joining hands: Exploiting monolingual treebanks for parsing of code-mixing data (2017). arXiv:1703.10772
74. Ghosh, U., Sharma, D.M., Khanuja, S.: Dependency parser for bengali-english code-mixed data enhanced with a synthetic treebank. In: Proceedings of the 18th International Workshop on Treebanks and Linguistic Theories (TLT, SyntaxFest 2019), pp. 91–99 (2019)
75. Artetxe, M., Labaka, G., Agirre, E.: Learning principled bilingual mappings of word embeddings while preserving monolingual invariance. In: Proceedings of the 2016 Conference on Empirical Methods in Natural Language Processing, pp. 2289–2294 (2016)
76. Partanen, N., Lim, K., Rießler, M., Poibeau, T.: Dependency parsing of code-switching data with cross-lingual feature representations. In: International Workshop on Computational Linguistics for Uralic Languages, pp. 1–17. ACL (2018)
77. Zhang, M., Zhang, Y., Fu, G.: Cross-lingual dependency parsing using code-mixed treebank. In: Proceedings of the 2019 Conference on Empirical Methods in Natural Language Processing and the 9th International Joint Conference on Natural Language Processing (EMNLP-IJCNLP), pp. 997–1006 (2019)
78. Dozat, T., Manning, C.D.: Deep biaffine attention for neural dependency parsing (2016). arXiv:1611.01734

Convolutional and Recurrent Neural Networks for Opinion Mining on Drug Reviews

Nesma Settouti and Fatiha Youbi

Abstract Online health forums are places where patients share their experiences about their disease(s), treatment(s), etc. Under the cover of anonymity, they express their personal experiences very freely. These forums are therefore a very useful source of information for healthcare professionals to better identify and understand the problems, behaviors, and sentiments of their patients. In this study, our goal is to unveil the secrets of sentiment analysis on drug reviews using deep learning-based approaches. More specifically, we focus our research on genericity, to reduce human intervention in the excavation process. We, therefore, explore the problem of opinion mining using methods based on deep convolutional and recurrent neuronal networks like CNN, LSTM (Long Short-Term Memory), GRU (Gated Recurrent Unit), and hybrid models to reach complementarity, instead of standard methods which require a priori resources such as vocabulary, and sentence structure. In this work, we attempt to extend the genericity of our process by reducing this need for resources a priori.

Keywords Deep learning · Convolutional neuronal networks · Recurrent neuronal networks · Hybrid models · Drug reviews · Opinion manning · Sentiment analysis

1 Introduction

Social media is the most popular online activity. In 2020, the world's social media registered more than 3.6 billion users, an estimation that is expected to be close to 4.41 billion by 2025 [32]. Consequently, the availability of text documents expressing opinions or feelings has exploded, for example, newsgroups, blogs, forums, and other product review sites.

Many large hospitals and healthcare organizations have also joined the social media movement and are followed by many users that comment on news and other

N. Settouti (✉) · F. Youbi
Biomedical Engineering Laboratory, University of Tlemcen, Tlemcen, Algeria
e-mail: nesma.settouti@univ-tlemcen.dz

F. Youbi
e-mail: fatiha.youbi@univ-tlemcen.dz

© The Author(s), under exclusive license to Springer Nature Switzerland AG 2022
T.-P. Hong et al. (eds.), *Deep Learning for Social Media Data Analytics*,
Studies in Big Data 113, https://doi.org/10.1007/978-3-031-10869-3_4

topics related to these large hospitals. Pharmaceutical companies and healthcare institutions are taking this information into account and opinion analysis has long been an important component in their decision-making.

Currently, patients connected to social media can search for health information, participate in a debate on a health topic, appreciate a service or product from a health organization; they can even get in touch with health experts. The data generated by these users and the opinions they express via social media are widely considered in areas such as e-commerce or politics. More specifically, with social media, patients are increasingly informed and engaged in making choices about their healthcare.

These forums are spaces where patients, under the cover of anonymity, freely relate their personal experiences. Hancock et al. [12] have shown that communication and anonymity behind a computer facilitate the expression of emotions and opinions more than traditional communications.

Since patients are actors at the very heart of healthcare delivery, their conversations and the data they generate are important indicators for measuring opinions on the quality of care; these resources are therefore very rich for healthcare professionals who have access to exchanges between patients, between patients and professionals and even between professionals that can help improve their image and obtain valuable decision-making information.

The need to automatically process opinions is therefore strongly felt. Automatic opinion analysis (opinion mining), is about extracting sentiment from a source such as a text without a predefined structure. Recognized sentiments are classified as positive, negative or in more finely defined classes. Opinions Mining/Sentiment Analysis involves several disciplines, namely Natural Language Processing on the one hand and Machine Learning on the other.

In this study, our goal is to unlock the secrets of sentiment analysis using a machine learning approach. More precisely, we focus our research on genericity and specificity. To do so, we use examples of drugs reviews annotated according to the associate sentiment (positive, neutral, or negative). We place ourselves in a supervised learning framework and explore the opinion mining problem using deep learning, recurrent learning methods, and hybrid models instead of standard methods that require a priori resources such as vocabulary and sentence structure. In this report, we attempt to push the genericity of our process by decreasing the need for a priori resources by studying the application of convolutional and recurrent networks and their hybridization.

The proposed work is organized as follows: In Sect. 2, some generalities over sentiment analysis are presented. Section 3, is an overview of the related works in drug reviews sentiment analysis. Next, the theoretical details of the compared DL architectures (CNN, LSTM, BiLSTM, GRU, BiGRU) and hybrid models are given in Sect. 4. The comparison results and discussion are reported in Sect. 5, followed by the conclusion and perspectives in Sect. 6.

2 Opinions Mining/Sentiment Analysis Generality

Sentiment analysis tools permit the identification of the subjectivity and polarity of a given statement [27], to create structured and exploitable knowledge that can be used by a decision support system. It can be applied at the level of the document, the sentence, or a group of words.

The categorization of sentiments can be objective or subjective. When a sentence is objective, no other basic task is required. When a sentence is subjective, its polarities (positive, negative, or neutral) must be estimated.

The first step when applying sentiment analysis is to specify the text that will be analyzed for a given study. In general, there are three analysis level [7]: for the entire text (*document level*), for each sentence in a text (*sentence level*) and a more refined analysis with sentiment and polarity *aspect level*.

Sentiment Analysis Approaches

In the literature, there are many methods and algorithms for implementing sentiment analysis systems, which can be classified as follows [20, 27]:

Automatic Approach
The text is represented in the space of descriptors (e.g., terms) so that it can be processed by statistical approaches that rely on machine learning methods.

Rule-based Approach
Determines an ensemble of rules that identifies the subjectivity, polarity, or subject of an opinion. This approach can use various inputs, such as classical NLP techniques, rooting, tokenization, etc. Other rule-based operations use the dictionary of sentiments (opinion words) and match them to data to identify polarity.

Hybrid Approach
The concept of hybrid methods is very intuitive: simply combine the best of the two approaches, the rule-based and the automatic. Generally, by combining the two approaches, methods can improve accuracy.

3 Existing Works on Drug Reviews

The importance of sentiment analysis is present in many fields and many applications have emerged in this context [21]. The subject problem was mainly approached from an automatic and/or linguistic point of view [20]. It is usually modeled as a classification problem in which a classifier is supplied with sentence/text and returns the corresponding polarity, in the case of polarity analysis, e.g., [26]. These methods are known for their genericity (good recall). On the other hand, linguistic methods, also called rule-based methods, have been widely deployed for opinion analysis, e.g., [16, 23]. These methods are known for their specificity (good accuracy). Finally, other works have attempted to mix the genericity of the automatic approach and the specificity of linguistics to propose methods that are both robust and accurate (hybrid methods that use both lexicons and machine learning algorithms.), e.g., [6, 34]. Until recently, support vector machines SVMs [15] and naive Bayes NB classifiers [28] represented the most widespread classifiers in this field. Following the current trend, recent work employs deep learning [1, 3, 39].

In this study, we focus on drug reviews sentiment analysis, where few works have been done. This is due to the small number of resources developed. We cite some existing works according to their predicted output (sentiment polarity and drug rating).

Drug Rating

Grasser et al. [9] perform sentiment analysis and aspect-based analysis of patients sentiments by dragging the https://www.drugs.com and https://www.Druglib.com online review sites, and converting the reviews into ratings to formulate both approaches as a classification of side effects and drug effectiveness problems. The n-grams approach was applied to represent the user reviews and logistic regression (LR) for classification. The Transformer-Based Language models for classifying drug user ratings were proposed by [30]. The transfer learning and fine-tuning strategy have been applied with pre-trained transformer-based neural network models on unlabeled data to extract patterns in the language, then two classifiers (RF and NB) based on TF-IDF were evaluated. Garg [8] proposed a drug recommendation system that indicates the best drug for a given disease using some vectoring processes such as Bow, TF-IDF, Word2Vec, and a manual feature model with several supervised classifiers trained on Drug Review Dataset (https://www.Drugs.com).

Sentiment Polarity

Gurdin et al. [10] compared the performance of Naive Bayes (NB), Random Forests (RF), Support Vector Machines (SVM), and Convolutional Neural Networks (CNN) models for drug reviews on https://www.WebMD.com. Han et al. [11] experiment a pre-training and Multi-task learning model based on Double BiGRU for aspect-level sentiment classification applied to SentiDrugs dataset. In [13], lexicon-based sentiment analysis is conducted on patient reviews of four drugs. At first, unsupervised ML (latent Dirichlet allocation, LDA) is applied to cluster similar documents (drugs) together, second, three supervised ML algorithms such as LR (regularized logistic regression), SVM, and ANN (artificial neural network) are developed to predict the review polarity. In [37] the authors proposed an analysis of classical and deep learning methods for opinion mining on drug reviews. On their part, Vineeta et al. [36] proposed to conduct various ML comparison to classify the best uterine fibroid treatment as formulate by the patient. In [5] three models (CNN, LSTM (Long Short Term Memory), and BERT (Bidirectional Encoder Representations from Transformers)) and their combination were experimented with various word embedding models for drug reviews classification. Min [22] proposed a Weakly Supervised hybrid CNN-LSTM to resolve the problem of a limited number of labeled comments on social media for Adverse Drug Reactions (ADR). The Hybrid CNN and Bi-directional LSTM building model is applied to train weakly labeled data with a small number of manually labeled data tests to improve ADR identification.

Other research studies both drug rating and sentiment polarity as [35] who aimed to study the Count Vectorizer and TF-IDF scores as text feature extraction combined with neural networks and regular machine learning algorithms such as SVM, Logistic Regression, and Random Forests to perform sentiment prediction of drug reviews and the reviewer rating. A CNN-LSTM and Regional CNN-LSTM models architectures are considered for drug rating and sentiment analysis by [33], the embedding algorithm used is GloVe, the singularity of this work is the use of sentiment dimensionality, which is much better than sentiment polarity for representing users' opinions.

The research progress of drug reviews sentiment analysis over the past two decades can be divided into three main phases. First, we note the use of classical machine learning approaches, including probabilistic models. The second period is marked by the use of deep neural networks. Their great success has prompted researchers in this field to move in this direction. Finally, the research community has recently taken an interest in knowledge transfer learning, which has triggered a remarkable paradigm shift by reaching the state of the art in opinion mining and sentiment analysis tasks.

In this proposal, we will try to look at deep learning (DL) based approaches as well as those used in the literature. We will study the contribution of several architectures (CNN, LSTM, BiLSTM, GRU, BiGRU) and hybrid models in our specific framework. We also tested and compared several classic machine learning algorithms with different types of continuous word representations. We analyze the errors of our system and the relevance of these representations for Drug reviews sentiment Analysis. The distinguished results obtained will be duly summarized at the end of this study.

4 Convolutional and Recurrent Neuronal Networks

To build our models, we used three families of deep neural networks: convolutional (CNN), recurrent (LSTM, BLSTM, GRU, and BiGRU), and hybrid combination (GRU-CNN, LSTM-CNN) for drugs sentiment classification.

4.1 CNN (Convolutional Neuronal Network)

A convolutional network is a special type of neural network introduced by LeCun and Bengio [18]. A CNN is composed of convolution layers (feature extractors), pooling layers (which combine the outputs of convolution layers to detect higher-level features), and fully connected layers (often with the ReLU function [25] as an activation function).

The convolution layers represent a set of kernels/filters that extract features directly from the data. Its eliminate the need to perform manual feature extraction, which is usually a tedious task. The characteristics used are not pre-trained, i.e. the network learns them itself by training on images. This automated feature extraction allows CNNs to achieve high performance in object classification.

In the case of NLP, the advantage of CNN is mainly to detect recurring patterns in a sentence. One could for example detect the intensification: "very good", "really bad", etc. We could also detect the negation: "I don't like", "not bad", etc. With the pooling layer, we can also detect n-grams at a distance: "as much ..., as much ..." or "I like ..., but".

4.2   RNN (Recurrent Neural Networks)

RNN are mainly used in the context of NLP problems. However, in practice, it is difficult to train them and reach their theoretical power. One of the reasons is when the sequence to be processed is too long, the back-propagation of the error gradient can either become much too large and explode or on the contrary become much too small. The network then no longer differentiates between information that it must take into account or not [2], which makes the RNNs unable to learn in the long term. The gradient descent adapted to RNNs, Back-propagation Through Time— BTT or BPTT [24], encounters numerical difficulties. The gradients explode or crash numerically - Vanishing and exploding gradients.

4.2.1   LSTM (Long Short-Term Memory)

RNNs models only take into account very short-term dependencies, the LSTM [14] proposes a solution to this problem by taking into account a memory vector via a system of gates and states. The LSTM cell in recurrent neural networks is much more complex than a traditional RNN cell or a traditional neuron. It is composed of a cell, a Forget Gate, an Input Gate, and an Output Gate and manages a dynamic memory that evolves according to the temporal data sequence (Fig. 1).

Given the more complex structure, in particular the presence of 3 weight matrices (Forget gate, Input gate, and memory calculation weights), it is logical that the learning phase of LSTMs requires more time than that of a conventional neural network or RNN. Nevertheless, an LSTM gives much better performance. Most of the interesting results obtained today by recurrent networks are obtained by LSTMs, because of their ability to handle long-term dependency.

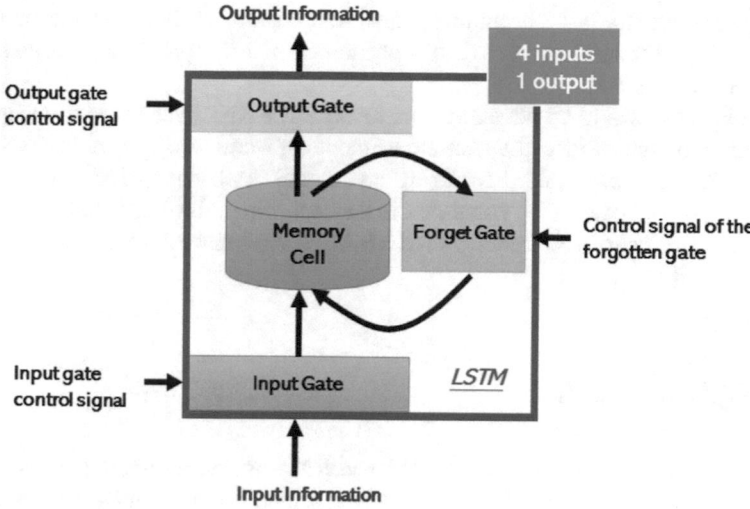

Fig. 1 LSTM model representation

4.2.2 GRU (Gated Recurrent Unit)

GRU [4] it a variant of LSTM; It structure is simpler than LSTM in the sense that it has fewer parameters. GRU with only three gates (no internal cell state), the information is instead integrated into the hidden state of the nearest unit and transmitted to the next closed recurring unit.

To make the parallel between these two models, GRU can be seen as an LSTM with no output gate and no cell memory. In practice, GRUs and LSTMs provide comparable results. The advantage of GRUs over LSTMs is that the execution time is faster since fewer parameters have to be calculated.

4.3 BiRNN (Bidirectional Recurrent Neuronal Network)

In most sequence analysis problems, it is interesting to have a memory of the past to make good decisions at the moment. But in some recognition problems, it is sometimes interesting to look at future observations, when available. The BiRNN [29] consists in traversing a one-dimensional signal along with its two directions.

Research on RNN architectures for translation tasks has shown that the classical reading direction of the sequences is not optimal. Indeed, they have shown empirically that reading the sequences backward significantly improves the performance of the model in such problems. The combination of the two reading directions (bidirectional RNN) gives even better results. The latter result is to be expected since a bidirectional representation allows the model to give as much importance to the first characters of the sequence as to the last. The unidirectional version, on the other hand, mechanically "dilutes" the first characters, even if this phenomenon is less pronounced in an LSTM than in a simple RNN.

Therefore, it would be advantageous to exploit RNNs over the entire sequence for each time step. Since the first experiments, several variants of BiRNN have emerged either with standard recurring layers, LSTM layers (BiLSTM), or GRU layers (BiGRU). It should be noted that in the case where the data is to be processed in real-time, the use of a bidirectional RNN is impossible since the end of the sequence is not accessible.

4.4 Hybrid Models

Today, the lines between the use of CNN and RNN are somewhat blurry, as we can combine these architectures to form a CRNN, which brings increased efficiency in solving specific tasks like video tagging or recognition of gestures.

Note that using the spatial modeling capabilities of a CNN to capture what is fundamentally a sequential phenomenon is not optimal by definition and requires a lot more effort and memory to accomplish the same task. Nevertheless, there are cases where the two models complete each other. For example, in recognizing the emotions of a text. Typically, a short snippet is sent to an LSTM which will extract the main ideas (the LSTM is good at analyzing text and making it appropriate). Then, this information is sent to a one-dimensional CNN (therefore which is suitable for working on time series) which will then predict whether the person who wrote the text has expressed anger, joy, or sadness. Also for the CNN-LSTM, we can consider that the CNN is the phase of preparation of the data for the LSTM. This phase is adaptive/intelligent, unlike a more classic pre-processing (with mathematical/statistical tools), since the CNN will work on each image differently depending on its content.

In short, a hybrid network is simply a chain of two networks. We use the notion of hybrid to emphasize that their work is complementary/intricate. Since we are talking about learning, in the case of CNN-LSTM networks example, the learning can be in parallel or pre-trained the CNN, it depends on the use case and the resources available.

5 Experiment and Results

In this study, we use UCI ML Drug Review dataset (https://www.Druglib.com) [9]. The dataset provides patient opinions on specific medications and associated conditions, with overall patient satisfaction indicated by a 10-star rating system. The data was obtained by exploring online pharmaceutical review sites. The dataset consists of over 4143 drug reviews conducted for over 700 different medical conditions, respectively split into 75% training set and 25% test set.

To develop our system and validate our proposal, we used the Python programming language and the Google Colab environment. For deep learning we used Google's library called TensorFlow, and also the Numpy library to manipulate matrices.

5.1 Hyper-Parameters and Training

Before text data is introduced into deep learning classifiers for Drug reviews sentiments analysis, the sentiment polarity scores are applied via the Vader sentiment function resulting in three labels positive, neutral, and negative. Intending to inject the least amount of linguistic information by hand, we perform very little preprocessing as:

- Tokenization: the text must be parsed to remove words;
- Deletion of punctuation; special characters;

For CNN architecture Fig. 2, in the following work, we use a small network with 3 paralleled convolution layers, with filters of size 512, 512, and 256 on the sentence length and of size 200 on the word-vector size (which corresponds in fact to the length of the vectors); after each convolution layer, a resizing is applied with a max-pooling layer, thereafter, to improve generalization a dropout layer of 0.2 is added. Finally, we add a layer with the softmax activation function which allows us to transform the output of the network into a probability distribution on the different possible categories (positive, negative, neutral).

During our experiments, we created two RNN models with different architectures for LSTM/BiLSTM and GRU/BiGRU and hybrid models. Their following architectures are presented as follows:

For LSTM architecture Fig. 3, the Input Layer includes the batch size (batch size = 128), the maximum length of the sequence (max length = 7484). Embeddings are the

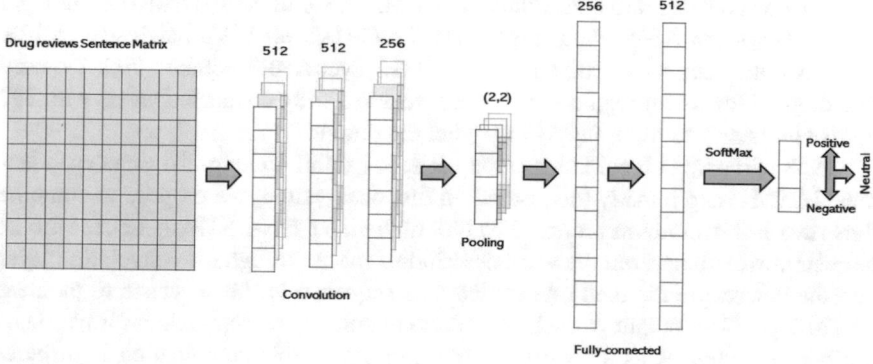

Fig. 2 CNN architecture

Fig. 3 LSTM architecture

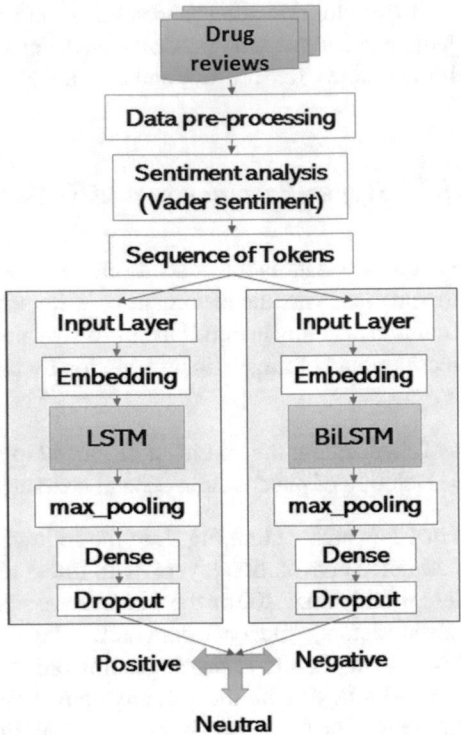

max length of the sequence the size of the embeddings is equal to 200. Three LSTM layers with 256 hidden nodes for each layer are applied followed by a max-pooling layer. In our model, three fully connected layers are used with the softmax activation function to get the desired output. Dropout is applied to allow faster learning. The same architecture is duplicated in BiLSTM architecture one for each direction.

The applied GRU integrates an embedded layer, which is necessary to convert the tokens in our dataset into embedding of size 50. The bidirectional GRU input layer (i.e. the output of the previous layer) like the GRU layer has a batch size of 128, the maximum length of 13000, and each GRU layer size is equal to 200. We used two dropout layers for regularization, and we add 3 fully connected layers with 200 hidden units and softmax supply to predict the output.

CNNs can extract local information but seem to fail on long-distance dependencies; LSTMs help remedy this. Based on this observation, we propose to combine these two architectures as proposed by [40] with their CNN-LSTM model, the model extracts correlations from the word description space through the convolution layer and the information is then transmitted as a sequence to the sequence to the next LSTM layer. The output of each convolution contains the semantic information of the whole sentences as a sequence. The purpose of this combination is to independently learn the best possible representation of emotions in Drugs review. It is important to have a module that is as light as possible in terms of parameters but which brings a performance of result as well as certain stability compared to the recurring module which follows it.

The CNN-LSTM architecture is built as follows: the Input Layer includes the batch size (batch size = 128), a dropout type regularization layer of 0.25, a convolution layer with 64 filters, a convolution filter of size 5, and activation of type ReLU, followed by a layer of max pooling aggregation, thereafter, a second convolution layer with 64 filters, a convolution filter of size 5 and activation of type ReLU. A max-pooling layer, a Dropout-type regularization layer of 0.4, beyond that, an LSTM layer of 70 connected to a FC layer. Learning is performed by the Adam [17] optimizer gradient descent.

For the LSTM-CNN model as proposed in [38], each feature sequence extracted by the LSTM model, is then injected into a convolution layer followed by a max-pooling operation and a FC layer. The parameters setting are batch size = 128, 100 hidden node in the LSTM, LR equal to 0.1 and Adam optimization strategy, filter size 5×64.

In order to aggregate the advantages from one model to another, we propose to combine CNN and GRU upstream and downstream. The structure of the CNN module contains 2 convolution layers followed by a max-pooling operation and a flatten operation.

In the CNN-GRU model, the CNN layers are used for spatial features extraction with 64 convolutions (with a kernel size of 3), extended to a dropout 0.2 and a flattening operation, where each convolution layer contains a convolution operation and a max-pooling operation, which then fed into GRU with 64, dropout = 0.1, recurrent dropout = 0.5. Finally, a dense layer of 3 outputs is applied.

The GRU-CNN model uses GRU to learn context features and then inject those context-aware vectors into CNN. The parameters setting are 60 units; 50 filters with kernel size equal to 2 and dropout 0.2.

5.2 Results and Discussion

For the test phase, we calculate the accuracy of a data test, if the accuracy is high (more than 70%) we fix the model otherwise, we repeat the treatment with other hyper-parameters. Our experimentation was applied to recurrent and convolutional methods and their combination as mentioned before, the evaluation was done in terms of precision metrics, as summarized in Table 1.

Table 1 describes the best results obtained for each of the models tested according to the parameters mentioned previously and the post-processing sets. The results obtained are satisfactory with a strong overall performance.

First, it has been shown that a simple structure (such as LSTM or GRU) can provide very good precision. This is not a general rule, as the results mainly depend on the

Table 1 Performance results for Drugs reviews classification

Model	Precision
CNN	0.8133
LSTM	0.8644
GRU	0.8313
BiLSTM	0.8633
BiGRU	0.8555
CNN-LSTM	0.8103
LSTM-CNN	**0.8844**
CNN-GRU	0.8390
GRU-CNN	0.8580

amount of the dataset used as well as the quality of the data. In our experiments, LSTM performs better than GRU, its lower performance is due to the three-gate structure. However, the interest and superiority of GRU can be observed in the case of reduced and noisy data.

Secondly, the same observation can be made concerning the BiLSTM and BiGRU models, an improvement is recorded for the simple models. Nonetheless, to be able to carry out the gradient backpropagation in the bidirectional GRUs architecture, it is necessary to keep in memory all the internal activations of the RNNs during the forward and backward passes of the MLP output. Depending on the size of the sequence and/or RNNs this can have a significant impact on the amount of memory used during learning.

From Table 1, we can remark that the best predictions of the proposed hybrid models are achieved with LSTM-CNN and GRU-CNN architectures. LSTM and GRU are applied for extracting word-level representation, the learning work is conducted by CNN. In our experiments, GRU provides less information than LSTM which explains why the results of GRU-CNN are much less efficient than the LSTM-CNN model.

At the level of the convolution module of our CNN-LSTM and CNN-GRU network, the drop in performance can be explained by three hypotheses: Connect raw data from the input layer led to a lot of noise in the CNN layer, moreover, the application of pooling layers between convolution layers. Indeed, even if this type of layer allows to decrease the number of parameters and thus calculation in the network, it quite brutally reduces the size of the representations, in particular temporal, and therefore, there is an associated loss of information. Finally, CNNs are designed for getting different size features in computer vision tasks, it is not fit for our drugs review sequence classification task. The precision results show the genericity of the approaches tested. Indeed, despite a slight post-treatment on drugs reviews, the test results are satisfactory. Moreover, these results show that the hybrid models maximize the precision rates (88.44%) and confirm the complementarity of the two approaches.

Our preliminary approaches have many limitations which we detail in this section. Regarding the classification phase, the main limitation is that we only used precision as a metric to assess the performance of classifiers. Recall and precision are two important measures in assessing the quality of a classifier and should be taken into consideration. Another limitation relates to the interpretation of deep learning is that these later do not provide any transparency or interpretation even though the model is performing well [31]. Nevertheless, despite the small manually annotated messages, this works reveal a clear superiority of deep classifiers performances compared to classical approaches. Advanced DL algorithms are explored to enable the interpretation of these models [19] and might be useful in this context. Finally, the last limitation of our study is linked to the generalization of our results. Indeed, the task is very specific to the study subject, the small datasets, as well as the type of texts, make it difficult to generalize our approach. Our results show variability in the quality of classification according to the datasets. This would tend to say that the analysis is specific to them and consequently that the networks only adapt without generalizing their knowledge. However, our proposal and results may be used as a benchmark for further works on the automatic identification of categories from textual data.

6 Conclusion

In this work, we have explored the field of sentiment analysis which, like all other fields of natural language processing, has undergone a major evolution since the 2000s and achieved a major evolution and great interest since the discovery of deep learning. In this work, we shared our contribution to the sentiment analysis problem for drug reviews, representing the tools and datasets used, as well as the steps we have taken to obtain the results through the methods of CNN, RNN as well as hybrid models. Our approach highlighted the effectiveness of deep learning architectures for predicting categories on health forum data. The applicability of our study could serve patients and healthcare professionals through social media platform. For future perspectives and approaches that may improve our sentiment classification system, we are working on the application of Transformers for better analysis and explainable tools for more transparency.

References

1. Ain, Q.T., Ali, M., Riaz, A., Noureen, A., Kamran, M., Hayat, B., Rehman, A.: Sentiment analysis using deep learning techniques: a review. Int. J. Adv. Comput. Sci. Appl. **8**(6) (2017). http://dx.doi.org/10.14569/IJACSA.2017.080657
2. Bengio, Y., Simard, P., Frasconi, P.: Learning long-term dependencies with gradient descent is difficult. Trans. Neur. Netw. **5**(2), 157–166 (1994). https://doi.org/10.1109/72.279181

3. Chandra, R., Krishna, A.: Covid-19 sentiment analysis via deep learning during the rise of novel cases. PLOS ONE **16**(8), 1–26 (2021). https://doi.org/10.1371/journal.pone.0255615
4. Cho, K., van Merrienboer, B., Gülçehre, Ç., Bougares, F., Schwenk, H., Bengio, Y.: Learning phrase representations using RNN encoder-decoder for statistical machine translation. CoRR (2014). http://arxiv.org/abs/1406.1078
5. Colon-Ruiz, C., Segura-Bedmar, I.: Comparing deep learning architectures for sentiment analysis on drug reviews. J. Biomed. Inf. **110**, 103539 (2020). https://www.sciencedirect.com/science/article/pii/S1532046420301672
6. Dermouche, M., Khouas, L., Velcin, J., Loudcher, S.: How to learn with Naive Bayes and Prior knowledge: an application to sentiment analysis. In: International Workshop on Semantic Evaluation. Atlanta, USA (2013). https://halshs.archives-ouvertes.fr/halshs-01100027
7. Fersini, E.: Chapter 6—sentiment analysis in social networks: a machine learning perspective. In: Pozzi, F.A., Fersini, E., Messina, E., Liu, B. (eds.) Sentiment Analysis in Social Networks, pp. 91–111. Morgan Kaufmann, Boston (2017). https://www.sciencedirect.com/science/article/pii/B9780128044124000061
8. Garg, S.: Drug recommendation system based on sentiment analysis of drug reviews using machine learning. CoRR (2021), https://arxiv.org/abs/2104.01113
9. Gräßer, F., Kallumadi, S., Malberg, H., Zaunseder, S.: Aspect-based sentiment analysis of drug reviews applying cross-domain and cross-data learning. In: Proceedings of the 2018 International Conference on Digital Health, pp. 121–125 (2018). https://doi.org/10.1145/3194658.3194677
10. Gurdin, G., Vargas, J.A., Maffey, L.G., Olex, A.L., Lewinski, N.A., McInnes, B.T.: Analysis of inter-domain and cross-domain drug review polarity classification. AMIA Joint Summits on Translational Science proceedings. AMIA Joint Summits on Translational Science, vol. 2020, pp. 201–210, May 2020. https://pubmed.ncbi.nlm.nih.gov/32477639
11. Han, Y., Liu, M., Jing, W.: Aspect-level drug reviews sentiment analysis based on double bigru and knowledge transfer. IEEE Access **8**, 21314–21325 (2020). https://doi.org/10.1109/ACCESS.2020.2969473
12. Hancock, K., Clayton, J.M., Parker, S.M., der, S.W., Butow, P.N., Carrick, S., Currow, D., Ghersi, D., Glare, P., Hagerty, R., Tattersall, M.H.: Truth-telling in discussing prognosis in advanced life-limiting illnesses: a systematic review. Palliat. Med. **21**(6), 507–517 (2007). https://doi.org/10.1177/0269216307080823
13. Harrison, C.J., Sidey-Gibbons, C.J.: Machine learning in medicine: a practical introduction to natural language processing. BMC Med. Res. Methodol. **21**(1), 158– (2021). https://doi.org/10.1186/s12874-021-01347-1
14. Hochreiter, S., Schmidhuber, J.: Long short-term memory. Neural Comput. **9**(8), 1735–1780 (1997). https://doi.org/10.1162/neco.1997.9.8.1735
15. Imamah, Husni, Rachman, E.M., Suzanti, I.O., Mufarroha, F.A.: Text mining and support vector machine for sentiment analysis of tourist reviews in bangkalan regency **1477**, 022023 (2020). https://doi.org/10.1088/1742-6596/1477/2/022023
16. Kennedy, A., Inkpen, D.: Sentiment classification of movie reviews using contextual valence shifters. Comput. Intell. **22**(2), 110–125 (2006). https://onlinelibrary.wiley.com/doi/abs/10.1111/j.1467-8640.2006.00277.x
17. Kingma, D.P., Ba, J.: Adam: A method for stochastic optimization (2014). arXiv:1412.6980
18. Lecun, Y., Bengio, Y.: Convolutional Networks for Images, Speech and Time Series, The Handbook of Brain Theory and Neural Networks, pp. 255–258. The MIT Press (1998) https://dl.acm.org/doi/10.5555/303568.303704
19. Lipton, Z.C.: The mythos of model interpretability. CoRR (2016). http://arxiv.org/abs/1606.03490
20. Liu, B.: Sentiment Analysis and Opinion Mining. Morgan & Claypool Publishers (2012) https://doi.org/10.2200/S00416ED1V01Y201204HLT016
21. Liu, B., Zhang, L.: A Survey of Opinion Mining and Sentiment Analysis, pp. 415–463. Springer US, Boston, MA (2012). https://doi.org/10.1007/978-1-4614-3223-4_13

22. Min, Z.: Drugs reviews sentiment analysis using weakly supervised model*. In: 2019 IEEE International Conference on Artificial Intelligence and Computer Applications (ICAICA). pp. 332–336 (2019). https://doi.org/10.1109/ICAICA.2019.8873466
23. Morsy, S.A., Rafea, A.: Improving document-level sentiment classification using contextual valence shifters. In: Proceedings of the 17th international conference on Applications of Natural Language Processing and Information Systems NLDB'12, pp. 253–258 (2012) https://doi.org/10.1007/978-3-642-31178-9_30
24. Muller, B., Reinhardt, J., Strickland, M.T.: Btt: Back-propagation through time. In: Muller, B., Reinhardt, J., Strickland, M.T. (eds.) Neural Networks: An Introduction, pp. 296–302. Springer, Berlin, Heidelberg (1995). https://doi.org/10.1007/978-3-642-57760-4_28
25. Nair, V., Hinton, G.E.: Rectified linear units improve restricted boltzmann machines. In: Proceedings of the 27th International Conference on International Conference on Machine Learning, pp. 807–814. ICML'10, Omnipress, Madison, WI, USA (2010) https://dl.acm.org/doi/10.5555/3104322.3104425
26. Pak, A., Paroubek, P.: Twitter as a corpus for sentiment analysis and opinion mining. In Proceedings of the Seventh International Conference on Language Resources and Evaluation (LREC'10), Valletta, Malta. European Language Resources Association (ELRA). https://aclanthology.org/volumes/L10-1/
27. Pang, B., Lee, L.: Opinion mining and sentiment analysis. Found. Trends Inf. Retr. 2(1–2), 1–135 (2008)
28. Pratama, Y., Tampubolon, A.R., Sianturi, L.D., Manalu, R.D., Pangaribuan, D.F.: Implementation of sentiment analysis on twitter using naive Bayes algorithm to know the people responses to debate of DKI jakarta governor election. J. Phys.: Conf. Ser. IOP Publishing 1175, 012102 (2019). https://doi.org/10.1088/1742-6596/1175/1/012102
29. Schuster, M., Paliwal, K.: Bidirectional recurrent neural networks. IEEE Trans. Signal Proc. 45(11), 2673–2681 (1997). https://doi.org/10.1109/78.650093
30. Shiju, A., He, Z.: Classifying drug ratings using user reviews with transformer-based language models. medRxiv (2021). https://www.medrxiv.org/content/early/2021/04/20/2021.04.15.21255573
31. Shwartz-Ziv, R., Tishby, N.: Opening the black box of deep neural networks via information. CoRR (2017). http://arxiv.org/abs/1703.00810
32. Statista_Research_Department: Number of social network users worldwide from 2017 to 2025. Statista, 10 Sep 2021. https://www.statista.com/statistics/278414/number-of-worldwide-social-network-users/
33. Thoomkuzhy, A.M.: Drug reviews: cross-condition and cross-source analysis by review quantification using regional CNN-LSTM models. Masters Dissertation. Master's thesis, Technological University Dublin (2020). https://arrow.tudublin.ie/scschcomdis/202/
34. Valle-Cruz, D., López-Chau, A., Sandoval-Almazán, R.: Impression analysis of trending topics in twitter with classification algorithms. In : Proceedings of the 13th International Conference on Theory and Practice of Electronic Governance, pp. 430–441. ICEGOV 2020. https://doi.org/10.1145/3428502.3428570
35. Vijayaraghavan, S., Basu, D.: Sentiment Analysis in Drug Reviews using Supervised Machine Learning Algorithms. CoRR (2020). https://arxiv.org/abs/2003.11643
36. Vineeta, Manek, A.S., Mishra, P.: UFMDRA: Uterine fibroid medicinal drugs review analysis 1110(1), 012006 (2021). https://doi.org/10.1088/1757-899x/1110/1/012006
37. Youbi, F., Settouti, N.: Analysis of machine learning and deep learning frameworks for opinion mining on drug reviews. Comput. J. (2021). https://doi.org/10.1093/comjnl/bxab084
38. Zhang, J., Li, Y., Tian, J., Li, T.: Lstm-cnn hybrid model for text classification. In: 2018 IEEE 3rd Advanced Information Technology, Electronic and Automation Control Conference (IAEAC), pp. 1675–1680 (2018). https://doi.org/10.1109/IAEAC.2018.8577620
39. Zhang, L., Wang, S., Liu, B.: Deep learning for sentiment analysis: a survey. WIREs Data Mining Knowl. Dis. 8(4), e1253 (2018). https://wires.onlinelibrary.wiley.com/doi/abs/10.1002/widm.1253
40. Zhou, C., Sun, C., Liu, Z., Lau, F.C.M.: A C-LSTM neural network for text classification. CoRR (2015). http://arxiv.org/abs/1511.08630

Text-Based Sentiment Analysis Using Deep Learning Techniques

Siddhi Kadu and Bharti Joshi

Abstract Today's world is the word of the internet and word of information. In this modern area of development, technology has contributed a lot to the global platform called Web. Reviews and emotions play a vital role in our day to day lives as they help in learning communication, decision making, product evaluation, election prediction. Artificial Intelligence (AI) is the branch of computer science that has worked on the analysis of the reviews as well as opinions generated by the people, and helps the media in order to cope with the situation. Currently to improve the marketing strategy and product advertisement traditional web-based survey methods have been replaced with the Sentiment Analysis which improves customer service. Therefore, various approaches such as machine learning, lexicon-based, hybrid, and other approaches were used to analyze these sentiments/opinions in the past. With the current advancements in deep neural networks, deep learning-based methods are becoming very popular due to their accuracy enhancement in recent times. Various methods like Convolutional Neural Network (CNN), Recurrent Neural Network (RNN), Long Short-Term Memory (LSTM), Gated Recurrent Unit (GRU), and Bi-LSTM (Bidirectional LSTM) are used for sentiment analysis. This work highlights different deep learning techniques used for text -based sentiment analysis for reviews generated by users.

Keywords Artificial intelligence · Sentiment Analysis · Deep learning

1 Introduction

Today because of tremendous growth in correspondence media and progression of innovation, 4G and 5G web has become a significant piece of society and number

S. Kadu (✉) · B. Joshi
Ramrao Adik Institute of Technology, D.Y. Patil Deemed to be University,
Nerul, Navi Mumbai, India
e-mail: siddhi.kadu1989@gmail.com

B. Joshi
e-mail: bharti.joshi@rait.ac.in

of applications, for example, Sentiment Analysis, Electronic voice assistant, online searching, and Machine Translation that relate the people and computers communication using Natural Language are been used as our everyday activities [1]. Presently in this advanced time, the most significant thing is information, or say data which is produced each day via web-based media as text or discourse. Automation to completely examine this content and discourse information productively is the basic errand and thus, Natural Language Processing (NLP), a branch of Artificial Intelligence (AI), is used to get semantics from text or speech and encode into a structured format and also it is used for different kinds of computer processing [2, 3]. The present machines can examine more language-based data than people, without exhaustion and in a steady, fair-minded way to understand text, hear speech, interpret it, measure sentiment and sort out what parts are important.

Online verbal exchange has become a vital resource. It is seen that individuals are making use of the internet for different activities such as Online Shopping, Social Networking, Blog posting, and so forth. A tremendous amount of information or opinion is available easily on the internet related to specific services, products or concerned individuals are openly available in the reviews form. These reviews are nothing but the experience of the customer using particular products and services, which can be used as a useful source of information for decision making [4]. The Reviews are useful for both the customer as well as the manufacturers. A customer has an interest in such reviews to figure out which product or a particular brand to purchase, estimates the pros and cons of products. Manufacturers can find out the improvement region in their item from such obstinate reviews. All these feelings, attitudes, views, and opinions which express the behavior of people are nothing but Sentiment.

Sentiment Analysis (SA) is an important, detailed study, and a tough process in the area of NLP. Many purchasers are dependent on customer reviews about a product, they read and analyze thousands of reviews and measure the degree of customer fulfillment for deciding whether to purchase a specific product or not. With the help of these reviews companies can gather information about their product services and can improve accordingly which can be further useful for the product development. Analyzing this massive amount of data manually is a tedious task, which utilizes human resources as well as costs financially high. Many researchers in this field tried to extract the summary or important part from the information in terms of reviews analysis, feedback analysis, and question and answering rating using various techniques but, only a little part of this information is saturated and maximum information is left behind. With the help of Sentiment Analysis, the major part of this information can be extracted from the reviews available, which will be helpful in any domain of the area to make the application [4, 15].

Sentiment analysis is a booming research area and can be useful in many fields, which engrossed many researchers to propose, evaluate, and compare different methods continuously, such as lexicon-based, machine learning (ML), hybrid approaches, and other approaches. ML has shown progress in various advanced learning algorithms, and pre-processing techniques. Due to the progression of computer technology, the upsurge of data size and the proliferation of neural networks has given rise to deep neural networks (DNNs). To enhance the accuracy and produce better results DNN has influenced many researchers and has been used in recent years for sentiment analysis for supervised and unsupervised learning. DNN has achieved huge achievement in ML, especially in NLP and computer vision [1, 5, 12, 13].

The key contributions of the chapter are as follows: The goal of the chapter is analyzing the customer reviews and then classifying them to positive or negative; by using different deep learning algorithms like Convolutional Neural Network (CNN), Recurrent Neural Network (RNN), Long Short-Term Memory (LSTM), Gated Recurrent Unit (GRU) and, Bi-LSTM (Bidirectional LSTM) for sentiment analysis based on text data. The comparative analysis of the above algorithms based on different hyperparameters using evaluation metrics is shown in the chapter. The remaining chapter is planned in the following sections. Various papers including studies associated with sentiment analysis using different deep learning algorithms are discussed in Sect. 2. Section 3 explores different existing approaches of sentiment analysis for text data. Section 4 describes the five selected deep learning architectures for analyzing sentiments. In Sect. 5 experimental work is performed which includes design, dataset and hyperparameters considered. The results of the experiments are shown in Sect. 6. The chapter is concluded in Sect. 7.

2 Related Work

The research work elaborated by different researchers in the area of sentiment analysis of specifying opinions about the products and services that have been extensively used by the people is presented in the section. Previously, methods such as lexicon-based approaches and non-neural network classifier approaches were used for analyzing sentiments by extracting emotional values and constructing an emotional dictionary, etc. which is an inefficient and tedious procedure also, the main process of extracting features is the most laborious one. Recently various deep learning algorithms made amazing progress in areas like pattern recognition, computer vision, and also according to recent research it has also shown remarkable growth in the field of NLP. Therefore, we investigate and discuss recent related papers based on deep learning methods.

Jain et al. [6] have presented a hybrid system combining CNN- LSTM system (CNN Followed by LSTM) for sentiment analysis, considering batch normalization, dropout, and max pooling hyperparameters to get outcomes on Twitter airline and Airline quality datasets using Keras embedding for the conversion of a word in the reviews into vectors. Priyadarshini et al. [7] have proposed an ensemble approach considering LSTM-CNN grid search method for sentiment analysis. To reduce predefined losses a grid search hyperparameter optimization method is used which also helps to increase the accuracy of the system on the datasets Amazon reviews and IMDB Dataset of 50K Movie Reviews. Basiri et al. [8] proposed a Model which considers long as well as short tweets for performing sentiment analysis using Bi-LSTM and GRU layers by applying attention mechanism outputs of bidirectional layers. To decrease the dimensionality of features convolution and pooling mechanisms are been used. The GloVe is used for feature extraction in text data. Ali et al. [9] explore different deep learning algorithms (Multi-Layer Perceptron (MLP), CNN, and LSTM) and proposed a CNN-LSTM system using an IMDB dataset of 50K movies reviews and Word2vector method for word embedding. To improve and lessen the results obtained by the convolutional layer, the Maxpooling mask is applied.

Ligthart et al. [10] perform a study of the important topics and the various approaches for different tasks in sentiment analysis. Various features, algorithms, datasets, Challenges, and open problems are identified. Additionally, hundred and twelve recent deep learning algorithms are categorized on sentiment analysis. Based on this analysis, it is observed that LSTM and CNN algorithms are the most utilized. Sivakumar et al. [11] presented a long-term short memory (LSTM) with fuzzy logic a multi-class classification system that considers reviews generated by the customer at the sentence level. The labels categorized are positive, negative, highly positive, and highly negative on datasets of Amazon video games reviews and consumer reviews of amazon products. For feature extraction word embedding methods such as a bag of words have been used. Xu et al. [12] introduced an advanced word representation, which incorporates the traditional TF-IDF algorithm for generating word vectors. Further, weighted word vectors are inserted into the (BiLSTM) to obtain contextual information, and vectors are represented effectively. The sentiment of comments is classified using a feedforward neural network classifier. Colon-Ruiz et al. [13] introduced comparative analysis of several deep learning methods like CNN, LSTM, RNN, and their combinations and discovers that a combination of Bidirectional Encoder Representations from Transformers (BERT) and Bi-LSTM shows remarkable results on the dataset of drug reviews for sentiment analysis. Further the results of various pre-trained word embedding models are also studied. Gandhi et al. [14] present how CNN and LSTM architectures work with stop words and Word2vec preprocessing methods on the IMDB dataset containing 50,000 movie reviews. Nemes et al. [15] developed an RNN model to predict the emotional tweets and classifies into positive, negative weakly positive or negative, and strongly positive or negative.

Beseiso et al. [16] presented a text classification model using CNN which uses character level for analysis and RNN stacked on top of one another which works on a dataset of Arabic tweets and as the data is limited author proposed a data augmentation method for the problem. ONAN et al. [17] performed the analysis on MOOC reviews to identify sentiments by applying supervised learning approaches, ensemble learning approaches, deep learning approaches and evaluate the efficiency, and concludes that deep learning-based methods outperform. The author also evaluates the effectiveness of text representation structures and word-embedding structures. A comparative analysis is performed and maximum analytical performance has been attained by LSTM in aggregation with GloVe and scheme-based representation. Wadawadagi et al. [18] perform a study experiment on different deep learning methods for sentiment This article aims to provide an empirical study on various deep neural networks (DNN) used for classifying sentiments and also discuss various applications. Further, the performance evaluation of various deep learning algorithms is performed by fine-tuning different hyperparameters and by using data visualization methods various results have employed. Salur et al. [19] proposed an innovative hybrid deep learning structure with a combination of various word embedding like Word2Vec, Fast-Text, character-level embedding with different deep learning approaches such as LSTM, GRU, BiLSTM, CNN for classifying sentiments and compare the model with existing various basic models by performing experiments to prove the efficiency of the proposed structure. Kula et al. [20] presented a novel approach for the detection of fake news by using various deep learning approaches such as RNN, GRU, and LSTM for classification and use GloVe word embedding method by using Flair library which provides promising outcomes. Patel et al. [21] applied deep learning algorithm RNN for classifying text and measures the efficiency of the classifier based on pre-processing of data, and obtained the accuracy of 94.61%. Ni et al. [22] present a method for analyzing sentiments at sentence level based on the GloVe word model and RNN. Also, LSTM and GRU models are combined as RNN cannot work for long-term data, and at last the comparative analysis of models is performed it is found that LSTM- GRU model achieves the greatest efficiency. Santur et al. [23] implement the system for performing sentiment analysis using GRU on Turkish e-commercial platform datasets and achieve an accuracy of 95%.

Hence, deep learning has become an area of interest for researchers to study because of its benefits over other traditional approaches as it decreases the problem of designing features as it can be trained on huge volume with improved performance and are therefore adapted to other domains like Medical Image Segmentation, Image Processing, Speech Recognition, Internet of Things.

3 Taxonomy of Sentiment Analysis Methods

The current approaches for sentiment analysis are divided into ML methods, Hybrid methods, and Lexicon-based methods. The ML approach classifies sentiments using various machine learning techniques and linguistic features. The lexicon-based method classifies sentiments into positive and negative using a list of phrases and words. Hybrid methods use machine learning and lexicon-based methods to enhance performance in the area of SA. Figure 1 discusses the taxonomy of sentiment analysis approaches [24].

3.1 Machine Learning Methods

Machine Learning (ML) methods are useful for classifying sentiments into positive, negative, and neutral polarity based on train and test data, which are further divided into supervised, unsupervised, semi-supervised learning (SSL) [10, 24, 25].

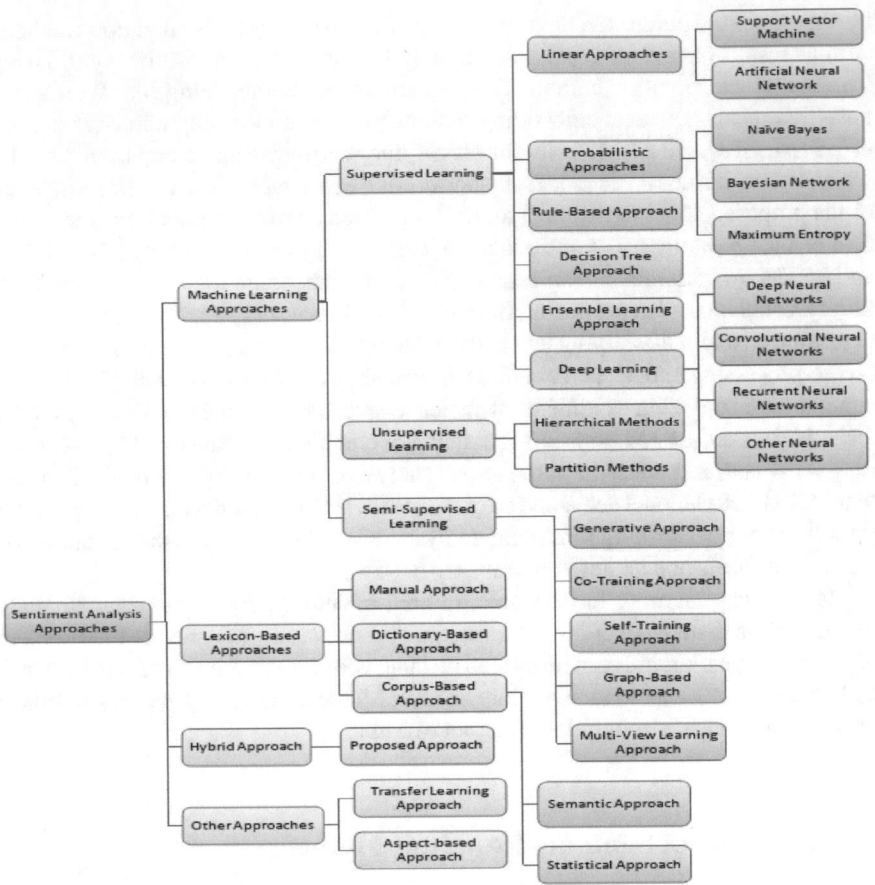

Fig. 1 Sentiment analysis methods [24]

3.2   Lexicon Based Methods

It is the traditional method that has been used for sentiment analysis and is further divided into Manual, Dictionary-based, and Corpus-based methods [10, 24, 25].

3.3   Hybrid Methods

The hybrid method integrates dictionary and machine learning methods. The hybrid method inherits the rise in accuracy from the ML approach and constancy from the Dictionary-based approach. It incorporates all the techniques from the previous two ways to overcome their limitations and use their advantages [10, 24, 25].

3.4 Other Methods

The firstly aspect-based method is a well-balanced activity for analyzing sentiments by predicting the polarities of the sentiment of a specific feature or words focused on texts. Features, characteristics, or attributes are the different aspects that need to be considered. Transfer learning uses the data distribution, similarity of data, modeling function, etc. to use information that has already been studied in a domain. This approach needs no time to train the algorithm from the beginning [10, 24, 25].

Thus, different approaches are been considered for the text-based sentiment analysis and it is observed that deep learning approaches have been extensively used for the analysis as it has provided a remarkable result in terms of accuracy. Many other approaches such as transfer learning and other hybrid approaches have been also used.

4 Deep Learning Architectures for Sentiment Analysis

Deep learning models have become a trending topic in recent years in the area of NLP. They are made up of Artificial Neural Networks (ANN) which have the input, output, and hidden layers between them. This section discusses some common deep learning architectures like Convolutional Neural Network (CNN), Recurrent Neural Network (RNN), Long Short-Term Memory (LSTM), Gated Recurrent Unit (GRU), and Bi-LSTM (Bidirectional LSTM).

4.1 Convolutional Neural Networks (CNN)

CNN which falls under the category of ANN has shown great success and development in the field of image processing and computer vision. CNN's basic architecture is shown in Fig. 2 which consists of the input layer, pooling layer, convolutional layer, and fully connected layer [6].

4.2 Recurrent Neural Networks (RNN)

RNN is an ANN class used to create a sequential model. Sequential data is used in various applications for example, in the translation of language. The primary function of the RNN is the sequential processing of information based on internal memory taken from targeted cycles. RNNs maintain the internal conditions of inputs that process each word in a sentence repeatedly as shown in Fig. 3. So, it helps to predict

Fig. 2 General architecture
of CNN [6]

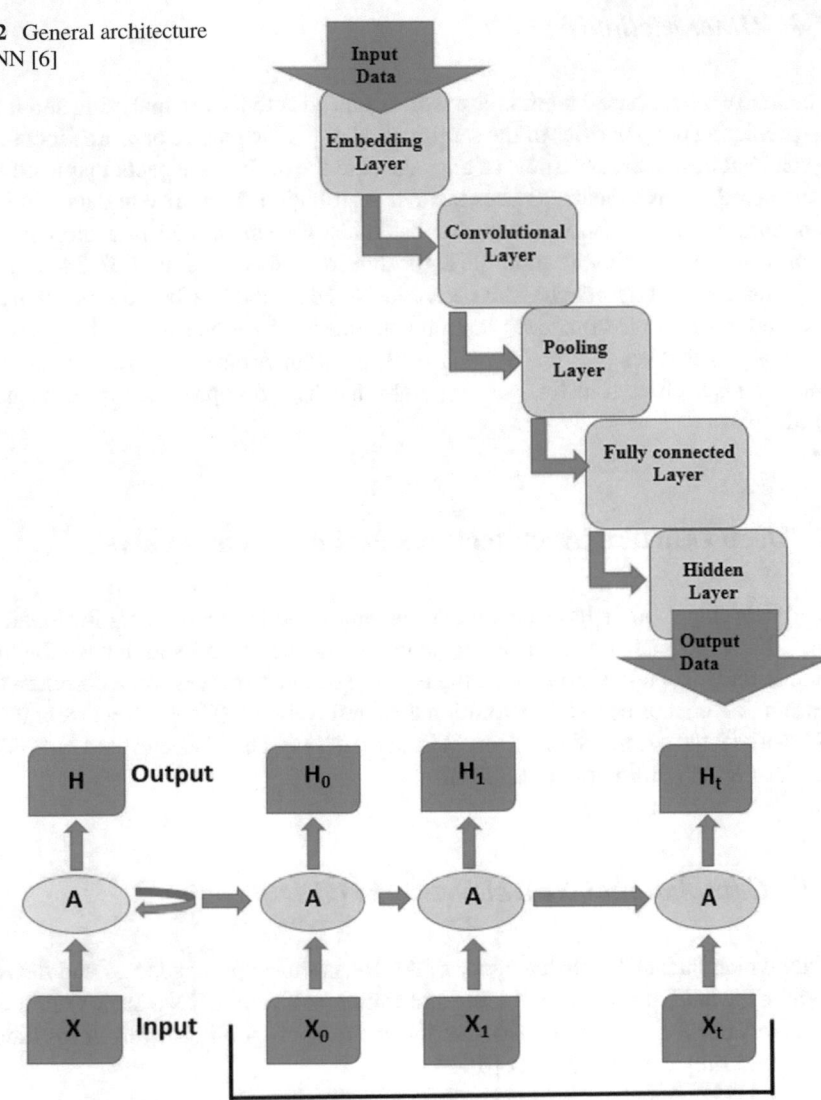

Fig. 3 General architecture of RNN [15]

the next word in a sentence by keeping all past words and relationships between them
[15].

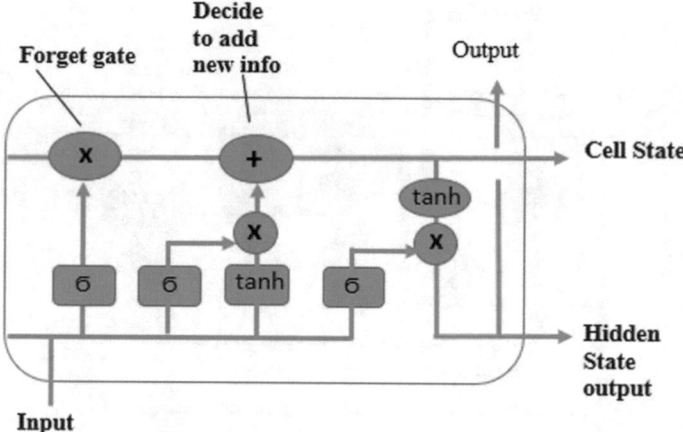

Fig. 4 General architecture of LSTM [7]

4.3 Long Short-Term Memory (LSTM)

The Long Short-Term Memory Model (LSTM) is an RNN special form, with the help of this network tasks such as prediction or classification are done. The problems such as exploding and vanishing gradient, and long-term dependencies are overcome by LSTM. As shown in Fig. 4 The standard LSTM consists of a cell, an input gate, an output gate and a forget gate. The output gate manages the data sent to the next layer and a forget gate manages the memory loss rate and determines which data can be reduced to the cell state [7, 8].

4.4 Gated Recurrent Unit (GRU)

The Gated Recurrent Unit (GRU) is a basic system of the LSTM model. The GRU accurately captures the long-term dependencies despite limited parameters. Hence, GRU is measured based on efficiency and performance in comparison with LSTM. The flow of data for each hidden unit is controlled with the help of update and reset gates [22, 23] as shown in Fig. 5.

4.5 Bi-LSTM (Bidirectional LSTM)

The Bi-LSTM neural network is made up of LSTM units that operate in both directions to capture past and future contextual information. Bi-LSTM can learn long-term dependencies without having to store duplicate context information. Therefore, it has

Reset Gate

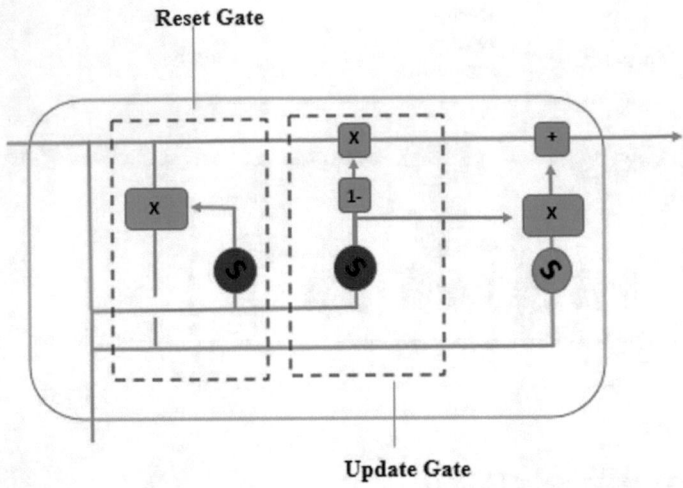

Update Gate

Fig. 5 General architecture of GRU [23]

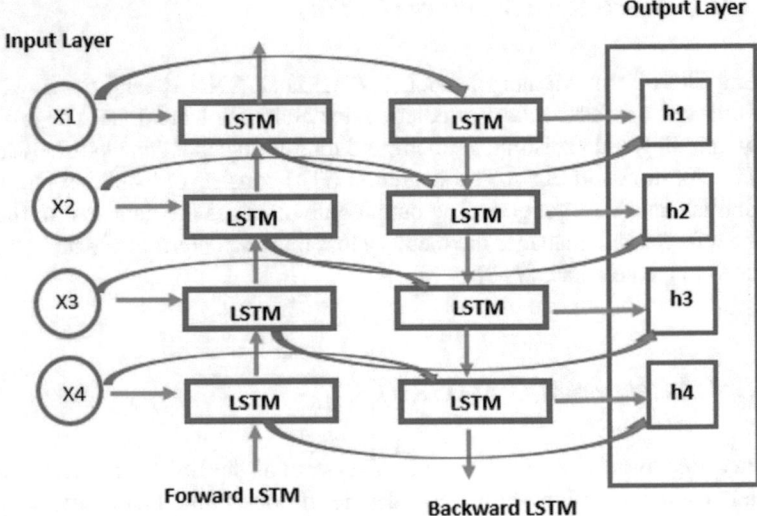

Fig. 6 General architecture of Bi-LSTM [12]

shown excellent effectiveness in sequential modeling problems and is widely used in text classification. The Bi-LSTM network has two parallel layers that are spread out on both sides with a forward and reverse pass to capture dependencies on two scenarios [12] as shown in Fig. 6.

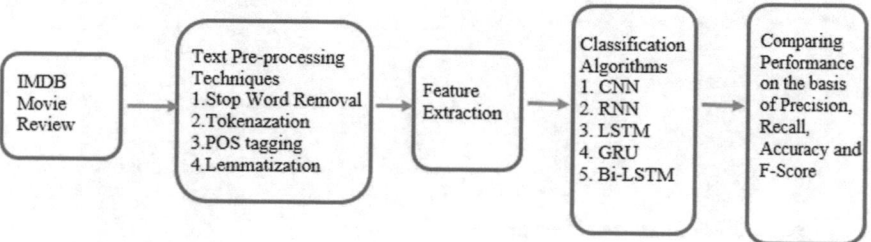

Fig. 7 Proposed framework

5   Experimental Work

The proposed framework as shown in Fig. 7 considers the IMDB movie reviews dataset then we applied text pre-processing techniques on the reviews and further extracted features using N-gram and Word2Vec techniques. Further, we implemented five classification deep learning algorithms and evaluated them based on four metrics.

5.1   Dataset Description

For the analysis purpose, we used an IMDB dataset which includes 50K movie reviews which we have divided into 40K training and 10K testing. In 40K rows, we had split data into 70:30 ratio for training and testing respectively for model Accuracy i.e., is the training accuracy and accuracy on 10k rows is the testing accuracy [7].

5.2   Data Pre-processing

Information obtained from various sources mainly from social media is in the form of raw data which contains a lot of noise such as spelling and grammar errors. Therefore, it becomes essential to process, clean, edit the text, and then perform analysis. The purpose of the pre-processing phase is to get better analysis by reducing the size of input data by deleting meaningless text. In the Fig. 8 the pre-processing techniques such as removing HTML, removing non-letters, converting to lowercase tokenization, stop words removal, part-of-Speech (Post) tagging, lemmatization were applied to the dataset [4, 7, 8].

```
def preprocess(review: str) -> list:

    # if show_progress:
    #     global counter
    #     counter += 1
    #     print('Processing... %6i/%6i'% (counter, total), end='\r')
    # 1. Clean text
    review = clean_review(review)
    # 2. Split into individual words
    tokens = word_tokenize(review)
    # 3. Lemmatize
    lemmas = lemmatize(tokens)
    # 4. Join the words back into one string separated by space,
    # and return the result.
    return lemmas

def clean_review(raw_review: str) -> str:
    # 1. Remove HTML
    review_text = BeautifulSoup(raw_review, "lxml").get_text()
    # 2. Remove non-letters
    letters_only = REPLACE_WITH_SPACE.sub(" ", review_text)
    # 3. Convert to lower case
    lowercase_letters = letters_only.lower()
    return lowercase_letters



def lemmatize(tokens: list) -> list:
    # 1. Lemmatize
    tokens = list(map(lemmatizer.lemmatize, tokens))
    lemmatized_tokens = list(map(lambda x: lemmatizer.lemmatize(x, "v"), tokens))
    # 2. Remove stop words
    meaningful_words = list(filter(lambda x: not x in stop_words, lemmatized_tokens))
    return meaningful_words
```

Fig. 8 Data pre-processing steps

5.3 Feature Extraction

Feature Extraction (FE) is a vital function because it directly affects the functioning of the sentimental phase in the process of sentiment analysis. The purpose is the extraction of important data (e.g., sentiment expressive words) that defines significant text data aspects. Some fundamental features that are considered for analysis of sentiments are Bag-of-Words, Word Embedding, and TF-IDF which converts the word into vectors for finding the relationship between a word and its synonymous [24, 25]. For the experiment we used Word2vec word representation is a two-layered network that analyzes text by 'vectorizing' words, in which the input is in a text corpus, and the output is a vector set: the feature vectors signify the words in that corpus.

5.4   Classification Algorithms

Convolutional Neural Networks (CNN) Model The Fig. 9 shows a summary of the CNN model with all layers with their respective output shapes. The model is defined as a sequential model as text is passed as a sequence. The embedding layer encodes each word into a vector and then the GlobalMaxPooling1D layer is applied after which all the outputs are concatenated. A Dense layer following Dropout layer and final Dense layer is applied [6–8, 14].

Recurrent Neural Networks (RNN) Model The Fig. 10 shows a summary of a simple RNN model with all layers with their respective output shapes [8, 15, 21].

Long Short-Term Memory (LSTM) Model The Fig. 11 shows a summary of the LSTM model with all layers with their respective output shapes [6, 7, 13, 14, 22].

Gated Recurrent Unit (GRU) Model The Fig. 12 shows a summary of the GRU model with all layers with their respective output shapes [22, 26].

Bi-LSTM (Bidirectional LSTM) Model The Fig. 13 shows a summary of the Bi-LSTM model with all layers with their respective output shapes [5, 12, 19].

```
Model: "sequential"

_____
Layer (type)                    Output Shape              Param #
=================================================================
embedding (Embedding)           (None, 150, 256)          13507072
_____
conv1d (Conv1D)                 (None, 150, 64)           49216
_____
global_max_pooling1d (Global    (None, 64)                0
_____
dense (Dense)                   (None, 128)               8320
_____
dropout (Dropout)               (None, 128)               0
_____
dense_1 (Dense)                 (None, 1)                 129
=================================================================
Total params: 13,564,737
Trainable params: 57,665
Non-trainable params: 13,507,072
```

Fig. 9   Model summary of CNN algorithm

```
Model: "sequential_4"
_____
Layer (type)                    Output Shape              Param #
=================================================================
embedding_4 (Embedding)         (None, 150, 256)          13507072
_____
simple_rnn_4 (SimpleRNN)        (None, 256)               131328
_____
dense_8 (Dense)                 (None, 8)                 2056
_____
dense_9 (Dense)                 (None, 1)                 9
=================================================================
Total params: 13,640,465
Trainable params: 133,393
Non-trainable params: 13,507,072
```

Fig. 10 Model summary of RNN algorithm

```
Model: "sequential"
_____
Layer (type)                    Output Shape              Param #
=================================================================
embedding (Embedding)           (None, 150, 256)          13507072

lstm (LSTM)                     (None, 256)               525312

dropout (Dropout)               (None, 256)               0

dense (Dense)                   (None, 64)                16448

dropout_1 (Dropout)             (None, 64)                0

dense_1 (Dense)                 (None, 1)                 65

=================================================================
Total params: 14,048,897
Trainable params: 541,825
Non-trainable params: 13,507,072
```

Fig. 11 Model summary of LSTM algorithm

5.5 *Evaluation Measures*

The performance of the models are evaluated using four metrics Precision (Pr), Accuracy (Acc), Recall (Re), F1-score (F 1) [24]. These metrics are widely used for tasks such as classifying text and sentiment analysis and are calculated as follows:

$$Precision = \frac{TP}{TP + FP} \tag{1}$$

```
Model: "sequential"
```

Layer (type)	Output Shape	Param #
embedding (Embedding)	(None, 150, 256)	13507072
gru (GRU)	(None, 150, 128)	148224
dense (Dense)	(None, 150, 1)	129

```
Total params: 13,655,425
Trainable params: 148,353
Non-trainable params: 13,507,072
```

Fig. 12 Model summary of GRU algorithm

```
Model: "sequential"
```

Layer (type)	Output Shape	Param #
embedding (Embedding)	(None, 150, 256)	13507072
bidirectional (Bidirectional	(None, 512)	1050624
dropout (Dropout)	(None, 512)	0
dense (Dense)	(None, 64)	32832
dropout_1 (Dropout)	(None, 64)	0
dense_1 (Dense)	(None, 1)	65

```
Total params: 14,590,593
Trainable params: 1,083,521
Non-trainable params: 13,507,072
```

Fig. 13 Model summary of Bi-LSTM algorithm

$$Recall = \frac{TP}{TP + FN} \qquad (2)$$

$$F1 - Score = 2 * \frac{Precision * Recall}{Precision + Recall} \qquad (3)$$

$$Accuracy = \frac{TP + TN}{TP + TN + FP + FN} \qquad (4)$$

Table 1 The hyperparameters of deep learning models [26]

Hyperparameters	Value
The nodes of hidden layer network	128, 256
Loss function	Categorical cross-entropy
Optimizer	Adam
Dimensions of word vector	256
Epoch	0, 10, 20, 30, 40
Batch size	100, 200

5.6 Hyperparameters of Deep Learning Models

Hyperparameters are variables that can control the learning process. Tuning hyper-parameters confirms that the model can solve the problem appropriately by lessening the predicted losses and providing precise results [26]. The hyperparameters of deep learning models are shown in Table 1.

6 Results and Discussion

In this section, the results are discussed which includes the comparative analysis of selected deep learning models like CNN, RNN, LSTM, GRU, Bi-LSTM based on different hyperparameters using evaluation metrics. The hyperparameters that are considered are the number of nodes of the hidden layer, epoch, and batch size. It is found that changing the above parameters has affected the evaluation parameters. The following figures explain the variation in accuracy of different deep learning models by changing the above-considered hyperparameters. Figure 14 shows a change in the accuracy of CNN, RNN, LSTM, GRU, and Bi-LSTM models, as the number of epochs changes from the range of 0–40 when the number of nodes is 128 and the batch size is 100 on the training dataset. It is observed that as the epoch changes the accuracy is increased.

Figure 15 shows the change in the accuracy of CNN, RNN, LSTM, GRU, and Bi-LSTM models, as the number of epochs changes from the range of 0 to 40 when the number of nodes is 128 and the batch size is 200 on the training dataset. It is observed that as the epoch changes the performance is increased but the only change in the batch size shows a minor change in the accuracy.

Figure 16 shows the change in the accuracy of CNN, RNN, LSTM, GRU, and Bi- LSTM models, as the number of epochs changes from the range of 0 to 40 when the number of nodes is 256 and the batch size is 100 on the training dataset. It is observed that as the epoch changes the performance is increased. The change in the number of nodes shows a major change and increase in the accuracy of the models as compared to previous parameters values.

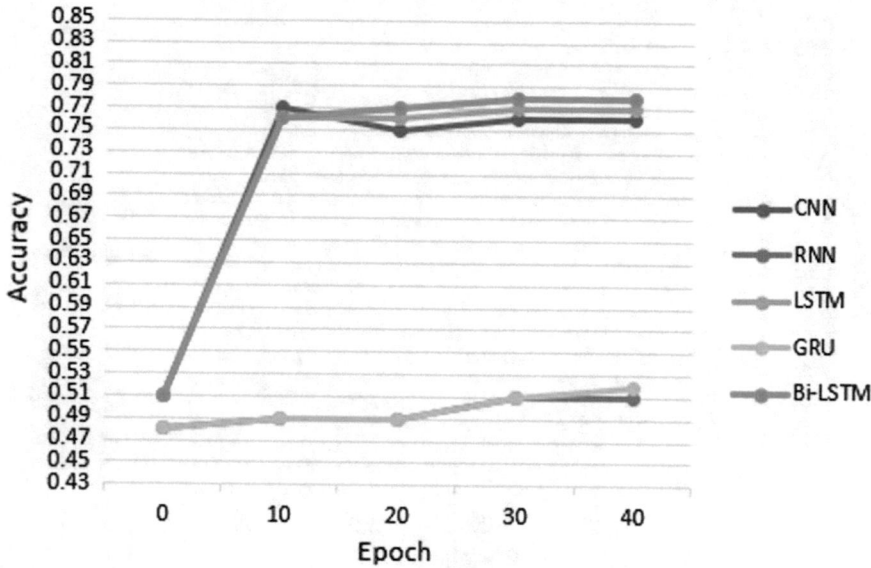

Fig. 14 Model accuracy across epoch considering number of nodes 128 and batch size 100

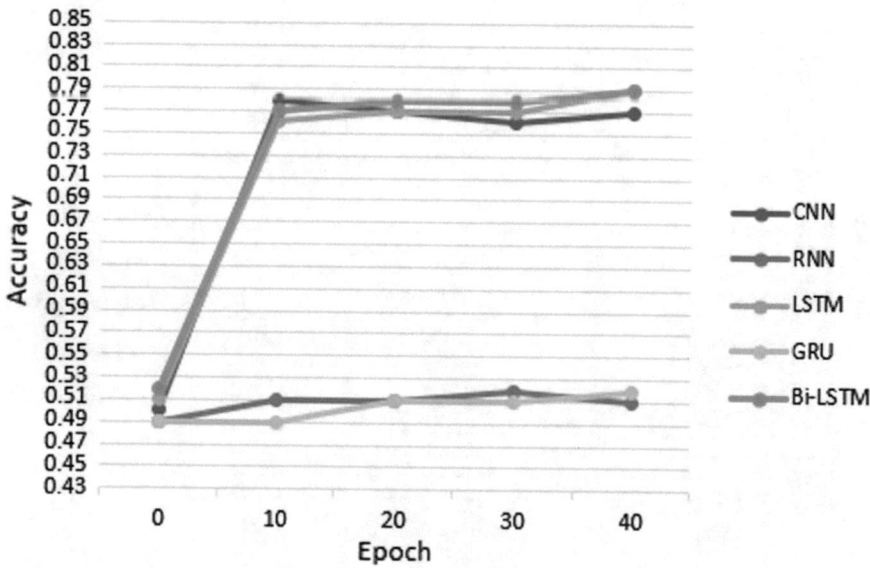

Fig. 15 Model accuracy across epoch considering number of nodes 128 and batch size 200

Figure 17 shows the change in the accuracy of CNN, RNN, LSTM, GRU, and Bi-LSTM models, as the number of epochs changes from the range of 0 to 40 when the number of nodes is 256 and the batch size is 200 on the training dataset. It is observed that as the epoch changes the performance is increase but the only change

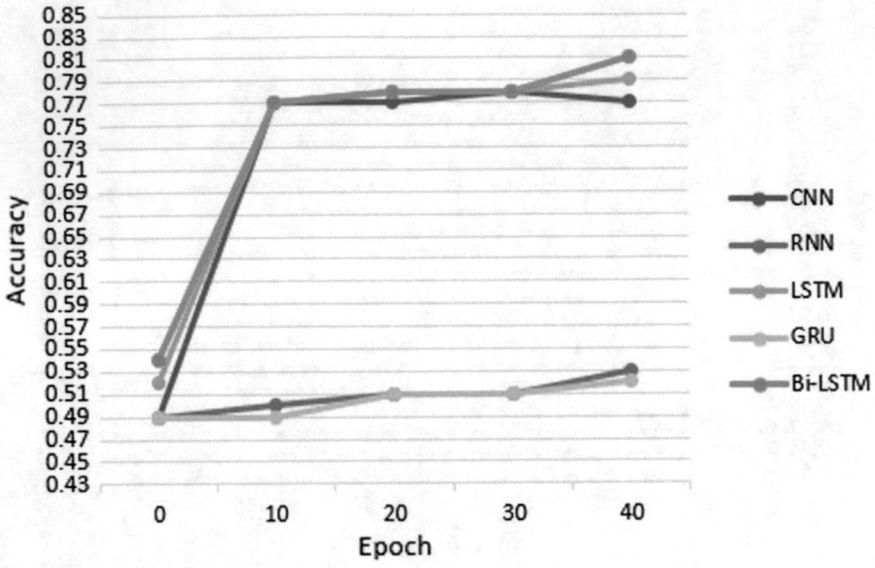

Fig. 16 Model accuracy across epoch considering number of nodes 256 and batch size 100

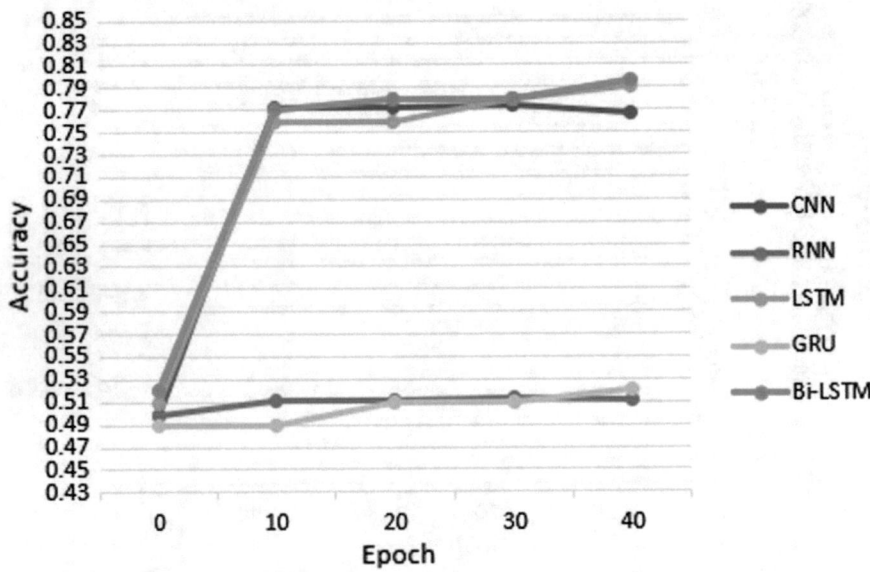

Fig. 17 Model accuracy across epoch considering number of nodes 256 and batch size 200

Table 2 Comparisons of accuracy, precision, recall, F1-score on IMDB movie reviews dataset

Algorithm	Accuracy	Precision	Recall	F1-score
CNN	77.2	79.0	74.0	76.4
RNN	51.4	51.0	92.2	66.0
LSTM	79.2	81.0	76.0	78.4
GRU	52.0	51.0	99.0	67.1
Bi-LSTM	81.0	78.0	80.4	79.0

in the batch size shows a minor change in the performance. The following Table 2 discusses the comparison of different models based on evaluation metrics such as Acc, Pre, Re, F1 on the IMDB movie reviews dataset and it is observed that LSTM and Bi-LSTM show better performance as compared to other models in terms of accuracy on considering hyperparameters, the number of nodes as 256, batch size as 100 on epoch 40.

7 Conclusion

Sentiment Analysis plays a vital role in natural language processing, which is used in many applications. Previous research focused on conventional methods which show a decrease in performance when the data size increases and therefore non-traditional methods which include artificial intelligence is used for better performance. In this chapter, comparative experimentation is shown using different deep-learning architectures like CNN, RNN, LSTM, GRU, and Bi-LSTM for the classification of sentiments using the IMDB Movie Reviews Dataset. The accuracy results of the selected models are compared based on changing hyperparameters such as batch size, epochs, and the number of nodes, and it is been observed that models such as LSTM and Bi-LSTM performs better than other models and shows an accuracy of about 79 and 81% on considering hyperparameters of the number of nodes as 256, batch size as 100 on epoch 40 on training data. Further, hybridization of various models with algorithms can be performed to gain better accuracy. Hyperparameters other than epoch, batch size, and number of hidden layers can be considered to obtain better performance. multiple datasets and applications can be evaluated on proposed models.

References

1. Tembhurne, J.V., Diwan, T.: Sentiment analysis in textual, visual and multimodal inputs using recurrent neural networks. Multimed. Tools Appl. **80**(5), 6871–6910 (2021)
2. Al-Moslmi, T., Ocaña, M.G., Opdahl, A.L., Veres, C.: Named entity extraction for knowledge graphs: a literature overview. IEEE Access **8**, 32862–32881 (2020)

3. Otter, D.W., Medina, J.R., Kalita, J.K.: A survey of the usages of deep learning for natural language processing. IEEE Trans. Neural Netw. Learn. Syst. **32**(2), 604–624 (2020)
4. Duong, H.T., Nguyen-Thi, T.A.: A review: preprocessing techniques and data augmentation for sentiment analysis. Comput. Soc. Netw. **8**(1), 1–16 (2021)
5. Yadav, A., Vishwakarma, D.K.: Sentiment analysis using deep learning architectures: a review. Artif. Intell. Rev. **53**(6), 4335–4385 (2020)
6. Jain, P.K., Saravanan, V., Pamula, R.: A hybrid CNN-LSTM: a deep learning approach for consumer sentiment analysis using qualitative user-generated contents. Trans. Asian Low-Resour. Lang. Inf. Process. **20**(5), 1–15 (2021)
7. Priyadarshini, I., Cotton, C.: A novel LSTM-CNN-grid search-based deep neural network for sentiment analysis. J. Supercomput. 1–22 (2021)
8. Basiri, M.E., Nemati, S., Abdar, M., Cambria, E., Acharya, U.R.: ABCDM: an attention-based bidirectional CNN-RNN deep model for sentiment analysis. Futur. Gener. Comput. Syst. **115**, 279–294 (2021)
9. Ali, N.M., Abd El Hamid, M.M., Youssif, A.: Sentiment analysis for movies reviews dataset using deep learning models. Int. J. Data Min. Knowl. Manag. Process. (IJDKP) **9** (2019)
10. Ligthart, A., Catal, C., Tekinerdogan, B.: Systematic reviews in sentiment analysis: a tertiary study. Artif. Intell. Rev. 1–57 (2021)
11. Sivakumar, M., Uyyala, S.R.: Aspect-based sentiment analysis of mobile phone reviews using LSTM and fuzzy logic. Int. J. Data Sci. Anal. **12**(4), 355–367 (2021)
12. Xu, G., Meng, Y., Qiu, X., Yu, Z., Wu, X.: Sentiment analysis of comment texts based on BiLSTM. IEEE Access **7**, 51522–51532 (2019)
13. Colón-Ruiz, C., Segura-Bedmar, I.: Comparing deep learning architectures for sentiment analysis on drug reviews. J. Biomed. Inform. **110**, 103539 (2020)
14. Gandhi, U.D., Kumar, P.M., Babu, G.C., Karthick, G.: Sentiment analysis on twitter data by using convolutional neural network (CNN) and long short term memory (LSTM). Wirel. Pers. Commun. 1–10 (2021)
15. Nemes, L., Kiss, A.: Social media sentiment analysis based on COVID-19. J. Inf. Telecommun. **5**(1), 1–15 (2021)
16. Beseiso, M., Elmousalami, H.: Subword attentive model for Arabic sentiment analysis: a deep learning approach. ACM Trans. Asian Low-Resour. Lang. (2020)
17. ONAN, A.: Sentiment analysis on massive open online course evaluations: a text mining and deep learning approach. Comput. Appl. Eng. Educ. **29**(3), 572–589 (2021)
18. Wadawadagi, R., Pagi, V.: Sentiment analysis with deep neural networks: comparative study and performance assessment. Artif. Intell. Rev. **53**, 6155–6195 (2020)
19. Salur, M.U., Aydin, I.: A novel hybrid deep learning model for sentiment classification. IEEE Access **8**, 58080–58093 (2020)
20. Kula, S., Choraś, M., Kozik, R., Ksieniewicz, P., Woźniak, M.: Sentiment analysis for fake news detection by means of neural networks. In: International Conference on Computational Science, pp. 653–666. Springer, Cham (2020)
21. Patel, P., Patel, D., Naik, C.: Sentiment analysis on movie review using deep learning RNN method. In: Intelligent Data Engineering and Analytics, pp. 155–163. Springer, Singapore (2021)
22. Ni, R., Cao, H.: Sentiment analysis based on GloVe and LSTM-GRU. In: 2020 39th Chinese Control Conference (CCC), pp. 7492–7497. IEEE (2020)
23. Santur, Y.: Sentiment analysis based on gated recurrent unit. In: 2019 International Artificial Intelligence and Data Processing Symposium (IDAP), pp. 1–5. IEEE (2019)
24. Birjali, M., Kasri, M., Beni-Hssane, A.: A comprehensive survey on sentiment analysis: approaches, challenges and trends. Knowl.-Based Syst. 107134 (2021)
25. Jain, P.K., Pamula, R., Srivastava, G.: A systematic literature review on machine learning applications for consumer sentiment analysis using online reviews. Comput. Sci. Rev. **41**, 100413 (2021)
26. Liu, Y., Lu, J., Yang, J., Mao, F.: Sentiment analysis for E-commerce product reviews by deep learning model of Bert-BiGRU-Softmax. Math. Biosci. Eng.: MBE **17**(6), 7819–7837 (2020)

Social Sentiment Analysis Using Features Based Intelligent Learning Techniques

Prasannavenkatesan Theerthagiri (ORCID)

Abstract In these days, sentiment analysis has become increasingly popular. It examines how people feel about organizations, individuals, products, and services. Opinions from customers are collected through various resources such as social media networks, surveys, and forms. Social sentiment analysis is quite useful in determining how the product or brand is regarded in comparison to the competitors. Business organizations make use of reviews given by the customer to enhance product quality. In this study, sentiment analysis is applied to a variety of datasets, including Twitter sentiment Analysis, Internet Movie Database (IMDB) movie reviews, and Amazon mobile reviews. Desired aspects are extracted from the datasets, and the sentiment polarity of the aspects is identified. The sentiment associated with the reviews can have a positive, negative, or neutral tone. Various intelligence learning techniques are considered for the classification of sentiment. Parameters such as recall, precision, and accuracy are considered in order to estimate performance of the classification algorithms.

Keywords Pre-processing · Aspect extraction · Sentiment orientation · Classification algorithms

1 Introduction

Now a days, the Social media has become an inextricable aspect of our lives. Social media tools provide a way to share ideas, experiences, and thoughts with everyone in an effective manner. Business organizations collect opinions about their products, restaurants, or hotel events from the customers to know about their business. Reviews given by the users will be filtered at the feature level, and overall opinions about the particular product will be identified. Relevant comments will be automatically routed to product teams to ensure quality maintenance and good customer support [1]. Online client surveys and feedback forms also have a tremendous impact on

P. Theerthagiri (✉)
Department of Computer Science and Engineering, GITAM School of Technology, GITAM University Bengaluru, Bengaluru, India
e-mail: prasannait91@gmail.com

© The Author(s), under exclusive license to Springer Nature Switzerland AG 2022
T.-P. Hong et al. (eds.), *Deep Learning for Social Media Data Analytics*,
Studies in Big Data 113, https://doi.org/10.1007/978-3-031-10869-3_6

the acquiring choices of different customers. With the rapid increase in online users and an enormous amount of data available on the internet, young researchers got motivated to pursue their work on text mining and sentiment analysis (SA). It is also called Opinion mining.

Natural Language Processing (NLP) and intelligence learning methods are combined into a sentiment classification analysis system for the transcript analysis. Data analysts at big corporations use sentiment analysis to assess conduct advanced market research, public opinion, product reputation, monitor brand, and understand consumer experiences [2]. It is a part of text mining that recognizes the enthusiastic tone behind various content. Sentiment classification can be applied at different levels namely sentence level, document level, and aspect level. In the topic of sentiment classification, there has been a substantial amount of study. Previously, the majority of the work was devoted to classifying the overall sentiment of a corpus [3].

In opinionated texts like regression algorithms, product reviews, are routinely adopted to predict the degree of positivity of opinions. Similar connections between classes that match to points on a scale, such as a reviewer's star rating, are explained by regression methods. Modeling discourse structure, twists and turns in a document, allows for more efficient sentiment labelling. Bo Pang and Lillian Lee attempted to tackle part of the problem by integrating location data in the feature set. They turned a related triple conversion into compound back-off characteristics that are more generalizable than regular features.

Another type of binary emotion categorization is agreement detection. When a pair of text documents are compared, agreement detection determines whether the sentiment-related labels should be the same or different. Following the identification of the polarity categorization, the system may offer the polarity degrees of positivity—that is, it may position the viewpoint on a scale of positive to negative. It can also categorize multimedia materials based on their mood and emotional content, which may be useful for troll filtering, cyber-issue identification, and affective human–machine interaction. New issues occur if the text does not express strong sentiments or if it covers more than one topic or item.

The three main approaches to sentiment analysis now accessible are keyword detection, lexical affinity, and statistical techniques. The most simple approach, keyword detection, is also the most common for the reason of its simplicity and low cost. Text is categorized into affect categories based on the existence of reasonably explicit affect terms such as 'sad,' 'happy', 'bored', and 'afraid'. This technique has two flaws: it ignores the effect of negation and relies on surface features. The first issue in the approach is that, while it may correctly classify "today was a nice day" as cheerful, it is more likely to fail with a statement like "today wasn't a happy day at all."

In fact, rather than utilising affect adjectives, many phrases portray emotion through underlying substance. The phrase "My spouse just filed for divorce, and he wants to take control of my children away from me" elicits powerful emotions, but it contains no effect keywords. As a result, a keyword detection approach will not be able to identify it. Lexical affinity is a step up from keyword indexing in that it provides arbitrary words a probabilistic 'affinity' for a certain emotion instead of only looking for evident affect phrases. For example, the word "accident" may be given a 75% chance of expressing a negative influence, as in "vehicle accident" or "injured by accident."

Typically, language datasets are used to train these probabilities. Although it frequently surpasses keyword recognition method. For starters, statements like "I averted an accident" (negation) and "I met my lover by accident" may readily fool lexical affinity, which operates only on a word level (other word senses). Second, due to lexical affinity probabilitie sand the source of the linguistic corpora are frequently skewed toward literature of a specific genre. It's tough to create a domain-independent, reusable model because of this [4].

Text impact categorization has traditionally relied on statistical methods namely support vector machines and Bayesian inference. The system can learn the effective and useful valence of affect keywords (keyword spotting approach) as well as by feeding a machine learning system a huge training corpus of affectively annotated texts, the valence of additional unordered keywords (such lexical affinity), punctuation, and word co-occurrence frequencies Traditional statistical techniques, on the other hand, are frequently semantically weak, which implies that other lexical or co-occurrence features in a statistical model, aside from clear influence keywords, have little predictive value on their own. As a consequence, statistical text classifiers can only perform accurately if the text input is large enough. As a result, while these algorithms are capable of categorising user material on a page or paragraph level, they struggle to do so on shorter text units such as sentences or clauses [5].

Sentiment analysis based on concepts focuses on a semantic analysis of text utilising online ontologies or semantic networks to gather conceptual and emotional data tied to natural language opinions. Such strategies avoid relying just on keywords and word co-occurrence counts in favour of concentrating on the implicit qualities associated with natural language ideas by relying on large semantic knowledge bases. Concept-based methods, unlike solely syntactical tools, may identify nuanced moods as well; for example, by examining ideas that do not express emotion openly but are indirectly related to other concepts that do [6–11].

The concept-level examination aims to infer the semantic and emotional data related to natural language views, allowing for a feature-based fine-grained sentiment analysis that is comparable. Users are more interested in examining various items based on individual characteristics (e.g., iPhone5's vs Galaxy S3's touchscreen) or even sub-features (e.g., the iPhone5's vs. Galaxy S3's touchscreen fragility) rather than gathering isolated remarks on a single topic (e.g., iPhone5). Building vast common and common-sense knowledge bases is critical for feature detection and polarity identification. Rational thinking is required to properly dismantle natural language text into sentiments, such as estimating the concept "small room" as negative for a resort review but "small queue" as positive for a postal office, or the concept "go read the book" as positive for a book review but negative for a movie review.

In this study, a classification problem, sentiment analysis classifier is fed a set of reviews that classifies the sentiment associated with the given text. This study considered only positive and negative sentiments for classification purposes. Various preprocessing techniques are applied to these reviews to transform unstructured information into structured facts. The aspects are mined from the reviews, and then a polarity score is allotted. Finally, classification algorithms are used to estimate the performance of the classifiers.

A model is presented in this work that describes the sequence of steps that are carried out in sentiment analysis of Amazon mobile reviews, Twitter sentiment analysis, and IMDB movie reviews. In this study, we focus on feature-based opinion mining. SA is performed on reviews given by the customers. The related work is described in Sect. 2. The specifics of the planned work are outlined in Sect. 3. Section 4 analyses and visualises the experimental data, and Sect. 5 closes with conclusions and recommendations for further study.

2   Related Work

SentiWordNet and SenticNet are two existing affective knowledge bases that are largely utilised for concept-level sentiment analysis. A two-step technique combining random walk and iterative regression with in-link normalization process is used to create a concept-level sentiment dictionary, for example. Affective Norms for English Words (ANEW) and SenticNet are used to propagate sentiment values, which are based on the assumption that semantically similar ideas have comparable sentiment. Moreover, polarity accuracy, average-maximum ratio, and Kendall distance are used to assess sentiment dictionaries rather than mean error.

A similar approach was reported for adding emotional information to SenticNet thoughts by labelling them with an emotion name. The authors employ distance-based metrics, namely emotional affinity, point-wise mutual information, and similarity measures based on polarity data supplied in SenticNet (using WordNet-Affect) [12]. It suggests re-evaluating objective terms in SentiWordNet by analysing sentimental importance of such words and related emotional sentences, which is another new effort that builds on a foundation of emotional knowledge that already exists. For sentiment classification, two sampling procedures are presented, which are then combined with support vector machines. The presented technique overcomes the typical sentiment mining approach in SentiWordNet, according to the various experiments. The key difficulties surrounding the creation of a corpus for sentiment analysis and opinion mining are explored. A case study project named sentiment Turin University Treebank Senti–TUT has been published, and it is establishing a corpus for the examination of irony concerning politics in social media.

Sentiment analysis analyses the opinions given by the users in the form of comments or reviews to identify the sentiment behind the content. The opinions uploaded by the users serve as a good resource for sentiment analysis [13]. Sentiment classification are performed with the support of supervised and unsupervised intelligence learning techniques. Yassine et al. [14] had suggested a novice model that categorizes the positive and negative perception of product comments, reviews, and feedback. It has been concluded that the Random Forest and Support Vector

Machine (RFSVM) algorithm is apt well to the dataset of product reviews provided by Amazon. Nádia et al. [15] had proposed that the aggregation of Support Vector Machine (SVM), Random Multinomial Naive Bayes, Logistic Regression (LR), and Random Forest (RF) would result in a classifier that, when applied to public tweet sentiment datasets, it shows the enhanced accuracy.

The results showed that feature hashing could be considered when it comes to tweet sentiment analysis, and in terms of accuracy, bag-of-words is the best choice. Hadi et al. [16] had proposed Natural language processing (NLP) methods that adopted the skip-gram models and bag of words. Numerous classifiers such as SVM, RF, and LR for binary classification and Recursive Neural Tensor Networks for datasets with multi-class labels are used to perform sentiment analysis. It was concluded that Recurrent Neural Network (RNN) serves as best when compared to others and achieved 1.5% accuracy improvement by ensemble-averaging wh.

Brian et al. [17] had introduced a Hybrid Scoring SVM that combines supervised methods (Naive Bayes, SVM) and an unsupervised method (the scoring algorithm) to analyse paper reviews of an international conference. The results revealed that the Hybrid Scoring SVM method was more robust than the others, relative to a baseline, by means of standard metrics, namely precision, accuracy, F1-score, and recall. Oscar et al. [18] had propounded a combination of existing traditional sentiment classifiers and sentiment-trained word embeddings on six public datasets culled from the domains of tweeting and movie reviews. Chetashri et al. [12] have discussed the implementation of various techniques for aspect classification followed by polarity classification of product reviews and concluded that 78% accuracy was achieved using SVM along with domain-specific lexicons.

DoaaMohey et al. [19] had presented a survey on challenges in sentimental analysis. The results showed that domain independence and accuracy rate are prominent factors to recognize the sentiment challenges. Michelle et al. [20] had presented a Support Vector Machines for finding sentiment polarity on widely accessible sets of film reviews. The results proved that the classification accuracy of an algorithm depends on the types of features chosen. Xing et al. [21] had dealt with the problem of categorization sentiment and various levels of sentiment categorization on the Amazon reviews dataset.

Many works investigate the knowledge bases and statistical approaches in an ensemble setting. For example, to deal with ambiguity and incorporate the context of sentiment phrases, [20] recommends a hybrid technique that combines lexical analysis and machine learning. The context-aware technique finds ambiguous phrases with different polarities based on the situation and encodes them in sentiment lexicons that context unaliased. When used in combination through semantic knowledge bases, these lexicons support in the identification of ambiguous sentiment phrases and the ideas that correlate to their polarity [22]. It provides a novel way for retrieving product characteristics and viewpoints gleaned from a collection of unstructured text consumer evaluations about a product or service, which is another machine-learning-based study. A language-modelling framework is used in this technique, which may be used to review any area and language from the set of opinion words. To estimate model characteristics for retrieval, this method combines a kernel-based model of opinion words (trained from a seed set of opinion words) with a statistical mapping between words [23].

In the area of concept-level sentiment analysis, challenges such as domain adaptation, opinion summarization, and multimodal sentiment analysis are examples of recent work. There are two separate demands in the problem of domain adaptation: instance adaptation and labelling adaptation [24]. The majority of current research, on the other hand, concentrates on the former attribute while ignoring the latter. A complete strategy known as feature ensemble plus sample selection (SS-FE) is provided in this paper. The SS-FE considers both adaptation methods: first, a feature ensemble (FE) model applied to learn a new labelling function, and then feature re-weighting is utilised as a supplement to FE [25].

Starlet, a unique method adopted for extractive multi-document summary for the evaluative of text. It uses feature grading distributions and language modelling as summarization characteristics concept-level summarization. Such aspects encourage the adoption of lines in the summary that reflect the general opinion distribution found in the original reviews and whose language is the most similar to that of the reviews. The proposed method outperforms the standard summarization technique on multi-document summarising evaluative language [26]. Multimodal sentiment analysis is a sub-area of sentiment classification that is growing in popularity. It studies multimodal sentiment analysis using language, visual information, audio etc. A collection of 105 Spanish films ranging in duration from 2 to 8 min, including 84 female and 21 male speakers, was gathered at random on the social media site YouTube and labelled for ternary emotion by two labels [27]. It resulted in 550 words and around 10,000 utterances. When all three feature types are combined, the result is far better than when each modality is used alone.

It is supported by the second set of English videos. Instead, the authors of [28] provide the Multi-Modal Movie Opinion database of individual movie evaluations culled from ExpoTV (78 clips) and YouTube (308 clips). The final collection comprises 370 of these 1–3 min English videos, each with one to two coders providing ternary emotion annotation. The 20 video features, 2 k audio features, and extra textual features constitute the feature foundation for selection. Then, several types of domain reliance are measured: cross-domain analysis, in-domain analysis using the 100 million written Metacritic movie review corpus for training, and usage of online data sources [29]. Sentiment classification analyisi gradually establishes itself as a distinct area, falling among natural language processing and natural language comprehension. Unlike normal syntactic NLP errands, namely auto-categorization, and summarization The suggested opinion mining concentrates on natural language semantic inferences and emotional information. It does not need a thorough understanding of the text. We envision sentiment analysis research moving toward content-, concept-, and context-based natural language text analysis, helped by time-saving parsing algorithms suitable for large-scale social data analysis [30].

There are a plethora of new sources for sentiment analysis and opinion mining. Users can offer their thoughts in video and audio format rather than text using webcams embedded in touchpads, smartphones, or other devices. Consider this: every minute, YouTube viewers upload the equivalent of 2 days' worth of video content to the site. The audiovisual format enables for the extraction of mood and viewpoints in addition to converting spoken language to written text for analysis [31]. Many new areas, including as facial expressions, body movement, and a video blogger's music or colour filters, might be useful in opinion mining. A similar area, affect analysis, deals with the utilisation of verbal, auditory, and (possibly) video data. This field is concerned with a larger range of sentiments or the assessment of constant emotion primitives, such as valence and sentiment. Researchers present updated investigations on vocal and transcribed-language-based analyses in one study, while investigators in another study look into more multimodal combinations. There is very little study on multimodal sentiment analysis [32].

Raaijmakers and their researchers integrate acoustic and linguistic data, but instead of employing automatic speech recognition output, they use a transcript of the spoken content. In addition to this study, Louis-Philippe Morency and his colleagues used auditory, textual, and video aspects to measure opinion polarity in 47 YouTube videos. Using Hidden Markov Models for classification, they show a considerable improvement in leave-one-video-out assessment. Polarized words, grins, gazes, pauses, and voice pitch were found as important aspects by the writers. Again, the researchers analysed the material using transcripts rather than the actual spoken word [33].

Multimodal sentiment analysis hasn't been completely studied yet, but it has a lot of potential as a tool. It might be really beneficial when a textual transcript is unavailable and we need a performance point of view for synergy effects and fail-safeness. In this approach, other modalities such as physiological and neurological signals, as well as the use of contextual knowledge, would be particularly exciting [34]. Then, in addition to the audiovisual confidence estimate, we'll need to look into assessments of robustness against disruptions in individual (or all) modalities.

3   Proposed Methodology

In this section, detailed architecture of the proposed sentiment classification approach has been presented (Fig. 1).

3.1   Dataset

We considered different datasets such as Twitter Sentiment Analysis, IMDB Movie reviews and Amazon mobile reviews for classification. Aspects are extracted from given datasets and provided as input to the system. The datasets considered for training and testing must be labelled data. We consider the comma-separated values (CSV) file format as input.

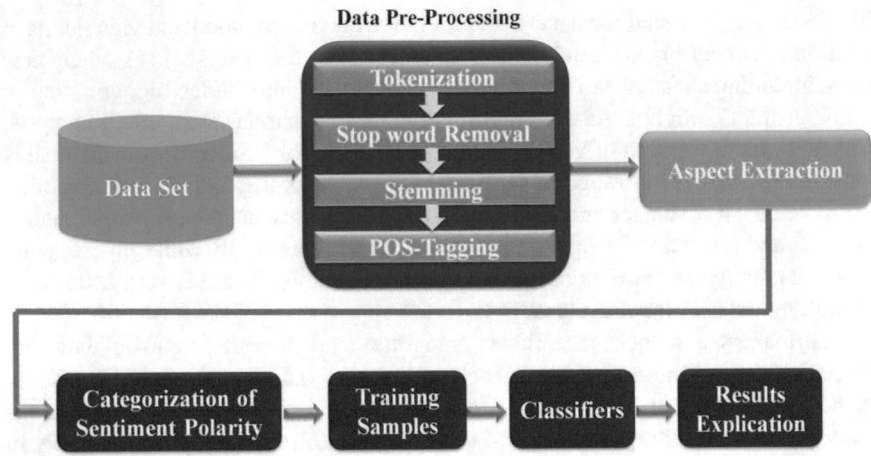

Fig. 1 Proposed System Model

3.2 Pre-processing

The data pre-processing is the fundamental stage for data mining and data analysis. The reviews are extracted from the web are always incomplete, inconsistent and may contain unwanted symbols or special characters. Hence, pre-processing techniques are applied to extracted reviews in order to remove special symbols, extra spaces, and lines. The stop words namely a, the, as, be, an', etc., are removed using the stop word removal technique as these words do not contain any significant centrality that should be used in search queries. Stemming is performed to diminish each word to its root word. POS (Parts of Speech) tagging should be done where the tag is assigned to each word. The common POS tags are nouns, verbs, adjectives etc.

Aspect Extraction: A bag of n-grams representation was used in this work for the evaluation and analysis.

Tokenization: It's a method for breaking down lengthy strings of text into smaller pieces. It breaks down the phrases into a series of tokens or words. A piece of text is generally handled further once it has been appropriately tokenized [35].

Counting: It determines the number of times the token appeared in the article.

Normalization: In normalization, weight is assigned to each word based on its occurrence in a corpus. Most occurring words, such as stop words, will be given less importance.

3.3 Categorization of Sentiment Polarity

Polarity in sentiment analysis defines the emotion expressed in a given sentence. It can be negative, positive, or neutral. This study considered only the positive and negative aspects of sentiment classification.

3.4 Classification Using Machine Learning Algorithms

Support Vector Machine: A supervised learning algorithm, Support Vector Machine (SVM) helps in categorizing data points with hyperplane in N-dimensional space, N denotes the number of features. Support vectors are used to build a hyperplane and influence its position [36]. The SVM uses a recursive training algorithm to decrease the error function to get an optimal hyperplane. Linear classifier for a binary classification problem with aspects x, weight parameter w and bias b is given as:

$$(x) = wx + b \tag{1}$$

As SVM requires input data to be in the form of numerical form, classified review text files are converted into numerical form. The tuning parameter values considered for SVM classifiers are C as 1.0, kernel as 'linear', degree as 3, and gamma as 'auto'.

Naive Bayes Classifier: Naïve Bayes Classifier (NB) is deduced from the Bayes theorem and is generally adapted for huge datasets. A certain point in the dataset is identified by considering the attributes of the Amazon, Twitter, and IMDB datasets [37]. In a supervised learning algorithm, an example set is iterated, whereas in a supervised algorithm, certain protocols are followed. The probability of predicting sentiment for a given aspect can be found by using the following rule.

$$P(Sentiment|Aspect) = P(Sentiment)P(Aspect|Sentiment)/P(Sentiment) \tag{2}$$

K-Nearest Neighbor (KNN): The KNN method is a supervised machine learning technique for solving classification and regression issues [28]. The dataset stored during the training phase classifies the new data into a category that suits the closest, which depends on certain measures like distance or proximity.

Decision-Tree: Decision-tree is a pictorial structure resembling a flowchart that comprises of nodes and branches. Classification rules are exhibited by the path from the root to the leaf; internal nodes are attributes, and leaf nodes denote class labels. Target variables take distinct values in classification trees and continuous values in regression trees [38].

Logistic Regression: Logistic regression (LR) is a classification algorithm that helps find the dependent variable's probability [39]. We go for Logistic Regression when the target variable is bilateral; that is, the range of value would be 0 to 1. The tuning parameter values considered for Logistic Regression: C as1.0, no_iter as 100.

Random Forests Algorithm: Random Forest (RF) is the most suitable algorithm these days. It is an update of the decision tree algorithm with added attributes like complexity and accuracy [40–45]. Random Forest analyses the data completely and identifies if any specific associations exist in the data, and then maps it to various forms of output obtained. The tuning parameter values considered for Random Forest are n_Rfestimators as100 and maximum depth as 30.

4   Implementation and Result Analysis

The proposed model has been developed in python version 3.7.0. This work adopted the Natural Language Toolkit for text processing which includes libraries such as word tokenizer, stop words, POS tagger, Text Blob, Scikit-Learn library etc., to build machine learning models. While assigning the polarity to a particular aspect, sentiment orientation is checked. If sentiment orientation is greater than zero, it is regarded positive, and if it is less than zero is taken as negative and finally fed into the classifiers.

The datasets are from the machine learning repository at the University of California, Irvine (UCI). All these datasets contain class labels and the number of features. These datasets include positive as well as negative opinions of customers about particular features. A detailed description of these datasets can be found at https://archive.ics.uci.edu/ml/.

4.1   Result Analysis

Some of the sample positive and negative terms are extracted from the Amazon mobile datasets are exposed in Tables 1 and 2. The performance of the classifiers is examined by considering distinctive datasets, and assessed results are contrasted using individual classifiers. Table 3 shows the values obtained by using the different classifiers. The figures (Figs. 2, 3 and 4) show the graphical representation of performance evaluation of various classifiers on Amazon mobile reviews, Twitter sentiment analysis and IMDB movie review datasets. In this experiment, 70% of the available reviews are utilized for training the system, while the remaining 30% are used to test it. After examining the results of the classifiers, it is concluded that the Random Forest classifier surpassed all other classifiers in terms of accuracy.

From the results given in Tables 1, 2, and 3, it is clear that the proposed random forest-based methodology produces better results as compared with other machine

Table 1 Sample positive terms

Good	Awesome	Great	Ok	Love
Cool	Happy	Thanks	High	Worth
Amazing	Enjoy	Amazing	Enjoy	Glad
Love	Clear	Honest	Cute	Easy
Avid	Magic	Fair	Superb	Superior
Marvelous	Reasonable	Special	Pretty	Nice
Sophisticated	Wise	Magic	Effective	Fantastic

Table 2 Sample negative terms

Blurred	Ill	Falsely	Trouble	Low
Slow	Worst	Frustrating	Depress	Minor
Poor	Horrible	Irritate	Ridiculous	Wrong
Missing	Wonky	Unfortunate	Dishonest	Hardly
Waste	Needless	Skeptical	Mess	Dead
Outdated	Weak	Broken	Sad	Needless

learning algorithms. In Table 3, the random forest methodology produces the accuracy, precision, recall, and F1 score as 92.5, 78.52, 95, 80.25% for Amazon mobile reviews dataset, 90.25, 80.65, 92, 88% for Twitter SA datasets, and 88.08, 78.52, 90, 86.05% for IMDB movie reviews dataset.

The performance of supervised machine learning classifiers can be measured by using a confusion matrix. This matrix has 4 diverse combinations of actual and predicted values such as True Positive (TP), False Positive (FP), True Negative (TN), and False Negative (FN) [46–51]. Performance evaluation metrics are demonstrated as follows.

$$Accuracy = \frac{TP_{Sentiment} + TN_{Sentiment}}{TP_{Sentiment} + TN_{Sentiment} + FP_{Sentiment} + FN_{Sentiment}} \quad (3)$$

$$Precision = \frac{TP_{Sentiment}}{TP_{Sentiment} + FP_{Sentiment}} \quad (4)$$

$$Recall = \frac{TP_{Sentiment}}{TP_{Sentiment} + FN_{Sentiment}} \quad (5)$$

$$F1Score = \frac{2 * Precison * Recall}{Precison + Recall} \quad (6)$$

Figures 2, 3 and 4 clearly depicts when compared to existing machine learning methods, it is obvious that the suggested random forest-based technique generates superior outcomes. The accuracy, precision, recall, and F1 score [49] for the Amazon

Table 3 Performance analysis of the different classifiers

Amazon mobile reviews dataset

ML classifier	Acc. (%)	Pre. (%)	Rec. (%)	FL_Sc. (%)
NB	75.71	95	61.36	76.05
SVM	76.82	90	82.93	78.85
KNN	84.28	80	86	88
Decision tree	87.3	85.45	90	92.15
LR	77.36	75.65	84.65	82.63
Random forest	92.5	78.52	95	80.25

Twitter SA dataset

NB	68	92.5	85	88.88
SVM	71.25	82.05	88	75.54
KNN	86.83	85	94	86
Decision tree	73.49	87.56	89	92.15
LR	74.64	82.25	90	85
Random forest	90.25	80.65	92	88

IMBD movie reviews dataset

NB	67.85	85	94	91
SVM	79.56	92	91	85
KNN	76	85	92	86
Decision tree	82.55	85.45	86.3	92.15
LR	78.75	75.65	80.75	82.63
Random forest	88.08	78.52	90	86.05

*Acc. = Accuracy, *Pre. = Precision, *Rec. = Recall, *F1 Sc. = F1_Score

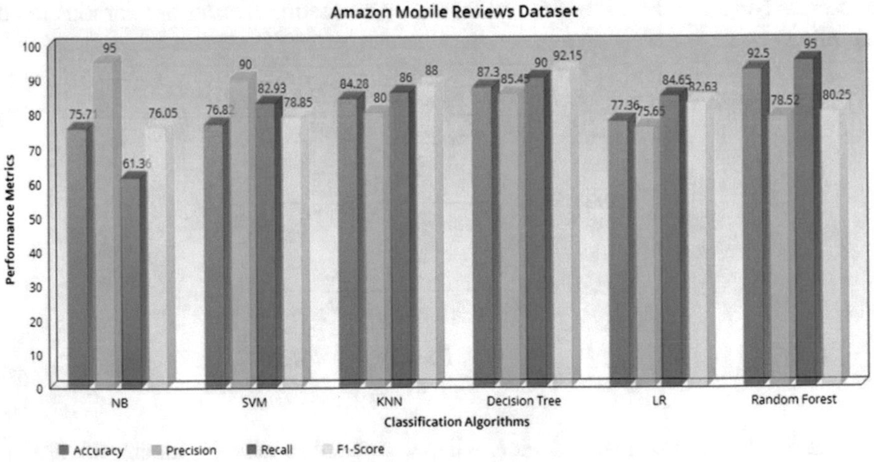

Fig. 2 Aspect statistics of Amazon Mobile Reviews dataset about classifiers

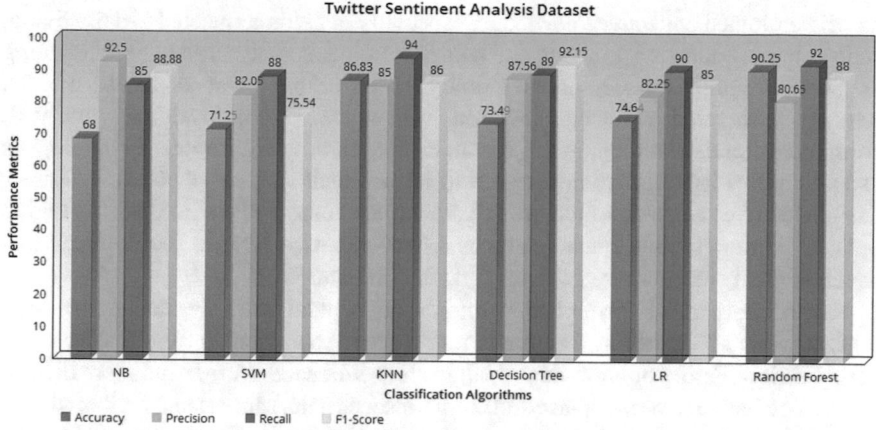

Fig. 3 Aspect statistics of Twitter Sentiment Analysis dataset about classifiers

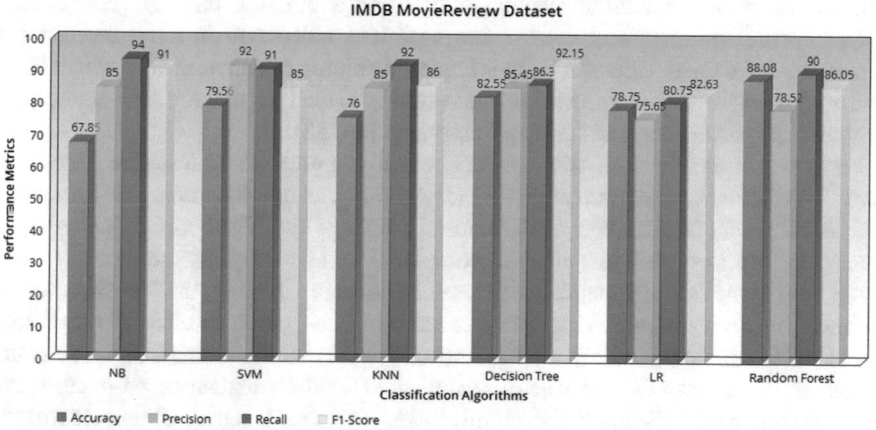

Fig. 4 Aspect statistics of IMDB Movie Reviews dataset about classifiers

mobile reviews dataset are 92.5, 78.52, 95, 80.25%, further, 90.25, 80.65, 92, 88% for Twitter SA datasets, and 88.08, 78.52, 90, 86.05% for the IMDB movie reviews dataset, according to the illustrated figures.

5 Discussions

The Web has evolved from a "read-only" to a "read-write" environment. This growth resulted in a community of passionate users who interacted and shared via online communities, social networks, wikis, blogs, and other collaborative media. The Web has disseminated collective knowledge, especially in fields connected to daily living, such as trade, education, tourism, and health. Sentiment analysis and opinion mining are still emerging as new multidisciplinary topics, despite great advances [4, 6, 52, 53]. For automated affect classification from voice, video, physiology, and text, computer scientists and engineers use machine-learning algorithms. Psychologists use a long history of emotion research to inform their rhetoric, theories, and practises. Affective sciences, which investigate human emotions, are inextricably tied to opinion mining and sentiment analysis. Affect-sensitive systems development and psychological emotion research must go hand in hand [22].

Recent study tries to better understand the conceptual principles that govern feelings, as well as suggestions that might aid people in transitioning from realisation to verbalization. Future opinion-mining algorithms will need a larger and more diverse knowledge base of commonplace and commonsense data. More extensive knowledge must be linked to reasoning techniques impacted by human psychology and cognition. It will aid academics in grasping natural language perspectives and bridging the gap between unstructured multimodal data and machine-processable data (structured). Combining scientific schemes of emotion with hands-on engineering goals of analysing sentiments in natural language text will result in more bioinspired approaches for the design of intelligent opinion-mining systems capable of handling learning new affective knowledge, semantic knowledge, making analogies, and perceiving, perceiving, and "feeling" emotions [54, 55].

The depth and breadth of the concept-based approach's knowledge bases are crucial. Without a large library of human knowledge, an opinion mining machine will fail to comprehend the semantics of natural language text. Furthermore, knowledge bases' ability to cope with semantic complications is limited by the fact that they only hold standard information associated to concepts. Finally, the fixed/flat shape of thoughts limits inferences of semantic and emotional qualities associated to them. Typically, statistical methods are semantically weak, which means that, aside from clear impact keywords, the rest of a statistical model's lexical or co-occurrence components have little predictive significance. As a result, statistical text classifiers are only effective when given a large volume of text to work with. As a result, while these algorithms may be able to accurately categorise a user's work on a page or paragraph level, they may struggle to do so on smaller text units such as phrases or clauses.

6 Conclusion

This work presented the process of constructing feature-based sentiment embeddings of reviews collected from the users through machine learning classifiers. This paper addresses the fundamental issue of sentiment analysis and sentiment categorization. Various datasets such as movie reviews, mobile reviews, and Twitter sentiment analysis datasets are used for this study. Sentence-level categorizations have been performed to identify the polarity of the aspects. This paper applies multiple machine learning classifiers such as KNN, Logistic regression, NB, SVM, Decision tree, and Random forest to the above-stated datasets. Results are evaluated and compared with other classifiers. Based on the experiment results, it can be determined that Random Forest outperformed amongst all other classification methods.

In many cases, customers can express their opinion using user ratings and emoji images. In future work, we may focus on these issues to improve the quality of sentiment analysis. Aspect extraction can be done by applying more specific feature selection techniques and using hybrid classification algorithms.

References

1. Jagdale, R S., Shirsat, V.S., Deshmukh, S.N.: Sentiment analysis on product reviews using machine learning techniques. In: Cognitive Informatics and Soft Computing, pp. 639–647. Springer, Singapore (2019)
2. Mäntylä, M.V., Graziotin, D., Kuutila, M.: The evolution of sentiment analysis—a review of research topics, venues, and top cited papers. Comput. Sci. Rev. **27**, 16–32 (2018)
3. Dina, N.Z., Juniarta, N.: Aspect based sentiment analysis of employee's review experience. J. Inf. Syst. Eng. Bus. Intell. **6**(1), 79–88 (2020)
4. Nasim, Z., Haider, S.: ABSA toolkit: An open source tool for aspect based sentiment analysis. Int. J. Artif. Intell. Tools **26**(06), 1750023 (2017)
5. Cambria, E.: An introduction to concept-level sentiment analysis, pp. 478–483. In Mexican International Conference on Artificial Intelligence, Springer, Berlin, Heidelberg (2013)
6. Birjali, M., Beni-Hssane, A., Erritali, M.: Machine learning and semantic sentiment analysis-based algorithms for suicide sentiment prediction in social networks. Proc. Comput. Sci. **113**, 65–72 (2017)
7. Ahmad, M., Aftab, S., Muhammad, S.S., Ahmad, S.: Machine learning techniques for sentiment analysis: a review. Int. J. Multidiscip. Sci. Eng. **8**(3), 27 (2017)
8. Hasan, A., Moin, S., Karim, A., Shamshirband, S.: Machine learning-based sentiment analysis for twitter accounts. Math. Comput. Appl. **23**(1), 11 (2018)
9. Tyagi, A., Sharma, N.: Sentiment analysis using logistic regression and effective word score heuristic. Int. J. Eng. Technol. (UAE) **7**, 20–23 (2018)
10. Elmurngi, E., Gherbi, A.: Detecting fake reviews through sentiment analysis using machine learning techniques. IARIA/data analytics, pp. 65–72 (2017)
11. Jain, A., Jain, V.: Sentiment classification of twitter data belonging to renewable energy using machine learning. J. Inf. Optim. Sci. **40**(2), 521–533 (2019)
12. Araque, O., Corcuera-Platas, I., Sánchez-Rada, J.F., Iglesias, C.A.: Enhancing deep learning sentiment analysis with ensemble techniques in social applications. Expert Syst. Appl. **77**, 236–246 (2017)
13. Wang, Y., Chen, Q., Shen, J., Hou, B., Ahmed, M., Li, Z.: Aspect-level sentiment analysis based on gradual machine learning. Knowl.-Based Syst. **212**, 106509 (2021)
14. AlAmrani, Y., Lazaar, M., El Kadiri, K.E.: Random forest and support vector machine based hybrid approach to sentiment analysis. Procedia Comput. Sci. **127**, 511–520 (2018)
15. Da Silva, Nadia, F.F., Eduardo, R.H., Estevam, R.H. Jr.: Tweet sentiment analysis with classifier ensembles. Dec. Supp. Syst. **66**, 170–179 (2014)
16. Saleena, N.: An ensemble classification system for twitter sentiment analysis. Proc. Comput. Sci. **132**, 937–946 (2018)
17. Pouransari, H., Ghili, S.: Deep learning for sentiment analysis of movie reviews. Technical Rreport, Stanford University, Technical report (2014)
18. Keith, B., Fuentes, E., Meneses, C.: A hybrid approach for sentiment analysis applied to paper. In: Proceedings of ACM SIGKDD Conference, Halifax, Nova Scotia, Canada, p. 10 (2017)

19. Bhadane, C., Dalal, H., Doshi, H.: Sentiment analysis: measuring opinions. Proc. Comput. Sci. **45**, 808–814 (2015)
20. Hussein, D.M.: A survey on sentiment analysis challenges. J. King Saud Univ.-Eng. Sci. **30**(4), 330–338 (2018)
21. Annett, M., Kondrak, G.: A comparison of sentiment analysis techniques: polarizing movie blogs. In: Conference of the Canadian Society for Computational Studies of Intelligence, pp. 25–35. Springer, Berlin, Heidelberg (2008)
22. Pa supa, K., Ayutthaya, T.S.N. Thai sentiment analysis with deep learning techniques: A comparative study based on word embedding, POS-tag, and sementic features. Sustain Cities Soc. **50**, 101615 (2019).
23. Shakhovska, N., Fedushko, S.: Data analysis of music preferences of web users based on social and demographic factors. Proc. Comput. Sci. **1**(198), 730–735 (2022)
24. Babu, N.V., Kanaga, E.: Sentiment analysis in social media data for depression detection using artificial intelligence: a review. SN Comput. Sci.. **3**(1), 1–20 (2022)
25. Garg, N., Sharma, K.: Text pre-processing of multilingual for sentiment analysis based on social network data. Int. J. Electrical Comput. Eng. **12**(1), 2088–8708 (2022)
26. Hossain, M.S., Cucchiara, R., Muhammad. G., Tobón, D.P., Saddik, A.E.: Special Section on AI-empowered Multimedia Data Analytics for Smart Healthcare. In: ACM Transactions on Multimedia Computing, Communications, and Applications (TOMM), vol. 18(1s), pp. 1–2, (2022)
27. Rawat, R., Mahor, V., Chirgaiya, S., Shaw, R.N., Ghosh, A.: Sentiment analysis at online social network for cyber-malicious post reviews using machine learning techniques, pp. 113–130. InComputationally Intelligent Systems and their Applications, Springer, Singapore (2021)
28. Bennett, K.P.: Decision tree construction via linear programming. University of Wisconsin-Madison Department of Computer Sciences (1992)
29. Chauhan, P., Sharma, N., Sikka, G.: The emergence of social media data and sentiment analysis in election prediction. J. Ambient. Intell. Humaniz. Comput. **12**(2), 2601–2627 (2021)
30. Naseem, U., Razzak, I., Musial, K., Imran, M.: Transformer based deep intelligent contextual embedding for twitter sentiment analysis. Futur. Gener. Comput. Syst. **113**, 58–69 (2020)
31. Alam, M., Abid, F., Guangpei, C., Yunrong, L.V.: Social media sentiment analysis through parallel dilated convolutional neural network for smart city applications. Comput. Commun. **154**, 129–137 (2020)
32. Sharma, A., Shekhar, H.: Intelligent learning based opinion mining model for governmental decision making. Proc. Comput. Sci. **173**, 216–224 (2020)
33. Alqaryouti, O., Siyam, N., Monem, A.A., Shaalan, K.: Aspect-based sentiment analysis using smart government review data. Appl. Comput. Inf. (2020)
34. Singh, M., Jakhar, A.K., Pandey, S.: Sentiment analysis on the impact of coronavirus in social life using the BERT model. Soc. Netw. Anal. Min. **11**(1), 1–1 (2021)
35. Fang, X., Zhan, J.: Sentiment analysis using product review data. J. Big Data **2**(1), 1–14 (2015)
36. Stefanovič, P., Kurasova, O., Štrimaitis, R.: The n-grams based text similarity detection approach using self-organizing maps and similarity measures. Appl. Sci. **9**(9), 1870 (2019)
37. Deng, Z., Zhu, X., Cheng, D., Zong, M., Zhang, S.: Efficient kNN classification algorithm for big data. Neurocomputing **195**, 143–148 (2016)
38. Friedman, J., Hastie, T., Tibshirani, R.: Additive logistic regression: a statistical view of boosting (with discussion and a rejoinder by the authors). Ann. Stat. **28**(2), 337–407 (2000)
39. Medhat, W., Hassan, A., Korashy, H.: Sentiment analysis algorithms and applications: a survey. Ain Shams Eng. J. **5**(4), 1093–1113 (2014)
40. Gupte, A., Joshi, S., Gadgul, P., Kadam, A., Gupte, A.: Comparative study of classification algorithms used in sentiment analysis. Int. J. Comput. Sci. Inf. Technol. **5**(5), 6261–6264 (2014)
41. Everingham, Y., Sexton, J., Skocaj, D., Inman-Bamber, G.: Accurate prediction of sugarcane yield using a random forest algorithm. Agron. Sustain. Dev. **36**(2), 27 (2016)
42. Catal, C., Nangir, M.: A sentiment classification model based on multiple classifiers. Appl. Soft Comput. **50**, 135–141 (2017)

43. Peng, H., Xu, L., Bing, L., Huang, F., Lu, W., Si, L.: Knowing what, how and why: A near complete solution for aspect-based sentiment analysis. Proc. AAAI Conf. Artif. Intell. **34**(05), 8600–8607 (2020)

44. Mowlaei, M.E., Abadeh, M.S., Keshavarz, H.: Aspect-based sentiment analysis using adaptive aspect-based lexicons. Expert Syst. Appl. **148**, 113234 (2020)

45. Huang, J., Meng, Y., Guo, F., Ji, H., Han, J.: Aspect-Based sentiment analysis by aspect-sentiment joint embedding. In: Proceedings of the 2020 Conference on Empirical Methods in Natural Language Processing (EMNLP), pp. 6989–6999 (2020)

46. Theerthagiri, P.: Forecasting hyponatremia in hospitalized patients using multilayer perceptron and multivariate linear regression techniques. Concur. Comput: Pract. Exp. **33**(16), e6248 (2021)

47. Prasannavenkatesan, T.: Probable forecasting of epidemic covid-19 in using cocude model. EAI Endorsed Transactions on Pervasive Health and Technology **7**(26), e3 (2021)

48. Kommina, L, Theerthagiri, P, Payyavula, Y, Vemula, PS, Reddy, G.D.: Post-Stroke readmission prediction model using machine learning algorithms. In: Emerging Trends in Data Driven Computing and Communications, pp. 53–65. Springer, Singapore (2021)

49. Theerthagiri, P., Jeena Jacob, I., Usha Ruby, A., Yendapalli, V.: Prediction of COVID-19 possibilities using K-Nearest neighbour classification algorithm. Int. J. Cur. Res. Rev. **13**, 156 (2021)

50. Prasannavenkatesan, T.: Prognostic analysis of hyponatremia for diseased patients using multi-layer perceptron classification technique. EAI Endorsed Trans. Pervasive Health Technol. **7**(26), e5 (2021)

51. Gopala Krishnan, C., Theerthagiri, P.: Extreme learning-based intellectual lung cancer classification using artificial intelligence. In: Tyagi, A.K., Abraham, A., Kaklauskas A. (eds.) Intelligent Interactive Multimedia Systems for e-Healthcare Applications, pp. 375–385. Springer, Singapore (2022)

52. Ikram, M.T., Afzal, M.T.: Aspect based citation sentiment analysis using linguistic patterns for better comprehension of scientific knowledge. Scientometrics **119**(1), 73–95 (2019)

53. Ahmad, M., Aftab, S., Ali, I.: Sentiment analysis of tweets using svm. Int. J. Comput. Appl **177**(5), 25–29 (2017)

54. Cambria E, Schuller B, Xia Y, Havasi C. New avenues in opinion mining and sentiment analysis. IEEE Intell. Ssyst. **21**, 28(2), 15–21 (2013)

55. Jiang, Q., Chen, L., Xu, R., Ao, X., Yang, M.: A challenge dataset and effective models for aspect-based sentiment analysis. In: Proceedings of the 2019 Conference on Empirical Methods in Natural Language Processing and the 9th International Joint Conference on Natural Language Processing (EMNLP-IJCNLP), pp. 6281–6286 (2019)

User Behaviour Analysis

Modified-PIP with Deep Neural Network (DNN) Architecture: A Coherent Recommendation Framework for Capturing User Behaviour

Bam Bahadur Sinha, Gurvinder Singh Yadav, and Sagar Badrish Kudkelwar

Abstract The use of online services such as e-commerce is rapidly growing, resulting in an exponential rise in the amount of information that is available. Recommender systems assist consumers in selecting relevant information from a vast pool of alternatives. Furthermore, varied recommendation contexts have their own challenges, necessitating the use of diverse recommendation approaches. Collaborative filtering techniques have been predominantly used owing to their high accuracy, simplistic approach, and low computational needs, whereas content-based filtering faces the problem of high computational needs. This chapter aims to capture the behaviour of users using a modified Proximity Impact Popularity (Modified-PIP) technique augmented with Deep Neural Network (DNN) to make efficient movie recommendations. The dominance of Modified-PIP over other similarity measures such as PIP similarity, Cosine similarity, and PCC (Pearson Correlation Coefficient) is tested on the MovieLens-100K, MovieLens-1M, and Jester datasets. A matching imputation technique is applied to reduce the sparsity of the dataset. The performance of the proposed Deep Neural Network (DNN) architecture in conjunction with the Modified-PIP similarity measure is evaluated using RMSE (root mean squared error), MAE (mean absolute error), R (recall), and P (precision). Low error scores obtained by the proposed model give justification for the high efficacy of the proposed model.

Keywords Collaborative filtering · Neural network · Similarity technique · Recommender system · MAE · RMSE

B. B. Sinha (✉)
Computer Science and Engineering, Indian Institute of Information Technology Ranchi, Ranchi 834010, JH, India
e-mail: bahadurbam43@gmail.com

G. S. Yadav
Data Science and Engineering, Indian Institute of Information Technology Dharwad, Dharwad 580009, KA, India

S. B. Kudkelwar
Department of Electrical Engineering, G.H. Raisoni Institute of Engineering and Technology, Nagpur 440016, MH, India

© The Author(s), under exclusive license to Springer Nature Switzerland AG 2022
T.-P. Hong et al. (eds.), *Deep Learning for Social Media Data Analytics*,
Studies in Big Data 113, https://doi.org/10.1007/978-3-031-10869-3_7

1 Introduction

Human beings are one of the most fragile organisms on the planet, yet they manage to be on top of the food chain and even manage to be more than just a survival process. As a species, if looked upon by humans, they are able to do so only because of the effective use of tools that they develop. These tools have taken us from the stone age, bronze age, iron age, and now the information age, where they have grown from being simplistic to developing highly sophisticated and advanced computer systems. The only reason we are able to do so is because of effective communication of information, which is hard to compile and even harder to share. But with the beginning of the new information age, the way information is shared has completely changed. Earlier, only a few had access to quality content, but now it is available to every single person on the planet who has access to the internet and a computer. Due to this exposure to the content, it has become even more important that everyone gets access to the content that is most relevant to them. In today's world, petabytes of data are being added to the internet every day, and making sense of this data for a single human being manually is nearly impossible. This bottleneck on how much content a person can consume and what makes a piece of content relevant for a person led to the introduction of recommendation systems. Netflix, which serves as a media service provider, makes use of a recommendation system to recommend movies and TV shows based on user profiles. User profiling includes the formulation of a set of users similar to the target user. The similarity among users is computed using different conventional similarity measures such as Cosine similarity, Pearson correlation, etc. These similarity measures show a deterrent effect on accuracy when the dataset being used is highly sparse in nature. This real-time recommendation challenge motivated us to test different existing similarity measures with a modified version of one such measure for effective similarity computation.

The recommender systems analyse the data fed into the system about the user and the content in order to generate recommendations for the content that is relevant for that particular user. This has huge implications as more and more businesses move to the online infrastructure. For example, there is a plethora of goods, services, and entertainment platform vendors in cyberspace, and to increase their profits, they use various techniques. The most important of these tools and techniques are the recommendation systems. These vendors use recommendation systems (RSs) to find the products they like the most and then recommend similar content so that the user spends more time and money on their platforms, but in order to do so, huge databases have to be maintained and analysed to produce accurate recommendations. The RSs are broadly divided into three fundamental filtering techniques, namely: CBF (content-based filtering), CF (collaborative filtering), and hybrid filtering [10]. CF uses user's information such as age, rating patterns, occupation, gender, and demographic information to generate recommendations, whereas CBF analyses the product on the parameters of shape, size, type, price, and furthermore, depending on the platform it generates the recommendation. The third approach is the hybrid approach, and as the name suggests, it is a mixture of both CF and CBS filtering techniques. This chapter focuses on building a hybrid recommendation system using the Deep Neural Network (DNN) architecture in conjunction with the Modified-PIP similarity measure to generate relevant recommendations. The highly accurate results obtained on the CF model using modified-PIP motivated us to make use of modified-PIP in our proposed deep learning-based hybrid model. The proposed model aims at generating an efficient recommendation list by recommending the items in which the end users might be interested.

Deep learning's foundation is widely used for the categorization of complex data in order to enhance the knowledge that can be obtained from it. Deep-structured learning is another name for deep learning. It is a rapidly developing area of machine learning research that aids in the modelling of data on many levels, as well as the discovery of relationships and correlations between various kinds of data from multiple sources, such as textual information, audio, pictures, etc., at a deeper level. Deep learning methods can anticipate and give better suggestions to users by building automated RSs based on monitoring interactions between consumers and products. Many businesses, like Amazon, Netflix, etc., utilise RSs to fulfil customer requirements and assist them in discovering what they need [1]. The conventional RS perceives user interests using CF techniques. The CF approach employs a matrix factorization (MF) approach to represent the interaction between consumers and products, capturing only a superficial degree of cooperation between consumers and products. Relying on the adoption of DL-based architecture, deep-learning techniques could capture interaction at a profound level and represent latent hidden characteristics at many layers. As a result, recommender systems based on deep learning may perform better and offer more customised suggestions.

Each recommendation context has its own set of problems, necessitating a variety of methods for developing recommendation systems. For example, news recommendations may place more emphasis on the freshness of news content, while movie recommendations may place greater prominence on the relatedness of movie content. Furthermore, user interests, particularly in the context of movies, evolve and change over time. It's conceivable that a user who often watches comedies may acquire an interest in romantic films for a variety of reasons. As a result, it's essential to take into account dynamic changes in preferences and provide better recommendations. While many current methods presume that the user interest remains constant, this assumption seems to be a stretch. This implies that dealing with temporal changes is necessary in the best interests of consumers. The following are the major contributions of this chapter:

- A modified-PIP is used to demonstrate its significance over other traditional similarity measures such as Cosine, Pearson, Jaccard, and Proximity Impact Popularity (PIP).
- The matching imputation technique is used to reduce the sparsity of the MovieLens-100K, MovieLens-1M and Jester dataset.
- The interaction between a consumer and a product (movies, news, etc.) is captured using a deep neural network.

- Stochastic Gradient Descent (SGD), RMSprop, Adam, and Adagrad are some of the optimization techniques used to make the DNN model more accurate and faster to converge.

The remaining section of the chapter is structured as follows: Sect. 2 discusses the related background. Section 3 highlights the different methodologies used in this chapter. Section 4 illustrates the proposed model. The experimental setup and results are discussed in Sect. 5. The closing Sect. 6 concludes the chapter with future direction.

2 Related Background

Despite plenty of research done in the domain of recommendation systems, the use of DNN in recommendation systems has gotten less attention. In this section, a representative collection of methods that are relevant to the proposed methodology is discussed.

2.1 Common Recommendation Approaches

The most extensively used strategies in recommendation generation are CBF-based and CF-based recommendation systems. The CBF approach primarily analyses the implicit rating provided by the text mining process while making a suggestion, whereas collaborative filtering considers user explicit evaluations [2]. Furthermore, the CF approach is classified into two techniques that are commonly employed in the recommendation systems, namely: model-based CF and memory-based CF. The model-based strategy creates a user model based on each user's ratings to determine the estimated value of unrated products [3]. The memory-based CF approach uses a similarity measure to identify neighbourhoods using explicit user rating data and then predicts which items should be recommended to the end user [4].

CBF is a personalised recommendation approach that is widely utilised in a variety of fields [5, 7]. CF, on the other hand, has a number of flaws, including the cold-start problem, over-specialisation and data sparsity [6, 8, 9].

The user experience is significantly affected as a result of these issues. Memory-based CF is further classified as: user-based and item-based filtering. The following paragraph discusses a collection of literature on user-based similarity techniques that was gathered for this study.

Since similarity measures play a key role in increasing the prediction accuracy of collaborative filtering systems, several studies have been conducted to introduce new similarity measures that can overcome the problems being faced by traditional similarity measures. The linear relationship (correlation) among different items/movies can be computed using the most widely used PCC technique. The similarity value

obtained using PCC $\in$ [-1 (negative relationship),+1 (positive or strong relationship)].
If the value turns out to be 0, it means there is zero correlation among the picked set
of users. The mathematical representation for PCC for computing similarity between
two users u_j and u_h is given by Eq. 1.

$$sim(u_j, u_h)^{PCC} = \frac{\sum_{i=1}^{m'}(r_{u_j,I_i} - \bar{r}_{u_j})(r_{u_h,I_i} - \bar{r}_{u_h})}{\sqrt{\sum_{i=1}^{m'}((r_{u_j,I_i} - \bar{r}_{u_j})^2}\sqrt{\sum_{i=1}^{m'}(r_{u_h,I_i} - \bar{r}_{u_h})^2}} \tag{1}$$

The Cosine similarity denotes a vector space model that is widely used for the retrieval
of information such that the angle between vectors of cosine is used to compute the
similarity among the set of users. The mathematical representation for computing
similarity between users is given by Eq. 2.

$$sim(u_j, u_h)^{COS} = \frac{\sum_{x=1}^{U'}(r_{u_j}, I_x) \times (r_{u_h}, I_x)}{\sqrt{\sum_{x=1}^{U'}(r_{u_j}, I_x)^2}\sqrt{\sum_{x=1}^{U'}(r_{u_h}, I_x)^2}} \tag{2}$$

Jaccard-based similarity is calculated by finding the ratio of cardinality of intersection
to the union of the rating values between two users. It ranges from 0 to 1, with
0 denoting no similar rating values, whereas 1 depicts all ratings as overlapping.
Equation 3 demonstrates the mathematical representation for computing similarity
among users u_j and u_h.

$$sim(u_j, u_h)^{JAC} = \frac{|I_{u_j} \cap I_{u_h}|}{|I_{u_j} \cup I_{u_h}|} \tag{3}$$

2.2 Recommendation System Based on Neural Network Approaches

In [11], a two-layer Restricted Boltzmann Machine (RBM) was utilised to simulate consumer's explicit item ratings. A pioneering study that employed the neural
network was carried out. Later, the technique was broadened to mimic the ordinal
nature of ratings [12]. Autoencoders have recently been an important component in
developing recommendation systems [13, 14]. The purpose of AutoRec [13] is to
discover hidden patterns that can rebuild the ratings of a user in light of their previous
ratings. This method has a similar context in terms of personalization as the item
[15] model, which represents a user as its rated item characteristics.

Wu et al. [16] presents a CDAE approach with implicit feedback. CDAE, unlike
the DAE-based CF [17], connects a user node to the autoencoder input for recreating
the user's ratings. Whenever the identity function is used to trigger the hidden layers
of CDAE, the authors demonstrate that it is similar to the SVD++ model [19]. Despite

the fact that CDAE is a CF model, it is based purely on item-item interactions, while the work presented here relies on user-item interaction. In contrast, the authors of [18] looked at DNN for recommender systems. They propose NCF (Neural Collaborative Filtering), a generic framework that substitutes the inner product with a deep network that can learn any function from the provided data. The user-item interaction is learned using a multi-layer perceptron (MLP). Matrix factorization (MF) may be expressed and generalised using NCF. They then employ a combination of MF linearity and DNNs non-linearity to describe user-item latent structures. This model is known as NeuMF(Neural MF).

3 Experimental Methodology

3.1 Similarity Measure

Shortcomings of Traditional Similarity Measures

The cosine and Pearson Correlation Coefficient are often utilised in widely used CF-based recommender systems. The following are the restrictions associated with Cosine and PCC: Single-value [20], Opposite value [20], Equal ratio [20], and Flat value [21] problem. Cosine is among the primary similarity metrics used in a broad range of research purposes. This method is commonly applied in the field of textual grouping to compute the correlation between two textual document [22], and is also utilised in the K-means approach to determine the resemblance among data items and their associated centroids. This resemblance metric outperforms traditional techniques in terms of accuracy [21]. Jaccard similarity is sometimes referred to as binary similarity. This is often used to calculate the similarity between categorical (binary) data sets. This metric yields a more optimal answer in evolutionary computation, like the fitness values for evolutionary algorithms (e.g., genetic algorithm) [23]. As previously stated, the study clearly demonstrates that cosine and Jaccard-based similarity metrics are widely utilised in real-world applications to address a variety of problems. The ordinal rating scale values are used in the context of CF-based systems. The COS and PCC assign equal weightage to the negatives and positives of a scale, resulting in erroneous similarity findings. This is one of the shortcomings of cosine-based and PCC-based similarity measures. The Jaccard similarity measure considers only the number of co-rated evaluations, not their severity.

3.1.1 Proximity Impact Popularity (PIP)

The PIP technique consists of two conditions (TRUE or FALSE) with respect to agreement. Three fundamental building blocks of PIP are: proximity (P), impact (I), and popularity (P) [24]. Equations 4 and 5 can be used to compute the similarity between user r_{u_j, I_x} and r_{u_h, I_y}.

$$PIP(r_{u_j,I_x}, r_{u_h,I_x}) = Proximity(r_{u_j,I_x}, r_{u_h,I_x})$$
$$\times Impact(r_{u_j,I_x}, r_{u_h,I_x}) \qquad (4)$$
$$\times Popularity(r_{u_j,I_x}, r_{u_h,I_x})$$

$$sim(u_j, u_h)^{PIP} = \sum_{x=1}^{U'} PIP(r_{u_j}, I_x) \qquad (5)$$

The PIP score is heavily influenced by the agreement criterion. It seems to be the only similarity metric used to quantify agreement criteria to distinguish positive from negative ratings, particularly when compared to other similarity metrics. This agreement criterion is used to differentiate evaluations based on the pattern of ratings provided by two different individuals. Considering the set of users ($U = u_1, u_2, ..., u_n$) and the set of objects ($I = X_1, X_2, ..., X_m$). The total users present in the system and the total number of items present in the system is represented by 'n' and 'm' respectively. They're linked to a utility matrix, which can be generalised as a *user $\times$ item* rating matrix.

To illustrate, consider the two ratings, r_1 and r_2, where r_1 is the rating provided by 'u_j' and u_h for item 'x' respectively. 'R_m' and 'R_n' denotes the maximum and minimum rating respectively for the given rating scale; additionally, let $R_d = \frac{R_n + R_m}{2}$; when both r_1, r_2 greater or less than R_d and $Agreement(r_1, r_2)$ is TRUE. Similarly, $Agreement(r_1, r_2)$ is FALSE, if the ratings provided by both users are in different directions. Equation 6 represent the functions defining agreement conditions.

$$Agreement(r_1, r_2) = True$$
$$if\ (r_1 < R_d\ and\ r_2 < R_d) or (r_1 > R_d\ and\ r_2 > R_d) \qquad (6)$$

$$Agreement(r_1, r_2) = False; otherwise \qquad (7)$$

Proximity

Usually Proximity signifies the closeness of any object but here Proximity Eq. 9 is utilised to define the absolute difference between rating values.

$$Abs.\ Diff(r_1, r_2) = |r_1 - r_2|; Agreement = TRUE$$
$$Abs.\ Diff(r_1, r_2) = 2 \times |r_1 - r_2|; Agreement = FALSE \qquad (8)$$

$$Proximity(r_1, r_2) = \{(2 \times (R_m - R_n) + 1) - Abs.Diff(r_1, r_2)^2\} \qquad (9)$$

The value of $Abs.Diff(r_1, r_2)$ can be computed using Eq. 8. The value of proximity ranges from 1 to 81; it ranges from 1 to 25 when Agreement is 'FALSE' and 49 to 81 when Agreement is 'TRUE'. The higher the value of this Proximity greater is the dissimilarity between the ratings.

Impact

Impact factor determines how strongly the users likes or dislikes a given item. It ranges from 0.11 to 0.25 when $Agreement(r_1, r_2) = False$ whereas its value ranges from 1 to 9 when $Agreement(r_1, r_2) = True$. Higher value of impact factor signifies stronger preference of item by that particular user.

$$I(r_1, r_2) = (|r_1 - R_d| + 1) \times (|r_1 - R_d| + 1); Agreement = TRUE$$

$$I(r_1, r_2) = \frac{1}{(|r_1 - R_d| + 1) \times (|r_1 - R_d| + 1)}; if \, Agreement(r_1, r_2) = FALSE$$

$$(10)$$

Popularity

Popularity signifies how popular is the given item when compared to its given counterparts. Its minimum value comes to 1 when $Agreement(r_1, r_2) = False$ and ranges from 1 to 5 when $Agreement(r_1, r_2) = True$.

$$Popularity(r_1, r_2) = 1 + ((\frac{r_1 + r_2}{2}) - \bar{r}_{I_i})^2$$

$$if \, Agreement(r_1, r_2) = True \tag{11}$$

$$Popularity(r_1, r_2) = 1$$

$$if \, Agreement(r_1, r_2) = False \tag{12}$$

3.1.2 Modified Proximity Impact Popularity

In PIP, every component is spread over a wide range and does not provide equal weightage to all the components. Considering the extreme situations when $u_i = 5$ and $u_j = 5$ the individual components (Proximity, Impact, and Popularity) come out to be 81, 9, 5 respectively, and the final PIP evaluates to 3645 comparing this with $u_1 = 5$ and $u_h = 1$, we see that individual component come to be 1,0.11,1 and the final PIP evaluates to 0.11 where Proximity and Popularity are given the same weightage. While each individual component provides vital information, some components are given greater preference in certain situations. To overcome this challenge, Modified Proximity Impact Popularity (MPIP) is proposed. In addition to this, MPIP gives a more normalized range of values. The upcoming section describes in further detail how to evaluate the MPIP.

Modified Proximity

The modified Proximity has a normalized range of values from 0 to 1. It is evaluated using the absolute deviation between the two ratings equation (13). R_{med} is the median rating of that particular item.

$$D(r_1, r_2) = |r_1 - r_2| \tag{13}$$

$$Proposed\ Proximity(r_1, r_2) = \left(\frac{D(r_1, r_2) - (\frac{med_+ + med_-}{2})}{R_{max} - R_{min}}\right)^2$$

$$if\ Agreement(r_1, r_2) = True \tag{14}$$

The proposed proximity can be computed using Eq. 14 where med_+ is the median of ratings equal to or above the median of the rating scale and med_- is the median of rating values less than the median of the rating scale as in Eq. 14.

$$Proposed\ Proximity(r_1, r_2) = \delta \times (\frac{\overline{D(r_1, r_2)}}{R_{max} - R_{min}})^2$$

$$if\ Agreement(r_1, r_2) = False \tag{15}$$

$$\delta = \begin{cases} 0.75 & if\ D(r_1, r_2) > R_{med} \\ 0.5 & if\ D(r_1, r_2) = R_{med} \\ 0.25 & otherwise \end{cases} \tag{16}$$

Delta acts as a weightage which penalises proximity for deviation between ratings. Equations 14 and 15 is divided by $R_{max} - R_{min}$ to get normalized value of Proximity ranging from 0 to 1.

Modified Impact

To normalize the impact, the expression is converted to an exponential expression when the $Agreement(r_1, r_2) = True$. The conversion of expression is discussed via Eq. 17.

$$Proposed\ Impact(r_1, r_2) = exp(-\frac{1}{(|r_1 - R_{mid}| + 1) \times (|r_1 - R_{mid}| + 1)})$$

$$if\ Agreement(r_1, r_2) = True \tag{17}$$

The larger the gap between r_1 and r_2 with respect to the median, the greater the 'Impact (I)' value. When both the scores are around the median, the 'I' has a lower value. This shows that both individuals agree on the median values, indicating that the rating has a negligible effect.

$$Proposed\ Impact(r_1, r_2) = \frac{1}{(|r_1 - R_{mid}| + 1) \times (|r_1 - R_{mid}| + 1)}$$

$$if\ Agreement(r_1, r_2) = False \tag{18}$$

Equation 18 indicates the proposed impact in case the 'Agreement' seems to be 'FALSE'. The expression is similar for $Agreement(r_1, r_2) = False$ as compared to Eq. 17 because its value already ranges between 0 to 1.

Modified Popularity

When two scores are in the same part of the scale and the mean of both ratings is very far from the item's average rating, the item is considered to be popular. Two individuals provide equivalent types of ratings for popular and unpopular items, and then they have a lot in common when it comes to rating behaviour. In the proposed method, when $Agreement(r_1, r_2) = False$, the value is set to 0.3010. The proposed popularity values ranges from 0.3010 to 0.778 when $Agreement(r_1, r_2) = True$. The mathematical representation for proposed popularity is given by Eqs. 19, and 20.

$$Proposed\ Popularity(r_1, r_2) = log_{10}(2 + (\frac{r_1 + r_2}{2} - \bar{r}_{I_i})^2)$$

$$if\ Agreement(r_1, r_2) = True \tag{19}$$

$$Proposed\ Popularity(r_1, r_2) = 0.3010$$

$$otherwise \tag{20}$$

$$sim(I_i, I_q)^{MPIP} = \sum_{j=1}^{n'} MPIP(r_{u_j, I_i}, r_{u_j, I_q}) \tag{21}$$

Equation 21 represents the final similarity measure that can be used to compute the similarity between different set of items.

3.2 Deep Neural Network (DNN)

A neural network (NN) is a kind of artificial intelligence that aims to mimic the process of learning that people use to acquire specific aspects of information. NN incorporates a wide range of neurons (artificial), similar to the real neurons found in the brain, and utilizes them to interpret and process information. A NN is made up of an input layer, hidden (one or more) layers, and output layers that convert the input into some meaningful output.

Figure 1 illustrates the workings of an artificial neuron. A DNN [26] is a collection of layered neural networks, or networks with many layers. DNNs with at least 2 hidden layers and movement only in the forward direction are also referred to as "feed-forward neural networks." Both categorization and prediction are possible with these neural networks. In this chapter, DNN is used to predict the ratings of unrated movies that might be of interest to the user. The DNN aims at extracting patterns from the incoming objects, and then use those patterns to predict output value. We devised a variety of hidden neurons with activation functions to do this. To create a network of active hidden neurons, non-linear activation functions such as Sigmoid, tanh, and ReLU are employed.

The weights (W) and biases (B) must be adjusted just so the activation function can trigger these hidden neurons. "W" and "B" are first set at random. The network

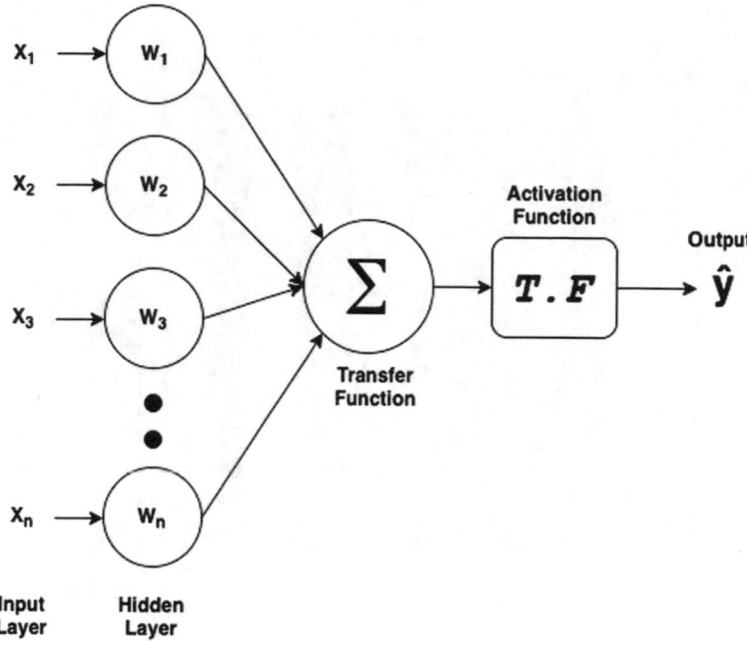

Fig. 1 Artificial Neuron [25]

is then trained using thousands of inputs. Back-propagation of error is used to adjust "W" and "B" such that the hidden neurons are activated at appropriate values. The feature set is a collection of "W" and "B" that will be used to identify a certain output. It is thought that DNN is the key solution when the pattern being utilised for discriminating is so complicated that conventional statistical and numerical methods fail. The working principle of DNN is illustrated via Fig. 2.

4 Proposed Model

The proposed flow consists of three phases as illustrated via Fig. 3. The first phase deals with the evaluation of different similarity measures with respect to collaborative filtering systems. The second phase deals with cutting back sparsity using MPIP and matching imputation. The third phase performs the key role of predicting the ratings of unrated items present in the system. The upcoming section explains the experimental outcomes of each phase of the proposed model.

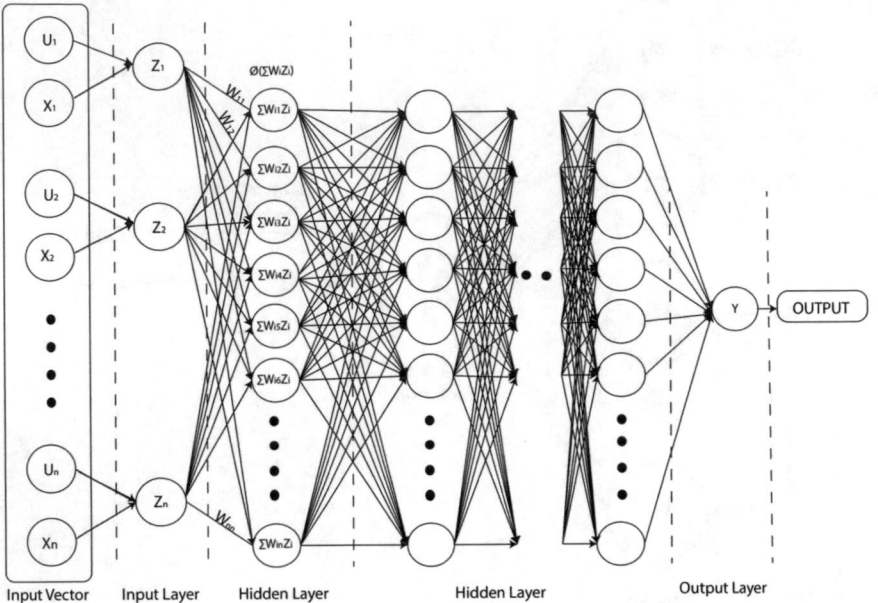

Fig. 2 Deep neural network [26]

5 Experimental Outcomes and Investigation

This chapter uses the MovieLens-100 K dataset to train and test the CF systems. Figure 4 shows the schematic structure of the u.data file. It has 4 columns: userId, itemId, rating, and timestamp. Figure 5 describes the schematic structure of the u.item file. It contains information about the following attributes: movieId, release date, IMDb URL, and genre information.

It can be clearly observed that most of the movies in the MovieLens-100K dataset are unrated. This inclines the dataset towards the problem of high data sparsity. The sparsity % of the MovieLens-100 K dataset is 93.7%. In this chapter, the sparsity issue is addressed by using matching imputation. The rating for any particular movie for any user is imputed based on the set of most similar users. Comparative analysis of different similarity measures is done in order to pick the one that performs best for computing the similarity among different users. Table 1 indicates the comparative analysis of different traditional similarity measures with our Modified-PIP similarity measure in the context of a collaborative filtering system. The Modified-PIP outperformed other similarity measures in terms of MAE, RMSE, Precision, Recall, and F1-Score.

The matching imputation approach helped in cutting back the sparsity to a great extent. The cutback level of sparsity obtained using matching imputation is specified in Table 2. The utility (user-movie rating) matrix formed after sparsity reduction is

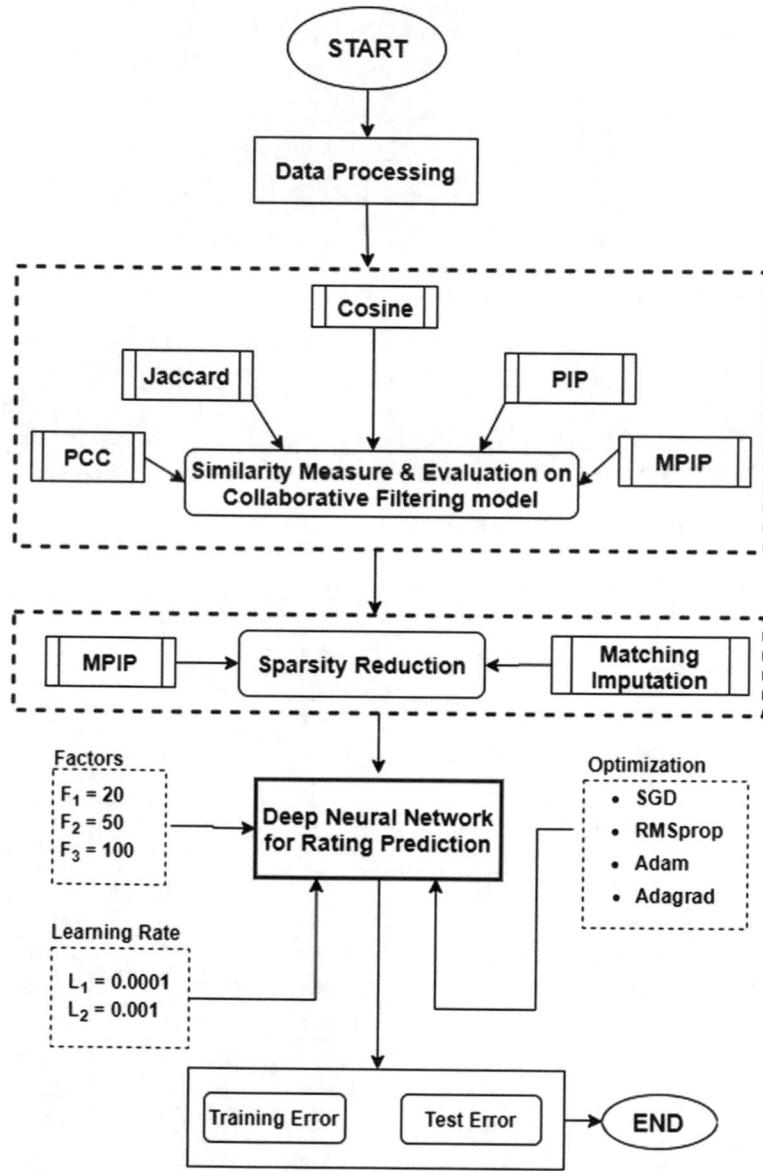

Fig. 3 The proposed architecture

	userId	itemId	rating	timestamp
0	196	242	3	881250949
1	186	302	3	891717742
2	22	377	1	878887116
3	244	51	2	880606923
4	166	346	1	886397596

Fig. 4 Dataset Description—(User:Rating Information)

Toy Story (1995)	01-Jan-1995	0	0	0	1	1	1	0	0	...	0	0	0	0	0	0	0	0	0	0
GoldenEye (1995)	01-Jan-1995	0	1	1	0	0	0	0	0	...	0	0	0	0	0	0	0	1	0	0
Four Rooms (1995)	01-Jan-1995	0	0	0	0	0	0	0	0	...	0	0	0	0	0	0	0	1	0	0
Get Shorty (1995)	01-Jan-1995	0	1	0	0	0	1	0	0	...	0	0	0	0	0	0	0	0	0	0
Copycat (1995)	01-Jan-1995	0	0	0	0	0	0	1	0	...	0	0	0	0	0	0	0	1	0	0

Fig. 5 Dataset Description—(Movie Information)

Table 1 Performance comparison of similarity measures

	MovieLens	−100 K	(1682 movies	And 943	Users)
Similarity	Precision	Recall	F1-Score	MAE	RMSE
Cosine	0.335744	0.273264	0.245345	1.370069	1.914075
Jaccard	0.243128	0.302515	0.217494	0.942697	1.205746
Pearson	0.299624	0.328616	0.270832	1.095055	1.084220
PIP	0.365533	0.331017	0.266618	0.887194	1.174997
MPIP	0.365495	0.342917	0.271720	0.849342	1.137702
MPIP-Scaled	0.352711	0.324872	0.277165	**0.844218**	**1.013293**
	MovieLens	−1 M	(3900 movies	And 6040	Users)
Similarity	Precision	Recall	F1-Score	MAE	RMSE
Cosine	0.195976	0.21426	0.182558	0.66259	0.73768
Jaccard	0.628424	0.58745	0.51613	0.76697	0.957043
Pearson	0.195976	0.21426	0.182558	0.74021	0.83768
PIP	0.720224	0.68344	0.584828	0.76453	1.27950
MPIP	0.82978	0.75433	0.737421	0.73432	0.928201
MPIP-Scaled	0.89558	0.79511	0.73768	**0.71426**	**0.825976**
	Jester	-data-1	(100 items	And 24983	Users)
Similarity	Precision	Recall	F1-Score	MAE	RMSE
Cosine	0.254888	0.41384	0.48458	1.25433	1.5686
Jaccard	0.138729	0.2684	0.30009	0.96259	1.24004
Pearson	0.347697	0.29326	0.31926	0.83697	1.08842
PIP	0.678336	0.79817	0.76253	**0.74210**	**0.88828**
MPIP	0.644869	0.56762	0.53018	0.76453	0.90128
MPIP-Scaled	0.644751	0.71511	0.58258	0.78124	0.93798

Table 2 Sparsity reduction using matching imputation

Dataset	Sparsity before imputation (%)	Sparsity after imputation (%)
MovieLens 100 K	93.69	43.45
MovieLens 1 M	95.8	56.71
Jesder-data-1	27.53	25.71

Table 3 DNN model MSE (Training Error :: Test Error)

Factors	SGD	RMSprop	Adam	Adagrad
	Dense-10	LR = 0.0001	Batch Size = 128	Epoch: 100
20	0.4234::0.2362	0.3241::0.2252	0.2778::0.1989	1.3757::1.3400
50	0.4663::02606	0.3008::0.2051	0.2786::0.1996	1.0675::1.0087
100	0.4423::0.2638	0.3022::0.2089	0.3071::0.1961	0.9793::0.8979
	Dense-10	LR = 0.001	Batch Size = 128	Epoch: 100
20	0.3484::0.2042	0.3141::0.2098	0.2833::0.2029	0.3232::0.1988
50	0.3043::0.2006	0.3536::0.2211	0.3102::0.2034	0.2871::0.1987
100	0.3145::0.2010	0.3088::0.2142	0.2852::0.2029	0.3440::0.2006
	Dense-20	LR = 0.0001	Batch Size = 128	Epoch: 100
20	0.3567::0.2285	0.2468::0.2039	0.2254::0.1939	1.0167::0.9529
50	0.4055::0.2535	0.2393::0.2017	0.2160::0.1945	0.8271::0.7491
100	0.3586::0.2240	0.2432::0.2020	0.2164::0.1947	0.7753::0.7086
	Dense-20	LR = 0.001	Batch Size = 128	Epoch: 100
20	0.2303::0.1973	0.2800::0.2125	0.2228::0.1982	0.2334::0.1954
50	0.2583::0.1972	0.2636::0.2125	0.2236::0.1981	0.2282::0.1946
100	0.2318::0.1972	0.6316::0.5646	0.2237::0.1995	0.2261::0.1954
	Dense-50	LR = 0.0001	Batch Size = 128	Epoch: 100
20	0.2962::0.2200	**0.2203::0.2018**	0.1884::0.1816	0.8280::0.7732
50	0.2969::0.2184	0.2208::0.2002	**0.1820::0.1826**	0.5069::0.4531
100	0.2996::0.2218	0.2153::0.1960	0.1829::0.1821	0.6355::0.5847
	Dense-50	LR = 0.001	Batch Size = 128	Epoch: 100
20	**0.2165::0.1944**	0.2660::0.2260	0.1941::0.1880	0.2091::0.1928
50	0.2162::0.1939	0.2291::0.2130	0.1915::0.1883	0.2104::0.1923
100	0.2169::0.1937	0.2406::0.2127	0.1925::0.1874	**0.2056::0.1914**

used by the DNN model to make predictions about unrated movies that might be of interest to the end user.

The formed dataset is divided into 60/40 ratios, such that training data comprises 60% of the data and test data comprises 40% of the data.

The Deep Neural Network model is tested on different parameters, as mentioned below:

Factors: 20, 50, 100; *Learning Rate*: 0.001, 0.0001; *Optimization*: Stochastic Gradient Descent (SGD), RMSprop, Adam, and Adagrad; *Batch Size:* 128; *Epoch:* 100.

The results (Train/Test error) obtained on different configurations are discussed via Table 3. The minimum error value obtained for the proposed DNN model is around 0.1820(Training error) and 0.1826 (Test Error). When the network is made even more dense, the generalisation gap between the training error and the test error increases. Thus, making the network more dense won't help it in converging towards the best solution; rather, it will make the system overfit. The best solution is obtained with the Dense-50 network, with a learning rate of 0.0001 and a batch size of 128. Adam optimization performs most efficiently as compared to other gradient descent optimization algorithms. The convergence of training and test error for a different number of epochs in different configurations possessing the worst performance is illustrated via Fig. 6a–d. Figure 7a–d illustrates the configuration setup possessing the best performance yielded by different optimization approaches implemented on top of DNN.

6 Conclusion and Future Work

In this chapter, Deep Neural Networks (DNN) have been used in collaboration with the Modified-PIP similarity measure for computing the similarity among the users and predicting the rating of unrated items. Testing of the DNN model has been performed on different configurations (factors, epoch, optimization technique, and learning rate (LR)). The proposed deep hybrid model yielded a low error, thus assuring the high performance of the model.

In the future, we will try to examine the impact of learning an error function rather than using the MPIP to compute the user-user similarity. We would also want to test our model against other recommendation contexts such as news recommendations, joke recommendations, song recommendations, etc. Apart from that, we would explore the potential of movie recommendation using reinforcement learning. The potential user implicit input may be utilised to model their preferences and provide recommendations to them.

Implementation Code: The python code for the implemented deep hybrid model can be found at the following GitHub link: https://github.com/gurvinder-yadav/mpip-dnn.

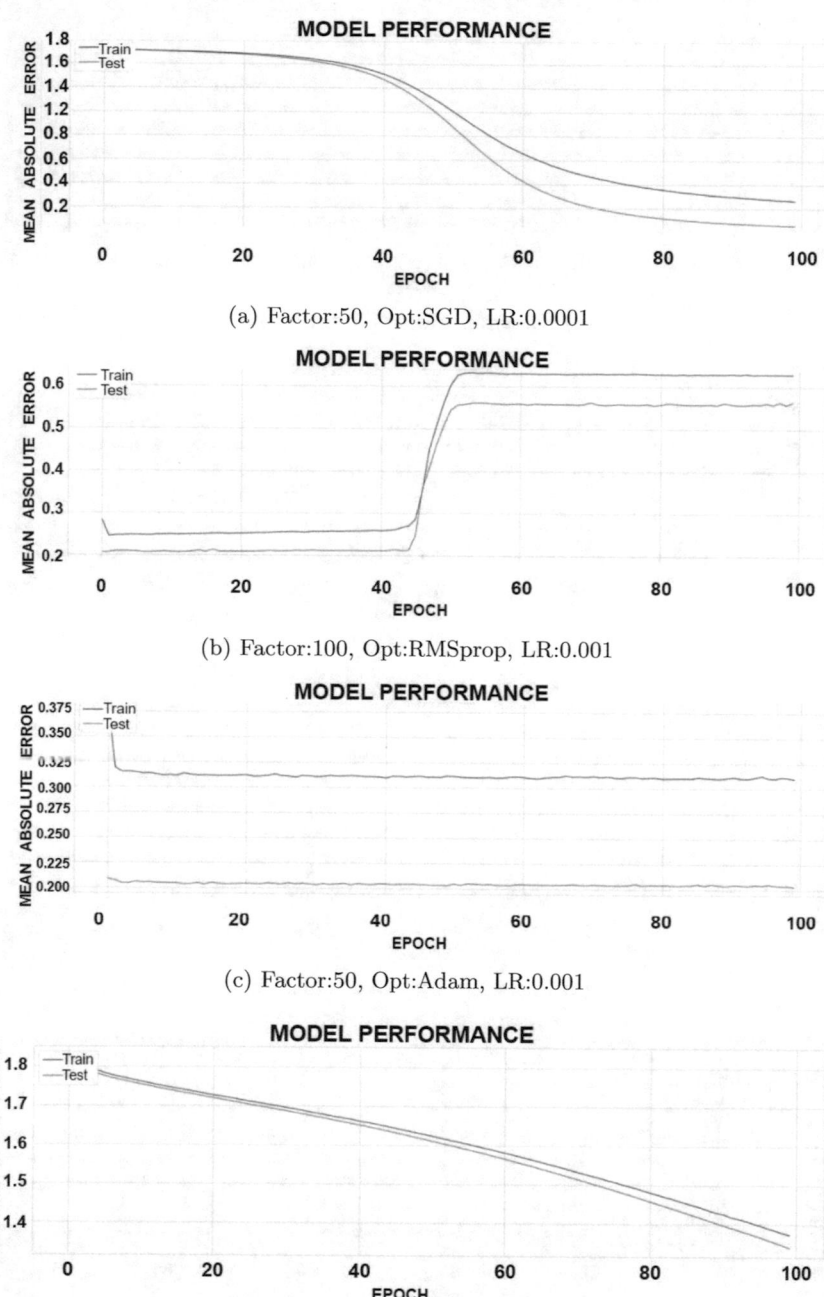

(a) Factor:50, Opt:SGD, LR:0.0001

(b) Factor:100, Opt:RMSprop, LR:0.001

(c) Factor:50, Opt:Adam, LR:0.001

(d) Factor:20, Opt:Adagrad, LR:0.0001

Fig. 6 Worst performing DNN architecture

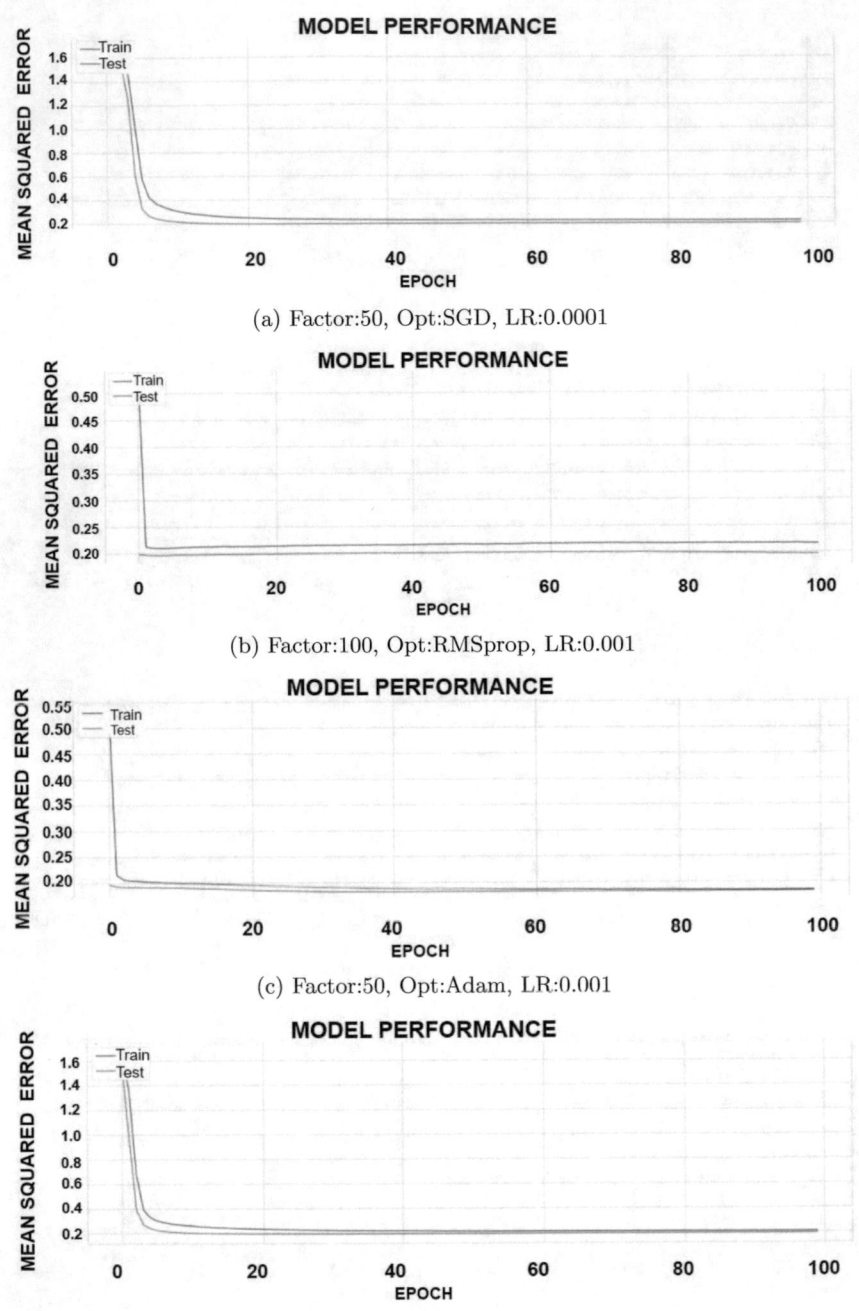

(a) Factor:50, Opt:SGD, LR:0.0001

(b) Factor:100, Opt:RMSprop, LR:0.001

(c) Factor:50, Opt:Adam, LR:0.001

(d) Factor:20, Opt:Adagrad, LR:0.0001

Fig. 7 Best performing DNN architecture

References

1. Lu, J., Wu, D., Mao, M., Wang, W., Zhang, G.: Recommender system application developments: a survey. Decis. Supp. Sys. **74**, 12–32 (2015)
2. Bag, S., Kumar, S.K., Tiwari, M.K.: An efficient recommendation generation using relevant Jaccard similarity. Inf. Sci. **483**, 53–64 (2019)
3. Liu, H., Hu, Z., Mian, A., Tian, H., Zhu, X.: A new user similarity model to improve the accuracy of collaborative filtering. Knowl.-Based Syst. **56**, 156–166 (2014)
4. Shi, X., Luo, X., Shang, M., Gu, L.: Long-term performance of collaborative filtering based recommenders in temporally evolving systems. Neurocomputing **267**, 635–643 (2017)
5. Bellogín, A., Sánchez, P.: Collaborative filtering based on subsequence matching: a new approach. Inf. Sci. **418**, 432–446 (2017)
6. Bobadilla, J., Ortega, F., Hernando, A., Gutiérrez, A.: Recommender systems survey. Knowl.-Based Syst. **46**, 109–132 (2013)
7. Sinha, B.B., Dhanalakshmi, R.: Evolution of recommender paradigm optimization over time. J. King Saud Univ.-Comput. Inf. Sci. (2019)
8. Luo, X., Xia, Y., Zhu, Q.: Incremental collaborative filtering recommender based on regularized matrix factorization. Knowl.-Based Syst. **27**, 271–280 (2012)
9. Bobadilla, J., Serradilla, F.: The effect of sparsity on collaborative filtering metrics. In: Proceedings of the Twentieth Australasian Conference on Australasian Database, vol. 92, pp. 9–18, Jan. 2009
10. Dhanalakshmi, R., Sinha, B.B.: Hybrid Cohort Rating Prediction Technique to leverage Recommender System (2019)
11. Salakhutdinov, R., Mnih, A., Hinton, G.: Restricted Boltzmann machines for collaborative filtering. In: Proceedings of the 24th International Conference on Machine Learning, pp. 791–798, June 2007
12. Yang, C., Bai, L., Zhang, C., Yuan, Q., Han, J.: Bridging collaborative filtering and semi-supervised learning: a neural approach for poi recommendation. In: Proceedings of the 23rd ACM SIGKDD International Conference on Knowledge Discovery and Data Mining, pp. 1245–1254, Aug. 2017
13. Sedhain, S., Menon, A. K., Sanner, S., Xie, L.: Autorec: autoencoders meet collaborative filtering. In: Proceedings of the 24th international conference on World Wide Web, pp. 111–112, May 2015
14. Strub, F., Mary, J.: Collaborative filtering with stacked denoising autoencoders and sparse inputs. In: NIPS Workshop on Machine Learning for eCommerce, Dec. 2015
15. Singh, P.K., Sinha, M., Das, S., Choudhury, P.: Enhancing recommendation accuracy of item-based collaborative filtering using Bhattacharyya coefficient and most similar item. Appl. Intell. **50**(12), 4708–4731 (2020)
16. Wu, Y., DuBois, C., Zheng, A. X., Ester, M.: Collaborative denoising auto-encoders for top-n recommender systems. In Proceedings of the Ninth ACM International Conference on Web Search and Data Mining, pp. 153–162, Feb. 2016
17. Yu, J.B.: Evolutionary manifold regularized stacked denoising autoencoders for gearbox fault diagnosis. Knowl.-Based Syst. **178**, 111–122 (2019)
18. Sinha, B.B., Dhanalakshmi, R.: DNN-MF: deep neural network matrix factorization approach for filtering information in multi-criteria recommender systems. Neural Comput. Appl. 1–15 (2022)
19. Rendle, S., Krichene, W., Zhang, L., Anderson, J.: Neural collaborative filtering versus matrix factorization revisited. In: Fourteenth ACM Conference on Recommender Systems, pp. 240–248, Sept. 2020
20. Tan, Z., He, L.: An efficient similarity measure for user-based collaborative filtering recommender systems inspired by the physical resonance principle. IEEE Access **5**, 27211–27228 (2017)
21. Guo, G., Zhang, J., Yorke-Smith, N.: A novel bayesian similarity measure for recommender systems. In: Twenty-third International Joint Conference on Artificial Intelligence, June 2013

22. Abualigah, L.M., Khader, A.T.: Unsupervised text feature selection technique based on hybrid particle swarm optimization algorithm with genetic operators for the text clustering. J. Super-comput. **73**(11), 4773–4795 (2017)
23. Abualigah, L.M.Q., Hanandeh, E.S.: Applying genetic algorithms to information retrieval using vector space model. Int. J. Comput. Sci. Eng. Appl. **5**(1), 19 (2015)
24. Ahn, H.J.: A new similarity measure for collaborative filtering to alleviate the new user cold-starting problem. Inf. Sci. **178**(1), 37–51 (2008)
25. Ahire, J.B.: The artificial neural networks handbook: Part 4. Medium (2018). https://medium.com/@jayeshbahire/the-artificial-neural-networks-handbook-part-4-d2087d1f583e
26. Sinha, B.B., Dhanalakshmi, R.: Building a fuzzy logic-based artificial neural network to uplift recommendation accuracy. Comput. J. **63**(11), 1624–1632 (2020)

A Survey on Graph Neural Network Based Video Recommendation System

Toshi Rawka and Mahipal Jadeja

Abstract Recommender systems helps the user to recommend something based on their interest. As online data is growing extensively, we require recommender systems more, and there has a lot of work going on in this field. Recommender system in online video streaming platforms plays a huge role. 80% of videos watched online are from recommendations. There are many techniques on which this system works. As most of the information is in graph structure and graph neural networks (GNNs) have a specialty in representation learning, the field of utilizing GNN in recommender systems is expanding. This chapter provides knowledge of GNN-based recommender systems and state-of-the-art models in developing this domain.

Keywords Recommender system · Graph neural network (GNN) · Deep learning · Tag ranking · GraphSage · GAT - Graph attention network · GCN - Graph convolution network

1 Introduction

A recommender system is a system that predicts the likeliness of an item for a user or a group. Users rely on recommender systems to help them deal with information overload and find what they're looking for among the immense sea of options (e.g., products, movies, news, or restaurants). Various platforms which use recommendations at their peak are Amazon, Flipkart, Youtube, Netflix, inshorts, and many more. An effective recommender system precisely predicts users' preferences based on their previous interactions (e.g., clicks, watches, reads, and purchases) to achieve this goal, as shown in Fig. 1. They indicate users' preferences by directly utilizing their previously interacted objects because of their ability to efficiently capture non-linear user-video relation and non-trivial user-video relation and effortlessly contain

T. Rawka · M. Jadeja (✉)
Malaviya National Institute of Technology Jaipur, Jaipur 302017, Rajasthan, India
e-mail: mahipaljadeja.cse@mnit.ac.in

T. Rawka
e-mail: 2020pcp5532@mnit.ac.in

© The Author(s), under exclusive license to Springer Nature Switzerland AG 2022
T.-P. Hong et al. (eds.), *Deep Learning for Social Media Data Analytics*,
Studies in Big Data 113, https://doi.org/10.1007/978-3-031-10869-3_8

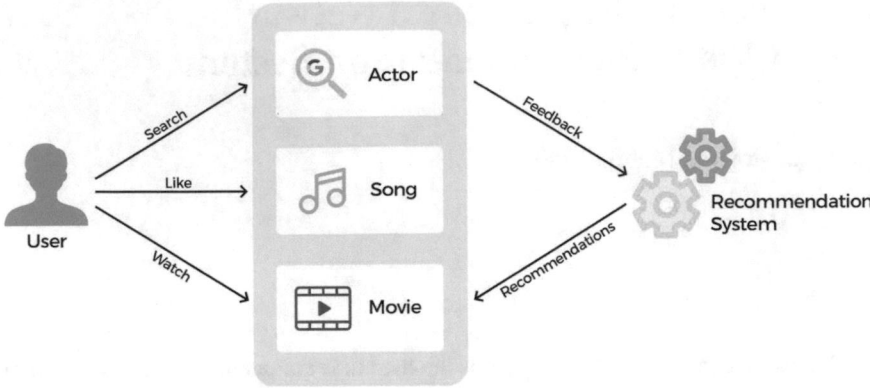

Fig. 1 Recommendation system

lots of data assets, including visible, textual, and contextual sort of data. Among all the recommendation categories, we focus on the recommendation of videos in this chapter. Unlike texts and images, videos often include more information that isn't immediately visible to the consumer. Although titles and cover images can assist in expressing crucial ideas in films, static abstracts can't capture all of the complexities of user preferences in videos.

There are many benefits of a Video recommender system to both user and the platform that uses the system. Some of the benefits are:

1. The Video Recommender system helps search the desired video of the user's interest.
2. The video consumption of videos on online video streaming platforms is mainly from recommendations, so a sound recommendation system for a video streaming company is vital.
3. As the number of videos available on the internet is multiplying recommender system is a must need.

Graph neural network is a combination of two words or two concepts: graph and neural network. The most crucial key in Graph neural network is a graph. A graph is a set of nodes and edges. A neural network helps the graph to think just like a human brain. As we all know, graph data is very complex to understand, and it creates a considerable challenge for existing machine learning and deep learning approaches. So, the solution that tries to minimize this challenge can be GNNs that are nothing but neural networks that can be directly applied to graphs. Existing machine learning and deep learning approaches are specialized for fixed-size and structure-like images, which we can say as fixed-size grid graphs. Also, represent a line graph to text and speech. We have more complex graphs, which have the variable size of unordered vertices, where vertices can have different neighbors. It does not help in existing machine learning algorithms because it has a core assumption that instances are independent of each other. This makes wrong for graph structure; each vertex is related to others by links/edges of different types.

The recommendation using Deep learning models is very widely used and rapidly growing in commercial application and research. Graph neural network (GNN) is undoubtedly the most appealing deep learning algorithm because of its greater capacity to train on complex graph data, which is essential for recommendation systems. A bipartite graph connecting user and item nodes, for example, can be used to describe interaction data in a recommendation application, which can observe interactions that are represented by edges. Item transitions according to users' behavior sequences can be represented in graphs as well. When including structured external information, like social relationships among users and knowledge graphs related to products, the usefulness of generating recommendations as a task on graphs becomes even more apparent. GNN offers a unifying perspective for modeling the numerous heterogeneous data in recommender systems in this way.

The rest of the chapter is organized as follows–Sect. 2 describes the preliminaries of recommendation system and associated challenges, Sect. 3 an in-depth finding for the underlying problem. Section 4 highlights the need for GNN based video recommendation system. Section 5 explains various state-of-the-art GNN based methods for video recommendation. A brief comparison of these methods is provided in Sect. 6. Section 7 briefly explain the potential application scenario, Sect. 8 describes the data privacy in the domain and Sect. 9 concludes the work. It also mentions the future scope in the same.

2 Preliminaries of Video Recommendation Systems and Challenges Associated with the Field

Video recommendation is a way of recommending videos by extracting users' profiles and history; after that, rank videos according to their preferences and watch time. Many giant companies implement video recommendations, including Netflix, Amazon Prime, and YouTube. Recommender systems use various techniques and algorithms to suggest relevant items to users. Video streaming services are dependent on the video recommender system to offer the videos which users would like to watch. Association between users and videos is the significant data source of video recommendation. Most of the existing recommendation systems calculate the relevance of video based on user feedback, e.g., watch and search activities. Traditionally, recommendations without using deep learning concepts are clustering, nearest neighbor, and matrix factorization. The problems of recommender systems without deep learning are to learn the effective user/video relations from their interactions, side information and to deal with the cold-start problem (whenever a new video is included to the existing video library), i.e., to bootstrap the video relevance score for the newly added video which is difficult mainly due to a very limited user behavior/interaction with respect to the newly added video. This chapter addresses this issue by discussing the content-based video recommendation approach by taking advantage of deep convolutional neural networks to cope up the cold-start problem. Deep learning-based recommender systems are better than traditional ones due to their capability to process non-linear data. The chapter discusses the key benefits of deep learning-based recommender systems, including non-linear transformation, representation learning, sequence modeling, and flexibility. The Graph neural network (GNN) techniques combine node information with the hidden topological structure. The great performance in graph data learning, GNN techniques have been widely applied in many fields, including but not limited to image recognition, natural language processing.

3 Findings

There are so many models are proposed by various researchers for video recommendation. In this section, we will discuss all the relevant models briefly.

First is with the use of content-based filtering, which uses the KNN model and cosine similarity; this similarity provides more accurate results than other similarity measures, proposed by Ramni Harbir Singh et al. in [1]; additionally, the running complexity in this similarity measure is comparatively low, in a movie recommendation system. Next is MOVREC, a movie recommendation system that works using a collaborative filtering approach presented by D. K. Yadav et al. in [2]; in this, explicit information is provided from the user this information is observed. The item is recommended, a recommendation is arranged in such a manner with the highest rating first. As both the proposed model, i.e., content-based and collaborative-based filtering, have their limitations, Luis M Capos et al. in [3] presented a model which uses collaborative filtering with a bayesian network. Harpreet Kaur et al. in [4] also discussed the hybrid recommendation, which is the integrative method of content-based and collaborative based. This model minimizes the limitation of both approaches. The user-item specific information makes a cluster by Utkarsh Gupta et al. using chameleon [5]. This is often an efficient technique based on Hierarchical clustering for the recommender system. The proposed approach has minimized the error and better clustered comparable items. Urszula Kuelewska et al. also suggested using the clustering concept to make a better recommendation; basically, they present and evaluate the combination of centroid-based solution and collaborative filtering, which occurs significant improvement in accuracy compared to centroid based method only. The user profile information, their watching history, and data involving items scored from other websites are used to aggregate similarity calculations to make a unique recommendation, ignoring the data specific to the user. To conclude, the models for video recommender systems are based upon the characteristics information, i.e., keywords, categories, preferences, profiles, ratings, likes, etc. According to that, mainly four types of algorithms are used in video recommender systems: simple, content, collaborative, and hybrid recommendation systems.

3.1 Simple Recommendation System

The Simple recommender gives recommendations to the user who is on trending video or is most popular. In this recommendation, it is assumed that the videos that are more popular or in trend have more ratings and will have a higher chance of being liked by the user. This recommender model fails to provide personalized recommendations based on user interest. This model implementation is very trivial. The videos need to be arranged in sorted order based on critical elements ratings and popularity and result from the highest-rated list of videos. This method is proper when users want a list of trending videos. It suffers from severe limitations; it gives the same recommendations to everyone, without considering the user profile and one's taste.

3.2 Content Based Recommendation System

In this system [6], it is assumed that if the user is interested in some item in the past, it may also be interested in the future. The main thing in this is past data of user profile with direct interaction with the user or indirectly asking from a user about their interests. It performs worse when the content doesn't contain enough information to classify the videos precisely. In this type of recommendation system, for making accurate recommendations, enough ratings are required. Still, if a new user is coming who has not rated any item yet, then this system is not working. Instead of the past watching history, the user gets a recommendation or suggestion for a similar item according to their rating or explicitly searches for. In this type, movie profiles are created, not the user profile. This is called item-based filtering, not user-based filtering. The suggestion in a content-based filtering system is not unexpected or surprising. It is also not able to judge the quality of a movie [7]. For example, it cannot differentiate between a good item and a lousy one if both items use similar words. It is difficult for a system to analyze the quality of a movie because it is the core features that depend on each user's interest, taste, framework, ideas, etc.

3.3 Collaborative Filtering Based Recommendation System

Till now, what we discussed was not providing good personalized recommendations. When any user queries, the system will receive a similar recommendation for a particular video; it ignores their data. This collaborative method [8] is quite different from the models we saw previously; it also involves other users. It assumes that similar users liked similar videos. This provides a better prediction of recommendations as compared to the content-based method. Youtube, Netflix, and Prime Video all use this model to make recommendations. So, now the question is how to identify two users are similar or not. For that, consider videos watched by both users and check how they rated the videos. Then if we have to predict the rating for unseen videos, look at similar user choices. Some important models of collaborative filtering are:

1. **KNN**: KNN is a class of machine learning approach which is used to find out the clusters of similar users according to their standard type of ratings, and make better suggestions as the name suggests KNN [9] means k nearest neighbor who makes use of the mathematical concept to average the rating and recommends top-k [10] nearest neighbors. It comes under the Collaborative filtering systems, which uses the behavior of users to suggest different videos. They could both be user-based or item-based.

2. **Principal Component Analysis (PCA)**: This is also a class of collaborative filtering methods used in recommendation systems that compute efficient similarity for different users and use the features for the video recommendations [11] system. This model employed the three metrics to calculate the errors in statistics, root mean square error (RMSE), mean absolute error (MAE), mean squared error (MSE), and for performance analysis recall, precision, and f-measure.

3. **Matrix factorization**: Matrix factorization [12] is a class of collaborative filtering models that are used in traditional recommender systems. It divides the user-item relationship matrix into the product of two low-dimensional matrices.

Disadvantage of collaborative filtering: The problem with Collaborative filtering is a cold start problem, i.e., when new users come into the system, it is difficult to make recommendations for users. So in those cases, the users have not given the rating to any items yet, so the recommender system cannot predict their interests. Also, this could face attacks like spam attacks; these attacks are mainly made by users interested in misguiding the system to recommend a particular product. One problem is the sparsity of the rating matrix; in most recommender systems, users do not rate each item, so most of the cells in the rating matrix are empty. In such cases, identifying similarities among different user or item groups creates challenges. Even in the presence of groups like these, it will still be a challenge to recommend end-users who do not give any feedback like affirmation or negation to any groups.

Singular Value Decomposition (SVD): As we see that sparsity of matrix is the problem in the collaborative filtering method, the model has to represent the expected behavior between users and items to handle this issue. The main idea is to convert the problems of recommendation systems to direct large data-set that change very frequently into a more straightforward problem. It can also be seen how better the model predicts the rating of items for every user. Commonly Root Mean Square Error (RMSE) is used. Better performance is observed whenever the RMSE is low. Latent factors are a promising approach that suggests a notion that a user or an item has. If talking about movies, the movie genre can generally be referred to by latent factor. Singular Value Decomposition extracts latent factors and thus decreasing the dimension of the utility matrix. Essentially, every user and item is mapped into a latent space with size. It helps in understanding the connection between end-users and items correctly because it can be compared one-to-one.

3.4 Hybrid Recommendation System

There was no user interference in a content-based recommendation system as any user would be recommended the same set of movies having watched similar movies. At the same time, in Collaborative filtering, there were no relations between movies watched by the user. If the user loves watching Action movies, the recommendation system will not capture his genre interest. Thus the idea of a Hybrid Movie Recommendation System came up as it tries to use both Content and Collaborative Filtering methods combining advantages and eliminating disadvantages of both to get a recommendation for users.

Only GNNs can be beneficial to the recommendation system as it focus on both content information (users, item features) and graph structures (user interactions with items i.e. video as example) [13]. In contrast, traditional models usually only use one of the two. The next section will see how and why we require a graph-based recommendation.

4   Why Do We Need GNN Based Video Recommendation System?

Graph neural networks (GNN) [14] came during the previous few years and have been very helpful for representing graph structures in many domains. Presently most of the data in the recommendation field are defined as a graph structure. So to make a better prediction, take advantage of the GNN technique. Earlier, the relationship between data can be represented in a matrix, but sometimes data may be sparse, creating a bipartite graph (as shown in Fig. 2 [13]) to illustrate the relationship of end-user and item vertices, where the links/edge indicates the relationship of the respective end-user and item. After that, a series of items be converted into a series

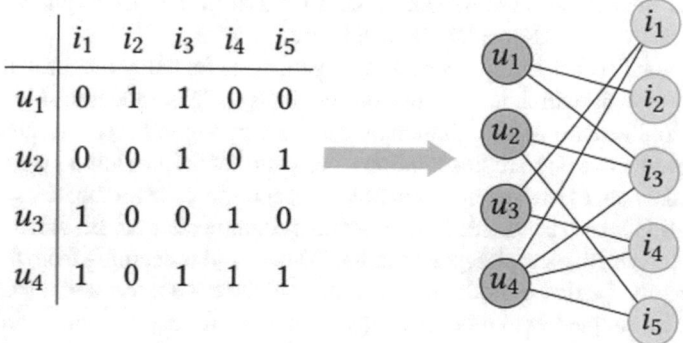

Fig. 2 User-item bipartite graph [13]

Fig. 3 Sequence graph [13]

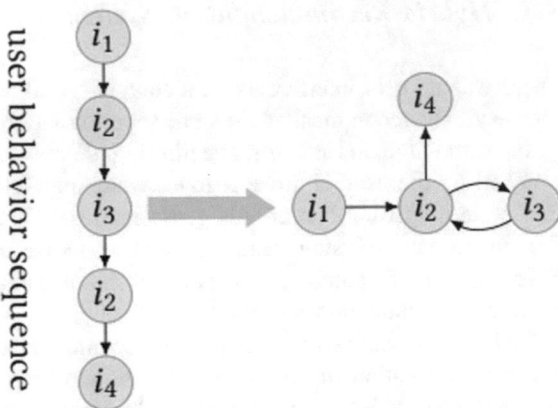

Fig. 4 Social relationship between users [13]

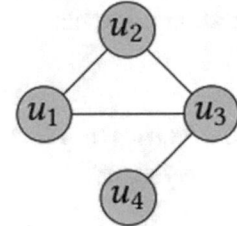

or sequence graph, where every item is linked to one or more following items, as shown in Fig. 3 [13].

Converting sequence data into a sequence graph makes it more versatile for item-to-item. Other information includes a social graph; a knowledge graph can also be represented as a graph, as illustrated in Fig. 4 [13].

Previously general recommendation was considered for paired interactions to understand user/item representation. Then for a sequential recommendation, conversion of sequence model to sequence pattern is deployed to understand the user's taste and predict the better recommendation because user preference might be changing according to their mood, locality, behavior, etc.

To effectively use that sequence pattern by learning for better results in the recommendation system, which is an enormous challenge, GNN model design came into existence to take care of this challenge. The role of general recommendation is to learn the productive node representations, i.e., user/item representations as user/item are denoted by nodes in the graph, and predict user preference for the items. Sequence representation is the reason for the sequential recommendation. In which the model learns the informative graph representation. The benefit of learning from the individual information (vertex or node information) and the edge information (relationship information), coupled with models taking advantage of the prevalent social network and knowledge graphs, makes GNNs a promising solution for the recommendation. This GNN technique is broadly divided into graph convolution network, graphSage, and Graph attention network. This further divides into the respective models: Light-GNN, DualGNN, Multimodal Graph Convolution Network (MMGCN), tag ranking based, Concept-Aware Denoising Graph Neural Network, Graph Refined Convolutional Network, Neighbor-Aware Graph Attention Network, Context-Dependent Propagating based Video recommendation. There are so many other models also proposed for this recommendation, we summarize with the help of some of the discussed models.

5 GNN Models for Video Recommendation

GCN MODELS:
Graph convolution network has many models which encounter in video recommendation systems. One of the models is LightGCN. LightGCN is come into existence to minimize the unnecessary design of GCN [15], mainly transformation of features and non-linear activation in a recommendation. In past times the critical idea behind the GCN approach is to learn the representation of nodes by using graph features. To reap this, LightGCN performs graph convolution repeatedly, summing the neighbor's features such as a new illustration of a destination node. Traditional method matrix factorization is directly mapping the unique ID to her embedding. The improvement in the quality of embedding can be made using the increased user ID, along with her interaction of history which has been given as the input. The key benefit of being simple is innumerable folds more illustratable, practically reliable to learn and maintain. In terms of technical, it is simple and easy to analyze the model behavior and amend it toward more effective instructions.

Another type of GCN model is MMGCN [16], which stands as Multimodal Graph Convolution Network. One of the concepts of neural graph networks is message passing, which is used in this type of framework, which is used to capture better user preferences. One can use model-specific depictions of users and videos. This constructs a user-item bipartite graph in each modality, along with it trying to enhance the representation of each node and topological feature and neighbors structure.

DualGNN is another model for recommending videos that focuses on two aspects: single modal and multimodal representations. To overcome the limitations of Dual-GNN, a brand new framework is proposed in [17] for the multimedia recommendation (dubbed dual Graph Neural network (DualGNN)). DualGNN works upon the user-video bipartite and user co-prevalence graphs, which strengthen the correlation among users to collaboratively mine the particular fusion sample for one and all users.

GraphSage Models:
One of the approaches to making video recommendations is using tags; the tag ranking model [18] has also been proposed in early work. At that time, videos can be seen or addressed by tags only ignoring the content, but with graph neural network, it is improved. As we know that the videos are attached with multiple tags and

other attributes, these tags are used to spotlight the contents of videos. This model presents the personalized list of tags for videos for the tag candidates. To improve tag click rate and watch time of the video, a model should entice users to click on extra tags, explicitly providing their interested tag channels, and watch more tag-related videos. The model should consider the user interest in the tags and videos. Basically, for improvement in video and tag, the neighbor similarity concept is used to encode the user choices in heterogeneous node representation. This will conclude that the model is proposed using a heterogeneous network which includes the dynamic or different relationship of video, tags, and users.

CONDE is a concept aware denoising graph neural network proposed in [19] for video Recommendation that addresses information redundancy challenges. In the video type of recommendation system, the system extracts or captures the concepts from captions and the comments linked or associated with the videos to make full use of the textual information for the recommendation model. Likewise, with the help of three attributes that are user, video, and concept nodes, a heterogeneous tripartite graph is constructed. Tripartite because a user can make a relationship or interact with multiple videos and every day it will receive lots of reviews, neither all videos nor all concepts of a video can make reflect users' preference accurately. Hence, a graph neural network is proposed to the personalized denoising concept to obtain the relationship between user and item. The outcome of GraphSage [20] presents homogeneous graphs with a direct connection of items and users, which can not make up for the negative or false impact of noisy information. At the same time, construction item to item and user to homogeneous user graph and the item-concept connectivity in information utilization and aggregation the hierarchical attention network comparatively to graph attention network (GAT) is more potent in performance improvement.

One recommendation model for multimedia and implicit feedback is the graph refined convolutional network proposed in [21], as we know that the one concept in representation learning is graphsage. This concept is useful when having rich node attributes, the trainable functions in the user-item graph, and the implicit feedback graphsage to predict the relationship between user and item. The model is trained and goes through the message along with the graph and gathers them to refresh each node's representation. Traditionally implicit feedback problem is faced in the GCN-based recommendation, which gives a solution in advanced user-item interaction graph for graph convolutional operations. In the interaction graph, this recognizes the wrong positive feedback and minimizes the corresponding noisy edges. It is also observed that the concept of message passing, node representation, and process of aggregation are not clear signals. Or in other words, we can say that it is not directly replicate the pattern of user preference in the local structure information in the graph construction with implicit feedback.

GAT Models:
Most of the GNN based recommendation models consider the complete collaborative multi-hop neighbors information to enhance the presentation but based on this

type of model, it doesn't account for relational information. Relational information may include differences in the expression for different neighbors explicitly in the neighborhood.

The effect of each neighboring user is addressing the representation of item preference by the correlation between the item and user neighboring items. For transformation in graph embedding aggregation, the internal correlations of neighbors are modeled. A neighbor-aware graph attention network named NGAT4Rec [22] is used to perform the recommendation task. In this model, one layer is called the neighbor-aware graph attention layer. This layer takes the coefficients of various neighbor conscious attention to different neighbors of a given node by calculating the pairwise attention among these neighbors. This model sums the neighbor's embeddings according to the respective coefficients of neighbor aware attention to engender subsequent layer embedding for every node. Moreover, it consolidates a more neighbor-aware graph attention layer to accumulate the powerful signals from multi-hop neighbors. It eliminates nonlinear activation and feature transformation that ended up pointless on collaborative filtering.

One video recommendation model is proposed [23], called CDPRec, a context-dependent propagation recommendation system. It is a neural network that includes HINs (Heterogeneous Information Networks) and multi-modal content. The word content is generally video; then, this neural network feeds into recommender systems naturally and effectively. This model tries to overcome the drawbacks of one of the methods. This path-based similarity introduces the context-dependent propagation, which means the model is automatically or self propagates the ability for preferences of the users and explores their hierarchical internal dependencies within the heterogeneous information networks. To make a straightforward prediction of the click-through rate, this model presents the concept of encoding; it encodes the context into the complete embedding of video. Encoding is done from video context into video embedding. Over the strong baselines, this method proved improvement and results experimentally on a real-world YouTube data illustration.

6 Comparison of Different Models

This section briefly compares the models, i.e., GraphSage, MMGCN, GAT. The dataset used for this comparison is Movielens (https://movielens.org/). This dataset is mainly used in terms of personalized recommendation, so we also take this dataset to compare our model. To measure the video recommendation system, the soundtracks from the video and keyframes from the video, and collection of the video features are extracted. With some initially trained deep learning models [24], the images, audio or sound, and text patterns are removed, and results are shown in Table 1.

GraphSage [21] type model is used to predict the relation and interaction between video and user. in which the user-item graph is created using implicit feedback.

The MMGCN [21] model is also used. It helps to store the relevant aggregate functions by which the message is passed into the graph structure to refresh the

Table 1 Effect of pruning operations on Movielens (Visual, Acoustic, and Textual)

Movielens		
Model	Precision	Recall
Visual	0.0633	0.2545
Acoustic	0.0621	0.2540
Textual	0.0611	0.2531

Table 2 Performance comparison between the GraphSAGE, MMGCN, GAT

Movielens		
Model	Precision	Recall
GraphSAGE	0.0496	0.1984
MMGCN	0.0581	0.2345
GAT	0.0569	0.2307

representation of each node. It also learns from the implicit feedback and guesses the prediction of users on watched videos and unwatched videos. This is generally based on model-specific user preference, in that the direct content information interchange among user and item is learned. This is somewhat content-based.

The third type of model is GAT. This type of model automatically trains and indicates or specifies different weights to every node's neighbors. Along with the trained consequences, this model removes the noise on the information taken from the neighbors to improve the personalized type of recommendation [21] effectively. As we know, precision and recall are the two main parts of deep learning. Precision is defined as the quality, and recall is the measure of quality. All the models in terms of these two elements are shown in Table 2.

6.1 Addressing Cold Start Problem

As discussed earlier, a cold start problem is any new thing that comes into the existing system which creates difficulty for the interaction between user and item. This problem can be understood in two ways first one is the user cold-start problem in which a new user is added into interaction or item cold start in that new item is added into interaction. For that explicit feedback is also needed so that whenever new item comes it is associated with some attributes by which user get attracted. As we can observe, many of the GNN models, including GCN, GraphSage, GAT, are discussed, which are used for video recommendation system and are built upon the user and item bipartite graph, these models are based on implicit feedback because these are ask-to-rate technique which do not apply to the real-world cold start scenario.

7 Application

This section briefly explained the potential application scenarios of the video recommendation system. Various applications of this domain are YouTube, Netflix, Prime-Video, Disney+Hotstar, Instagram Reels, Facebook. The main thing of these systems is to provide personalized recommendations that help users find their interested and relevant videos. The idea behind this is to update the recommendation engine regularly and keep track of the user's recent activity. In other words, the main aim of this system is to keep users entertained so that they can enjoy and engage in particular platforms. The system observed that most of the videos users watch are likely videos recommended by the system. A recommendation can make according to region, language, mood, and many more features.

8 Data Privacy

The recommendation system gathers the user's personalized information to suggest items based on their interest, so the data of individuals taken by the system is susceptible and should be taken care of. In other words, data that needs to be private is very important. So the main idea of a recommendation system is to use the data so that data is appropriately utilized without disclosing any identity. Dynamic privacy-preserving recommendation on graph date is discussed in one of the surveys [25]. Privacy is taken care of by the regulation, called general data protection regulation (GDPR).

9 Conclusions and Future Scope

As we see many models of GNN for recommendation due to which we observe that moving toward graph neural networks creates more remarkable improvement in this field due to the rich features. Train the model and predict the model in graph representation in collaborative signals and sequential patterns, this technique in recommender systems is very beneficial. The various methods, including content-based and collaborative-based methods, are discussed in this chapter with the GNN approach to make the recommendation system effective. Also, the comparison of the models is discussed where we can observe that MMGCN provides high precision compared to the other two models. In this chapter, a detailed discussion on the latest work on recommendation is done. Some categorization method to sort out the current existing result is also discussed.

As we all know, several users may change over time. Several items may also change or be added/removed, implied relationship between user and item may also change over time in real-world recommender systems scenarios. So to keep the record of the updated suggestions or recommendations, the recommender systems should continuously updated with the new arrival of information. From the view of the graph structures, the constantly updated information makes the graph dynamic instead of static. This dynamic graph is a challenge to the structure.

Also, if the users mistakenly click the items, social relationships might not capture. So, in that case, the intruder may also insert false information into the recommendation systems. Because of the vulnerability of GNN to noisy data, there are emerging efforts on the knowledge in the field of GNN. Some existing models based on GNN pay attention to adversarial learning, which is also a future direction to make a robust model. There is a colossal market dataset containing billions or trillions of vertices and edges. Each vertex has millions of features; the normal GNN is now challenging because of prominent memory uses and a long training time. We may reduce the graph size or make efficient GNN by keeping changing the model design. In recommendation, the graph contains various types of nodes and relationships between nodes, which is the main thing in the graph that we call heterogeneity. Furthermore, the methods that take heterogeneity into account always have more computational challenges, so that's why more studies are required for improvement in computational efficiency.

References

1. Singh, R., Maurya, S., Tripathi, T., Narula, T., Srivastav, G.: Movie recommendation system using cosine similarity and knn, pp. 2249–8958 (2020)
2. Kumar, M., Yadav, D., Singh, A., Gupta, V.K.: Article: a movie recommender system: Movrec. Int. J. Comput. Appl. **124**, 7–11 (2015)
3. de Campos, L.M., Fernández-Luna, J.M., Huete, J.F., Rueda-Morales, M.A.: Combining content-based and collaborative recommendations: a hybrid approach based on bayesian networks. Int. J. Approx. Reason. **51**(7), 785–799 (2010)
4. Harpreet Kaur Virk, E.A.S., Maninder Singh, Er.: Analysis and design of hybrid online movie recommender system. In: International Journal of Innovations in Engineering and Technology (2015)
5. Gupta, U., Patil, N.: Recommender system based on hierarchical clustering algorithm chameleon. In: International Advance Computing Conference (IACC) (2015)
6. Furtado, Singh, A.: Movie recommendation system usingmachine learning. Int. J. Res. Industrialeng. **9**(1), 84–98 (2020)
7. Uluyagmur, M., Cataltepe, Z., Tayfur, E.: Content-based movie recommendation using different feature sets. In: Proceedings of the World Congress on Engineering and Computerscience, vol. 1, pp. 17–24 (2012)
8. Subramaniyaswamy et al.: A personalised movie recommendation system based on collaborative filtering. Int. J. High Perform. Comput. Netw. **10**(1–2), 54–63 (2017)
9. Ahuja, R., Solanki, A., Nayyar, A.: Movierecommender system using K-Means clustering and K-NearestNeighbor. In: 2019 9th International Conference on Cloud Computing,Data Science & Engineering (Confluence). IEEE, pp. 263–268.6 (2019)
10. Hug, N.: Surprise: a python library for recommender systems. J. Open Source Softw. **5**, 2174 (2020)
11. Manimurugan, S., Almutairi, S.: A user-based video recommendation approach using CAC filtering, PCA with LDOS-CoMoDa. In: The Journal of Supercomputing, pp. 1–15 (2022)

12. Duan, R., Jiang, C., Jain, H.K.: Combining review-based collaborative filtering and matrix factorization: a solution to rating's sparsity problem. In: Decision Support Systems, p. 113748 (2022)
13. Wu, S., et al.: Graph neural networks in recommender systems: a survey (2020). arXiv:2011.02260
14. hen Gao et al.: Graph Neural Networks for Recommender Systems: Challenges, Methods, and Directions (2021). arXiv:2109.12843
15. He, X., et al.: Lightgcn: simplifying and powering graph convolution network for recommendation. In: Proceedings of the 43rd International ACM SIGIR Conference on Research and Development in Information Retrieval (2020)
16. Wei, Y., et al.: MMGCN: multi-modal graph convolution network for personalized recommendation of video. In: Proceedings of the 27th ACM International Conference on Multimedia (2019)
17. Wang, Q., et al.: DualGNN: dual graph neural network for multimedia recommendation. In: IEEE Transactions on Multimedia (2021)
18. Liu, Q., et al.: Graph neural network for tag ranking in tag-enhanced video recommendation. In: Proceedings of the 29th ACM International Conference on Information & Knowledge Management (2020)
19. Liu, Y., et al.: Concept-aware denoising graph neural network for video recommendation. In: Proceedings of the 30th ACM International Conference on Information & Knowledge Management (2021)
20. Hamilton, W., Ying Z., Leskovec J.: Inductive representation learning on large graphs. Adv. Neural Inf. Process. Syst. **30** (2017)
21. Wei, Y., et al.: Graph-refined convolutional network for multimedia recommendation with implicit feedback. In: Proceedings of the 28th ACM International Conference on Multimedia (2020)
22. Song, J., et al.: NGAT4Rec: Neighbor-Aware Graph Attention Network For Recommendation (2020). arXiv:2010.12256
23. Sang, L., et al.: Context-dependent propagating-based video recommendation in multimodal heterogeneous information networks. IEEE Trans. Multimed. **23**, 2019–2032 (2020)
24. Lund, J., Ng, Y.-K.: Movie recommendations using the deeplearning approach. In: 2018 IEEE International Conference on Information Reuse and Integration (IRI). IEEE, pp. 47–54 (2018)
25. Purificato, E., Wehnert, S., De Luca, E.W.: Dynamic privacy-preserving recommendations on academic graph data. Computers **10**(9), 107 (2021)

Characterisation of Mental Health Conditions in Social Media Using Deep Learning Techniques

Toshita Sharma, Rrubaa Panchendrarajan, and Akrati Saxena

Abstract Social media has become the easiest and most popular choice for online users to share their views, thoughts, opinions, and emotions with their friends and followers. The shared content is very helpful in making valuable conclusions about an individual's personality. Nowadays, researchers across the world are collecting social media data for understanding the mental health of social media users. Mental illness is a big concern in today's society, and in no time, it can turn into suicidal thoughts if efficacious methods are not taken into account. Early detection of such mental health conditions provides a potential way for effective social intervention. This has opened up opportunities to the research community to automatically determine various mental health conditions, such as anxiety, depression, and near-suicidal thoughts from user-generated content. The deep learning techniques have been extensively used in this research area to better understand the mental health of users on social networking platforms. In this chapter, we conduct a comprehensive review of the research works that use deep learning techniques to identify various mental health conditions from social media data. We discuss methods to detect different types of mental health conditions, such as anxiety, depression, stress, suicide, and anorexia. We also provide the details of the datasets used in these studies. The chapter is concluded with promising future directions.

T. Sharma
Institute of Technology, Nirma University, Ahmedabad, India

R. Panchendrarajan (✉)
Sri Lanka Institute of Information Technology, Malabe, Sri Lanka
e-mail: rrubaa.p@sliit.lk

A. Saxena
Department of Mathematics and Computer Science, Eindhoven University of Technology, Eindhoven, The Netherlands
e-mail: a.saxena@tue.nl

1 Introduction

Mental illness has been proved to be a leading cause of deteriorating mental health conditions worldwide. Globally, a near-about 264 million people are affected per year by the mental illness of depression alone [1]. It is even more perplexing to see that for lower to middle-income countries, around 76% to 85% people are unable to get any treatment for diseases in relation to mental illness [2]. Further studies [3] in this area show that the likelihood of having a major depressive outbreak within a period of one year is about 3–5% for males and a higher rate of about 8–10% for females. With the COVID-19 upheaval, the number of patients battling mental illness has worsened. Despite the gradual increase in the number of patients with such conditions, there is not enough care and treatments available worldwide.

With the advent of the internet, people are increasingly using social media platforms, such as Reddit, Twitter, and Facebook, to portray their feelings, thoughts, and opinions with their friends and family. These thoughts are mostly genuine, and hence they provide a means for excellent data that can be used to examine and further analyze users' mental health conditions. The negative emotions expressed by users might also provide insights about their mental health conditions, such as depression, anxiety, stress, suicidal intent, hints of anorexia, etc. Since mental health issues are proved to be some of the major illnesses that curb people, determining the mental health conditions from content actively shared on social media can be critically important. This data can be helpful in early detecting the mental illness of users and further providing required interventions. This has motivated the research community to develop solutions to automatically determine the mental health conditions of the users from their social media content.

The past few years have witnessed many research solutions to predict the mental health condition of users using their social media data [4]. They extract a large-scale dataset from online social media platforms and then use various techniques ranging from traditional machine learning algorithms to deep learning techniques to train a model for predicting different mental-health conditions. With the adoption of deep learning solutions to determine mental health conditions, state-of-the-art models have been produced, even using simple deep learning architectures. This chapter reviews state-of-the-art methods to identify mental health conditions by applying deep learning techniques to the data extracted from social media. We categorize the works based on mental illness, which includes the delineation of stress detection, anxiety detection, depression detection, anorexia detection, and suicidal prediction based on different deep learning techniques.

2 Mental-Health Detection Techniques

In the upcoming subsections, we discuss the literature on each mental illness detection using various deep learning techniques.

2.1 Stress Detection

Stress is one of the most found mental illnesses among people subsuming about 73% of the workforce where jobs are reported to be the leading cause of the illness. It is also reported that 45% of college students seek out counseling and therapy owing to stress.[1] For example, the following post made by a Reddit user expresses the stress she/he is undergoing due to the fear of a job and being neglected by the friends [5].

> I've had other things happening - lack of direction in life, nervousness about getting a job, feeling lonely by neglecting friends to do work and internships.

Stress can also lead to several other diseases, such as diabetes, increased risk of acid reflux, and heart disease. If stress can be detected at an earlier stage, then many health complications following that can be avoided or treated at the nascent stage.

Lin et al. [6] studied the detection of stress from three major social platforms, Tencent Weibo, Sina Weibo, and Twitter. They performed a series of analyses to present users' stress patterns, which can be shown through microblogs. The study was mainly focused on (i) content correlation - the differences in non-stressed and stressed microblogs' content, (ii) social engagement correlation - the differences between non-stressed and stressed users created via comments, likes and interactions, and (iii) behavioral correlation - the difference in the two types in terms of frequency and posting time. Using these three correlations as features, a convolutional neural network was developed with cross auto-encoders to predict stress from microblogs. The presented model was compared with the baseline models, including Naive Bayes, Random Forest, SVM, and Deep Neural Network. The authors concluded that the results are promising and accurate in detecting psychological stress using the data extracted from microblogs.

A similar analysis was performed by Hasan et al. [7] using Facebook and Twitter data. The authors defined two sets of features, tweet-level attributes and user-level attributes, to train a neural network. Tweet-level attributes included linguistic features, e.g., words that convey positive or negative sentiment [8] as well the other social attention factors, e.g., whether the post is liked or not. User-level attributes were extracted from the posting behavior, and social interactions were generated using the user's weekly posted content. The authors developed a hybrid model composed of a factor graph and CNN to determine whether a user is stressed or not. The results shown were very promising, with an improvement of 6-9% over state-of-the-art methods in terms of F1-score. The authors comprehensively studied the relation between the stress states and their social interactions behaviors. They observed that the social structures of stressed users tend to be sparsely connected and less complex than non-stressed users, and the number of sparsely connected social structures was 14% higher for stressed users than the non-stressed ones.

---

[1] https://medalerthelp.org/blog/stress-statistics/.

2.2 Anxiety Prediction

The terms *anxiety* and *stress* are often used interchangeably as they are both emotional responses. However, stress is a response to a threat in a situation, whereas anxiety is developed as a result of stress. There are several potential causes of anxiety disorders, including the use of excessive drugs or circumstances arising due to the work pressure or personal life. Family history and genetics also play a significant role in determining the amount of stress, and inadequate mechanisms to cope with that stress may further contribute to anxiety. In the United States alone, 40 million adults are affected by anxiety every year.[2] It is also astonishing to notice that given such a large number of patients, less than half of the affected people receive treatment for it.

One of the earlier lines of research in anxiety prediction using deep learning was proposed by Shen and Rudzicz [9]. The authors attempted to determine anxiety in users' posts on the Reddit platform. They generated multiple features, including (i) word embeddings using Continuous Bag of Words (CBOW) model, (ii) topics extracted using Latent Dirichlet Allocation (LDA), (iii) lexico-syntactic features (LIWC) [10], and (iv) likelihood of unigrams and bigrams obtained using a language model. The authors compared SCM and Logistic Regression (LR) as baseline models with a neural network developed with two hidden layers. By using the combination of N-gram probabilities and LIWC [10], the authors gained an accuracy of 98% for the neural network classifier. Their experiments concluded that anxiety has a high correlation with specific topics, such as alcohol consumption and school. The authors further highlighted potential future work by using larger datasets used along with diagnostic criteria or health records.

Tyshchenko et al. [11] also performed an anxiety disorder study in online forums using data scraped from *BlogSpot*. The authors experimented with a CNN architecture which is a derivative of the network proposed by Kim et al. [12]. Compared to social media posts which are generally short, forum posts may contain lengthy discussions. Therefore, the authors extended their study to investigate how does the length of a post in terms of word count could impact the performance of the classifiers. The experiment results revealed that the classifiers predict the illness better when the forum posts are short, and the accuracy decreases with the increase in the length. The authors proposed data-oriented improvements and the use of hierarchical attention networks to predict anxiety for future work.

2.3 Depression Detection

Depression is a mental condition in which one can have a long-lasting, deep feeling of despair or sadness. It can change an individual's feelings and affect the person's social behavior and sense of physical well-being. For example, the following tweet posted by a user expresses his/her depressed state due to loneliness [13].

---

[2] https://adaa.org/understanding-anxiety/facts-statistics.

lonely is not being alone, it's the feeling that no one cares. #alone #depressed

Depression affects a wide range of age groups, primarily affecting the group of 15 to 30 years of age. Next, we present state-of-the-art works on depression detection, and the works are categorized based on the used deep learning models.

Depression Detection using CNN The performance of convolutional neural networks (CNN) in classifying images has resulted in the adoption of the model for a wide range of tasks, including text classification. Kim et al. [14] used CNN for classifying whether the user's post belongs to depression or not. The CNN model architecture was laid down by describing the sequence of layers as including an embedding layer, convolution layer, max-pooling layer, dense layers, and the output, which is a result of binary classification. The dataset was collected from Reddit channels which was pre-processed using stop-word removal and tokenization. To address the imbalance in the stressed and non-stressed class, the authors used the synthetic minority over-sampling technique (SMOTE) [15]. The authors compared the CNN model with an XGBoost model using four metrics - F1-score, precision, recall, and accuracy. The CNN model approached an accuracy of 75.13 % and the XGBoost model with 71.69 %.

Similar research was performed by Yates et al. [16] using the Reddit platform. The authors created a large-scale dataset named Reddit Self-reported Depression Diagnosis (RSDD) that contains over 9000 Reddit users. The model for detecting depression was a two-layer CNN architecture, where the input to the first CNN layer is the users' posts, and the output was concluded using a softmax activation function. The model was compared against the FastText classifier [17] and SVM classifier [18]. The CNN model outperformed by a large margin when expressed in terms of recall and F1 on users who are diagnosed with depression, concluding that the model provides a strong approach to determinating posts that indicate the state of depression.

Trotzek et al. [19] studied the early detection of depression using Reddit posts dataset. The authors utilized the dataset published in the CLEF 2017 conference eRisk pilot task for early depression detection [20]. The CNN model followed by multiple fully connected layers was used to classify the posts. The authors experimented with two different word embedding models, fastText and Glove, to convert the text of the posts to vector representation to input to the classifier. The experimental results showed that the CNN-based architecture combined with fastText performed better compared to other combinations of proposed solutions and existing approaches. The authors proposed a possible improvement in the depression detection task by using recently published language modeling methods such as BERT, which can be used as an input to the network.

A fine-grained classification of depression was carried out by Mustafa et al. [21] using Twitter posts. The authors classified the Twitter posts into three different classes of depression, low, medium, and high, using CNN architecture. The class 'high' indicates the user feels guilty, worthless, and suicidal. The symptoms of anxiety, fatigue, and mood swings are considered as a medium level of depression. Seasonal affective disorder and feeling of someone harming you are categorized as low depression level.

The authors collected data of 179 depressive individuals with 200 to 300 average posts per user. The tweets were prepossessed using various techniques, including normalization, stop words removal followed by tokenization. The TF-IDF method is used for the computation of the weights of the tokens. The words were sorted in ascending order based on weights. Top 100 words with higher weights were used to train a CNN model, and they gained an accuracy of 91% in determining the depression level.

Depression Detection using RNN Recurrent Neural Networks (RNN) are quite popular for text classification tasks as they can effectively handle sequential data. One of the earlier lines of research that adopts RNN for depression detection was carried out by Paul et al. [22]. The authors aimed at the early detection of depression from posts and comments of Reddit users. The text classifier was trained using (i) *various textual features*, including Bag of Words, - generated using the number of occurrences of words, and (ii) *Unified Medical Language System* features - based on medical terms from the text.. The RNN model combined with fasttext embedding was used to perform the classification. They obtained a precision of 0.35 and recall of 0.15 as a result of fixing sentence length to 150 in the model, mainly due to the limited resources. However, the author insinuated to increase the sentence length for better performance.

A similar analysis of detecting early depression was performed by Shah et al. [23] using a Reddit dataset [20]. The authors explored various word embedding techniques, including Word2vec, Glove, Fasttext, and a trainable embedding layer as input to the neural model. They further extracted metadata indicating statistics of word usage as the input to a Bi-LSTM. The authors observed that the model performed better when words were converted to a vector representation using Word2vec embedding. Uban and Roso [24] used a similar set of features for early detection of depression. They used Glove embedding to convert words to vector representation, and extracted linguistic features, such as pronoun usage, the term related to mental disorder, and emotion and sentiment expressed. A Bi-LSTM combined with attention was used to detect early depression, and the results showed that considering the emotion of sentiment expressed in user posts improves the performance.

Following that observation of using emotion and sentiment of users for depression detection, Kholifah et al. [25] developed five criteria as a measure of depression in text. This includes emotion and sentiment of the author, pronouns used, negative terms used, and absolutist terms used. The five features were extracted using Natural Language Toolkit (NLTK) and lexicons developed by the authors. An LSTM model was used to determine the depression from text, and authors achieved a precision of 50.24% and a recall of 70.89%. The authors suggested that there can be an increase in the accuracy by optimizing hyperparameters and increasing the amount of training data.

Song et al. [26] focused on the next stage, where they performed the study on posts of Reddit users who are already diagnosed with depression. The authors developed a feature attention network that contains four feature networks. Each feature network was dedicated to different strong indications of depressive symptoms. This includes direct comments indicating depression, sentiments in posts reflecting negative thinking patterns, ruminative thinking capturing repeated expressions, and finally, the writing style mainly capturing the order of words and word types. The word sequence was initially converted to a one-hot vector and encoded to a feature vector using bi-directional RNN. This model outperformed all existing models except the state-of-the-art CNN model. The authors claimed this could be due to the differences in the size of the dataset used to train the model. A similar analysis was performed by Ahmad et al. [27] on Twitter data. The author used bi-LSTM model for classifying tweets as either *normal* or *depressed* and acquired an F1-score of 0.9. Their study emphasized the need for having balanced classes in the dataset to increase the accuracy of the existing approaches.

Depression Detection using CNN-RNN Several works [28–30] compared CNN and RNN models for detecting the depression due to their independent performance in detecting depression. Orabi et al. [28] compared the performance of CNN and RNN in determining whether a tweet posted by a user expresses a controlled mental state or depressed mental state or PTSD, which is a post-traumatic stress disorder. This comparison included CNN models with various convolutional layers and max-pooling sizes. The experimental results exhibited that the CNN model and its variations outperformed the RNN in determining depression. A similar comparison was performed by Wang et al. [29] on the Sina Weibo dataset. The authors compared the performance of CNN and RNN in determining the depression level of users out of four levels, and they observed a similar outcome, where the CNN model outperformed the RNN model. Shetty et al. [30] observed a similar result for binary classification performed to determine whether the user is depressed or not. This could be mainly due to the nature of CNN in extracting important keywords that can express the depression level of the users.

Independent performance of CNN and RNN models in detecting depression from text data drew the attention of the community to explore hybrid architectures [31–33] that include both CNN and RNN. [31] is one of the earliest research works that explored the hybrid architecture composed of CNN and RNN. The authors attempted to predict the granular level depression statuses classified into four classes - happy, response, low distress, and high distress. The proposed hybrid model consists of five layers - embedded layer, CNN, max-pooling layer, LSTM, and a dense layer at last. The author proposed an extension of the model with more linguistic feature extractions, such as POS tags, N-grams, as future work. A similar hybrid model was explored by Verma et al. [32] for classifying tweets as depressed or not, and the hybrid model outperformed the independent CNN and RNN as it utilizes the benefits of both models in the hybrid version. The model can be further extended by analyzing multimedia data to improve the accuracy.

As a timely solution, Ghosh et al. [33] investigated the impact of COVID-19 on the mental health of 482 users from four cities. The proposed hybrid model consists of four stages, an embedding layer, a bi-LSTM, a max-pooling layer, and multiple dense layers. The hybrid model achieved an accuracy of 99.42%. The study also revealed a significant increase in depressive tweets during the four months period of the pandemic. Besides this, during March 2020, the total number of tweets from the considered population was much higher, and more than 40% of the tweets conveyed depression.

Depression Detection using GRU Gated Recurrent Unit (GRU) was proposed as a solution to address the short-term memory issues in RNN, where they had a hard time carrying information of long sequences. This feature enabled many research works [34–36] in the literature to adopt GRU over RNN as a solution to determine depression from long text sequences. Sadeque et al. [34] experimented with a neural model composed of two layers of GRU that utilized various linguistic features. This included well-prepossessed vocabulary, uni-gram and bi-gram word count, a vectorized representation of the text obtained by an LSTM, and occurrences of concepts from the Unified Medical Language System (UMLS). While GRU achieved the best F1-score, the experiments displayed that the performance of the GRU model was boosted with the utilization of vector representation obtained for the user posts using an LSTM.

Sekulic and Strube [35] experimented a hierarchical attention networks that utilized GRU-based encoder to encode the entire sequence. This attention network was composed of four layers, a word encoder layer, a word-level attention layer to identify keywords, a sentence encoder made of GRU, and a sentence-level attention layer to identify key sentences. They modeled all the content posted by a user as a document which enabled them to predict the mental state of the user given all the posts published by the user. The evaluation performed on the Reddit platform dataset showed that the proposed model predicted various mental health illnesses of users successfully, including depression.

While the content posted by a user directly reflects the mental state of the user, the behavior of the user on social platforms, such as posting rate, amount of negative and positive posts, posting time, can also reveal the mental status. Zogan et al. [36] presented a novel framework called DepressionNet that utilizes both the content and user behavior to predict whether a user is depressed or not. The model was composed of two GRU-based sub-networks, each dedicated to analyzing user behavior and user posts. Users' behavior modeled as feature vector was passed to two layers of GRU to obtain its vector representation. Similarly, the summarized content of the user was passed to a convolution layer followed by a GRU layer to get a vector representation of user posts. Both vector representations were concatenated and provided to a dense layer to make the final prediction. An extensive evaluation showed that the DepressionNet achieved the best performance compared to existing approaches. This concluded the impact of utilizing users' behavior while predicting their mental state.

Other Deep Learning Techniques for Depression Detection Wongkoblap et al. [37] presented a hybrid model composed of a CNN layer, a GRU layer, and a multi-layer perceptron, that first classifies whether a user post is general or mental health-related and will use it further for classifying the user as depressed or not. The model achieved 72% accuracy, and the detailed analysis of the dataset also revealed that users who are depressed tend to post more mental health-related posts while the users who are not depressed post more general posts. The authors suggested that incorporating more features, such as interactions and comments, might enhance the model's performance.

Researchers also worked on identifying the depression state of the user with specific levels of the depression [38, 39]. Mann et al. [38] divided the depression states into four categories - intensity-severe, moderate, mild, and minimal. The authors collected posts with text and images from Instagram for this study. They compared three DL models and three feature engineering models, using only text, only images, and both text and images. For classifying the images, two DL models were used, (i) ResNet-18, 34, 50 [40] as it is a well-known public framework, and (ii) ResNeXt architecture [41], which was pre-trained on 940 million images. The text feature representation was learned using Bag of Words, FastText [42], and ELMo [43]. The experimental results exhibited that the combination of ELMo and ResNet-34 performed better in terms of F1 score, and visual models provided better precision scores while textual models provided higher recall scores.

A similar level of granularity in depression states was studied by Wang et al. [39] using Sina Weibo dataset. The authors classified users' posts into four depression levels ranging from 1 (mild) to 3 (severe). They used a pre-trained language model called BERT to train the classifier, and the performance was compared against CNN and LSTM. The BERT model is based on transformer-based architectures designed to capture the dependencies in long word sequences. While the BERT model outperformed other models in determining depression, this opens up research opportunities to adopt transformer-based architectures to analyze users' mental health from social media posts.

2.4 Anorexia Prediction

Anorexia is an eating disorder in which a person has an extreme fear of gaining weight and a distorted perception of weight. Retamero and Bedmar [44] worked on determining anorexia from social media posts using DL techniques. They used the erisk 2019 dataset as described in [45] with about 150 users' data. Users' posts were preprocessed by removing stopwords, punctuation, and numbers, and most representative words were retrieved using TF-IDF. The vector representation from the text was learned using three different embedding techniques, character embeddings, word embeddings, and writing embeddings. The authors experimented with various types of CNN and RNN models using different embedding representations and observed that the RNN trained with writing embedding outperformed the other combinations.

Ranganathan et al. [46] utilized the existing Neural machine translation (NMT) model composed of LSTM to predict anorexia from users' posts. NMT models are based on the encoder-decoder architecture, where the encoder encodes a sentence from one language to a vector representation, and then the decoder decodes the encoded vector to another language. In the anorexia prediction task, the NMT model was trained such that the decoder predicts whether the given post is a positive or negative indication of anorexia. The performance of the model was compared against SVM, and the results demonstrated that the NMT model outperformed the SVM model by a significant margin.

A similar study of early detection of symptoms of anorexia was performed by Amini and Kosseim [47] using eRisk 2019 dataset [48] containing Reddit posts. The authors used a neural classifier, CNN-ELmo, which was shown to be the best model in the eRisk shared tasks. The model was composed of ELmo word embedding layer, CNNs, and attention layer. Elmo word embedding layer generates vector representation of words using a bidirectional LSTM from the character-level information available. These vector representations were fed into the CNN to obtain a post-level vector representation, and all the post-level representations of a user were concatenated to obtain a user-level vector representation. Then, the attention layer was applied on user-level vector representation, which enabled the model to assign higher weights to users at risk of anorexia. The proposed model was shown to be performing well even with a limited number of posts available. The proposed model can be expanded for other binary classification mental health tasks in the field of NLP for clinical psychology.

2.5 Suicide Prediction

Suicidal thoughts and attempts are being a global concern contributing to 800,000 deaths and 25 million suicide attempts annually [49]. This massive count has promoted the research community to identify and prevent suicide using the content posted on social media. Sawhney et al. [50] scraped the anonymized data from online suicide forums, and manually annotated it to generate a lexicon of suicide-related phrases and words. Then they used these lexicon to collect Twitter dataset that was also manually labeled for the absence or presence of the suicidal intent. The authors trained multiple models to predict suicide, including an RNN, an LSTM, and a C-LSTM model composed of a CNN and LSTM stacked to make the prediction. Experiment results showed that combining CNN and LSTM provides the best results as CNNs are effective in spatially encoding the tweets into a vector representation, and LSTM has the capability of capturing long-term dependencies.

A similar study was done by Tadesse et al. [51] for detecting suicide ideation using a Reddit dataset containing 3549 suicide-intended and 3652 non-suicide intended posts. The authors used a unified LSTM-CNN to develop the classifier. First, the users' posts were converted to a vector representation using word embedding techniques, and input it to the LSTM followed by a CNN. The experimental results showed that the proposed model outperformed existing approaches mainly due to the nature of LSTM in processing long sequences, while CNN can identify key terms conveying suicidal thoughts.

Following the observation of combining CNN and LSTM for better performance in suicide prediction task, another hybrid model composed of multiple neural networks and Support Vector Machine (SVM) was presented by Mohammadi et al. [52] on a

Reddit dataset [53]. The proposed system is composed of eight sub-neural networks, each with a different type of word embedding as input and a hidden layer. Both ELmo and Glove were used as the word embedding layers, and Bi-LSTM, Bi-RNN, Bi-GRU, and CNN were used as the hidden layer of different sub-neural networks. The authors concluded that using the combination of neural features leads to a better performance.

Social network information can also be used to supplement textual features for suicide prediction. Mishra et al. proposed a model named SNAP-BATNET (Social Network Author Profiling - BiLSTM attention NETwork) [54] based on this concept. They created a dataset containing scraped tweets and social engagement data of all users, which were annotated by humans as suicidal intent present and not present. The authors used a wide range of textual and social graph features to train the model. Textual features include TF-IDF, POS tag, vector representation of words obtained using Glove word embedding, emotional words obtained using NRC lexicon, and topics extracted using the LDA topic model. The social graph features include embedding representations obtained for nodes of follower, mention, reply, and quotes graphs where existing node2vec technique [55] was used to convert node to a vector representation. Finally, the combined embedding representation for social graph features was obtained by training a Random Forest classifier. The model's performance was compared against several baseline models, including CNN, LSTM, and C-LSTM, and the results showed that using LSTM and SNAP-BATNET in combination with all feature sets outperforms. The results are in correlation with previous studies who also have showed the effectiveness of LSTM.

Following that Ophir et al. [56] utilized both textual features and social media activities of a user to develop neural network models for predicting suicide. The authors used ELMo, a contextualized word embeddings technique, to convert words to the vector representation of data extracted from Facebook. The overall textual activity of the user was represented as an average of its post vectors and was given as an input to two ANN (Artificial Neural Network) models. The output was a binary form (yes/no), which is for predicting the general risk of suicide. The first model was trained to only predict the risk of suicide and referred to as the Single Task Model (STM). In contrast, the second model was trained to predict a hierarchy of inter-related factors, including personality traits, psychiatric disorders, psychosocial risks, and risk of suicide, and it was referred to as a Multi-Task Model (MTM). The evaluations showed that MTM, i.e., based on multilayered and hierarchical sets of theory-driven risk factors, outperformed STM, as expected. The authors further elaborated on some of the limitations of the study, including the non-usage of external criteria for suicide risk assessment and the lack of usage of non-textual data, such as pictures.

A time-based analysis to determine suicide ideation was conducted by Sawhney et al. [57] using Twitter data. The collected historical tweets were annotated into two categories: suicidal intent present and suicidal intent absent. The authors proposed the STATENet model, i.e., a transformer-based architecture composed of two transformers dedicated to processing current and historical tweets. The first transformer processed the tweet to be assessed for suicide ideation and generated a vector representation of it. Similarly, the second transformer received past tweets of the same user and emotions expressed in them to generate an overall vector representation indicating the user's emotional state over time. Finally, both vector representations were concatenated and passed to a softmax layer for making the final prediction. In terms of model evaluation measured by metrics including macro F1, recall, and accuracy, the STATENet outperformed all mainly due to the utilization of time-aware emotional content posted by the user. One can further propose a priority-based suicide risk assessment model using live-streaming tweets for providing fast intervention.

Wang et al. [58] showed that latent features extracted in the form of word or document vectors are more useful compared to the handcrafted features for the detection of suicidal tendencies. They used CLPsych 2021 Twitter dataset [59] and extracted two types of features (i) latent features, using Doc2Vec to generate word and document embeddings, and (ii) handcrafted features, including emotions expressed in text, part-of-speech tags, and lexicon with words expressing suicide ideation. Using the latent features, a neural model called C-attention network was developed with a multi-head attention layer module followed by a convolutional layer and a softmax layer. In order to evaluate the impact of using handcrafted features, SVM, KNNs, and RF machine learning models were trained. The C-attention model outperformed other ML models trained using handcrafted features, and with these results, the authors concluded that latent features combined with neural models effectively predict the suicide ideation of users from the content posted on social platforms.

2.6   Predicting Multiple Mental-Health Conditions

The same factors could result in different mental health conditions, and a user may be facing multiple health conditions simultaneously. This enabled the research community to develop a single model to process content posted on social media and user activities to determine multiple mental-health conditions. Kshirsagar et al. [60] referred to multiple mental health conditions of a user, such as suicide, depression, self-harm, physical abuse, and eating disorders, as a *crisis* and attempted to predict the presence of a crisis on a dataset made available by Koko based on a clinical trial at MIT [61]. The authors developed an explainable model using the bidirectional GRU model with attention mechanism to predict the crisis and utilized the well-known LIME framework [62] to generate the seeds for the explanation of the predictions made by the model. They further proposed a method to generate the explanation. The results showed that the best models are proved to be effective for prediction, and at the same time, also produce explanations equivalent to human annotators.

Malhotra and Jindal [63] leveraged multimodal information from social media posts to perform binary classification of whether a post indicates depression/suicidal behavior or not. The authors collected data from Facebook, Twitter, and Instagram, and extracted images, videos, text, and emoticons. For the feature extraction, the authors used VGG-16 for images, word2vec for word embeddings, person detection in videos with faster-RNN, and a depression score based on the ratio of sad emoticons.

These feature representations were given as input to softmax layers, and the final decision was taken as a weighted average of predictions made by the softmax layers.

Benton et al. [64] focused on a more granular study to predict nine different mental states of the users, including depression, anxiety, and suicidal attempts, by leveraging multi-task learning. The authors evaluated both the single-task learning model (STL), that is a feed-forward two hidden layer network, and a multi-task learning model (MTL), in which the hidden layer from the bottom was shared among all tasks. The models were trained over multiple datasets extracted from Twitter. The results showed, the MLT model significantly outperformed the SLT model due to the benefit of sharing knowledge across multiple tasks through the shared hidden layer. Ive et al. [65] focused on classifying social media posts from Reddit further to 11 granular mental-health conditions. The authors proposed a hierarchical RNN architecture that processed word-level information to obtain a sentence-level vector representation which was further used to generate a document-level representation. Finally, the document-level representation was used to predict the presence of a mental health illness at a document level. The proposed model outperformed the CNN model in terms of predicting all 11 classes.

3 Datasets

In Table 1, we summarize social media datasets used for different mental health detection studies.

4 Conclusion and Future Directions

This chapter briefly explains the existing deep learning-based approaches to identify various mental-health conditions of users from content posted on social media. Many research works primarily used tweets and Reddit posts, though some works used blogs as their source of data. We observed that efficient preprocessing of the textual content plays a vital role in all mental-health prediction tasks, where word embedding models, such as Glove, word2vec, BERT, Fasttext, and ELmo, were widely used to convert text to vector representations. Further, the textual features, including TF-IDF, POS tag, the emotion expressed in text, and social network features, such as repost, mention, and like activities of users, are used to enhance the performance of the prediction models.

A range of neural models, CNN, LSTM, RNN, GRU, and transformer-based architectures, are shown to generate state-of-the-art solutions for various mental-health prediction tasks. This performance has been further enhanced by attention-based architectures, hierarchical models, and multi-tasking techniques. Further, it can be observed that the existing models' performance is largely affected by the data size and computational limitations in handling extremely large datasets. Especially,

Table 1 Datasets available for mental-health detection studies

Dataset	Size (users/posts)	Labels/Classes	Remarks
Reddit Anxiety Dataset [9]	22,808 posts	Anxiety, Control	Data collected from anxiety-related subreddits - r/panicparty, r/anxiety, r/socialanxiety, and r/healthanxiety; 9971 anxiety and 12,837 control posts
UMD Reddit Suicide Risk Dataset [66]	934 users	None, Low, Moderate, Severe (Risk)	Collected from /r/SuicideWatch
Higher Ed Depression Dataset [38]	221 students	severe, moderate, mild, minimal	Collected using Questionnaires and Instagram API; 136 females and 85 males students
Reddit Self-reported Depression Diagnosis (RSDD) [16]	116,484 users	Diagnosed and Control	Original data collected from https://files. pushshift.io/reddit/; 107,274 control users and 9,210 diagnosed users
Twitter Self-Reported Diagnoses Dataset [67]	2013 users	Anxiety, ADHD, Borderline, Bipolar, Depression, Schizophrenia, Eating, PTSD, OCD, Seasonal Affective	Self-reported diagnoses obtained using Twitter API followed by manual examination
eRisk Depression datasets [68]	1707 users	Depressed, Control	214 Depressed users and 1493 Control users
eRisk Anorexia Task [68]	1287 users	Anorexia, Control	134 Anorexia users and 1153 control users
eRisk Self-harm Task [69]	763 users	Self-harm, Control	145 self-harm and 618 control users
Posts of Mental-health-related Subreddits [14]	228,060 users with 488,472 posts	Depression, Anxiety, Bipolar, BPD, Schizophrenia, Autism	Collected from r/Anxiety, r/depression, r/schizophrenia, r/BPD, r/bipolar, r/autism, r/mentalhealth subreddits; Data is available at [70]
Suicide Attempt Exploratory Dataset [71]	554 users	Suicide	Collected based on users who mentioned on Twitter that they attempted to end their life
Twitter Stress DB-1 Dataset [6]	492,676 tweets, 23304 users	Stressed, Non-stressed	239,038 stressed tweets from 11,074 users

(continued)

Table 1 (continued)

Dataset	Size (users/posts)	Labels/Classes	Remarks
Sina Weibo Psychological Stress Dataset [6]	3,304 tweets, 210 users	Stressed, Non-stressed	1,459 stressed tweets from 98 users
Tencent Weibo Psychological Stress Dataset [6]	311155 tweets, 16084 users	Stressed, Non-stressed	Collected from Tencent Weibo; 138570 stressed tweets from 7845 users
CLPsych 2021 Shared Task Dataset [59]	136 users in Subtask 1, 194 users in Subtask 2	Suicidal or Non-suicidal	Data is available at [72]
#Suicidal-Twitter Dataset [73]	34,306 tweets and 32,558 users	Suicidal and Non-suicidal	3,984 suicidal tweets
Multiclass Twitter Depression Dataset [21]	179 users	Medium, Low, High Depression	Collected using Twitter API
Sina Weibo Depression Level Dataset [29]	13993 microblogs	Levels 0, 1, 2, 3	Collected from Sina Weibo
Twitter dataset [74]	1238 Users, 2986k tweets	Bipolar, Depression, PTSD, SAD	Users in different classes: 394 in Bipolar, 441 in depression, 244 in PTSD, 159 in Sad. Besides this 10k users in control group
Depressive-tweets-processed Dataset [75]	2345 tweets	Depressive tweets	Used in [33]
SMHD Dataset [76]	36948 Diagnosed users and 335952 control users	Anxiety, ADHD, Bipolar, Autism, Eating, Depression, PTSD, OCD, Schizophrenia	Collected from Reddit
CES-D [77]	614 users	Depression, Non-depression	466 depressed and 148 non-depressed users with Facebook profiles
Suicide Risk Facebook Dataset [56]	1002 users and 83,292 posts	Suicidal, Non-suicidal	361 users at risk and 641 users with no risk

the attention-based models are shown to perform well only with the presence of a larger dataset. As an alternative to this issue, traditional ML algorithms are used to generate accurate prediction models with limited data.

One can further work on developing models that caries the significance of emojis and emoticons to capture the user's emotional state more accurately. Moreover, deploying scalable methods for mass-public health tracking using several types of data, like EEG and speech signals, to detect these diseases better is required as a timely solution.

References

1. James, S.L., Abate, D., Abate, K.H., Abay, S.M., Abbafati, C., Abbasi, N., Abbastabar, H., Abd-Allah, F., Abdela, J., Abdelalim, A., et al.: Global, regional, and national incidence, prevalence, and years lived with disability for 354 diseases and injuries for 195 countries and territories, 1990–2017: a systematic analysis for the global burden of disease study 2017. The Lancet **392**(10159), 1789–1858 (2018)
2. Wang, P.S., Aguilar-Gaxiola, S., Alonso, J., Angermeyer, M.C., Borges, G., Bromet, E.J., Bruffaerts, R., De Girolamo, G., De Graaf, R., Gureje, O., et al.: Use of mental health services for anxiety, mood, and substance disorders in 17 countries in the who world mental health surveys. The Lancet **370**(9590), 841–850 (2007)
3. Andrade, L., Caraveo-Anduaga, J.J., Berglund, P., Bijl, R.V., De Graaf, R., Vollebergh, W., Dragomirecka, E., Kohn, R., Keller, M., Kessler, R.C., et al.: The epidemiology of major depressive episodes: results from the international consortium of psychiatric epidemiology (icpe) surveys. Int. J. Methods Psychiatr. Res. **12**(1), 3–21 (2003)
4. Briand, A., Almeida, H., Meurs, M.-J.: Analysis of social media posts for early detection of mental health conditions. In: Canadian Conference on Artificial Intelligence, pp. 133–143. Springer (2018)
5. Turcan, E., McKeown, K.: Dreaddit: a Reddit dataset for stress analysis in social media. In: Proceedings of the Tenth International Workshop on Health Text Mining and Information Analysis (LOUHI 2019), pp. 97–107, Hong Kong. Association for Computational Linguistics (2019)
6. Lin, H., Jia, J., Guo, Q., Xue, Y., Li, Q., Huang, J., Cai, L., Feng, L.: User-level psychological stress detection from social media using deep neural network. In: Proceedings of the 22nd ACM International Conference on Multimedia, pp. 507–516 (2014)
7. Ashfakul Hasan, Md., Anantha Raman, G.R., Kiran Kumar Reddy, P.: Deep learning cnn based model for detection of stress: a novel approach. In: Journal of Resource Management and Technology
8. Saxena, A., Reddy, H., Saxena, P.: Recent developments in sentiment analysis on social networks: techniques, datasets, and open issues. In: Principles of Social Networking, pp. 279–306. Springer (2022)
9. Hanwen Shen, J., Rudzicz, F.: Detecting anxiety through Reddit. In: Proceedings of the Fourth Workshop on Computational Linguistics and Clinical Psychology — From Linguistic Signal to Clinical Reality, pp. 58–65, Vancouver, BC. Association for Computational Linguistics (2017)
10. Pennebaker, J.W., Boyd, R.L., Jordan, K., Blackburn, K.: The development and psychometric properties of liwc2015. Technical report (2015)
11. Tyshchenko, Y.: Depression and anxiety detection from blog posts data. Nature Precis. Sci., Inst. Comput. Sci., Univ. Tartu, Tartu, Estonia (2018)
12. Kim, Y.: Convolutional neural networks for sentence classification (2014). arXiv:1408.5882

13. Coppersmith, G., Dredze, M., Harman, C., Hollingshead, K., Mitchell, M.: Clpsych 2015 shared task: depression and ptsd on twitter. In: Proceedings of the 2nd Workshop on Computational Linguistics and Clinical Psychology: From Linguistic Signal to Clinical Reality, pp. 31–39 (2015)

14. Kim, Jina, Lee, Jieon, Park, Eunil, Han, Jinyoung: A deep learning model for detecting mental illness from user content on social media. Sci. Rep. **10**(1), 1–6 (2020)

15. Chawla, N.V., Bowyer, K.W., Hall, L.O., Philip Kegelmeyer, W.: Smote: synthetic minority over-sampling technique. J. Artif. Intell. Res. **16**, 321–357 (2002)

16. Yates, A., Cohan, A., Goharian, N.: Depression and self-harm risk assessment in online forums (2017). arXiv:1709.01848

17. Joulin, A., Grave, E., Bojanowski, P., Mikolov, T.: Bag of tricks for efficient text classification (2016). arXiv:1607.01759

18. Wang, S.I., Manning, C.D.: Baselines and bigrams: simple, good sentiment and topic classification. In: Proceedings of the 50th Annual Meeting of the Association for Computational Linguistics (Volume 2: Short Papers), pp. 90–94 (2012)

19. Trotzek, M., Koitka, S., Friedrich, C.M.: Utilizing neural networks and linguistic metadata for early detection of depression indications in text sequences. IEEE Trans. Knowl. Data Eng. **32**(3), 588–601 (2018)

20. Losada, D.E., Crestani, F., Parapar, J.: erisk 2017: Clef lab on early risk prediction on the internet: experimental foundations. In: International Conference of the Cross-Language Evaluation Forum for European Languages, pp. 346–360. Springer (2017)

21. Ul Mustafa, R., Ashraf, N., Shabbir Ahmed, F., Ferzund, J., Shahzad, B., Gelbukh, A.: A multiclass depression detection in social media based on sentiment analysis. In: Proceedings of the 17th IEEE International Conference on Information Technology-New Generations, pp. 659–662. Springer (2020)

22. Paul, S., Kalyani Jandhyala, S., Basu, T.: Early detection of signs of anorexia and depression over social media using effective machine learning frameworks. In: CLEF (Working notes) (2018)

23. Muhammad Shah, F., Ahmed, F., Kumar Saha Joy, S., Ahmed, S., Sadek, S., Shil, R., Hasanul Kabir, Md.: Early depression detection from social network using deep learning techniques. In: 2020 IEEE Region 10 Symposium (TENSYMP), pp. 823–826. IEEE (2020)

24. Uban, A.-S., Rosso, P.: Deep learning architectures and strategies for early detection of self-harm and depression level prediction. In: CEUR Workshop Proceedings, vol. 2696, pp. 1–12. Sun SITE Central Europe (2020)

25. Kholifah, B., Syarif, I., Badriyah, T.: Mental disorder detection via social media mining using deep learning. Kinetik: Game Technology, Information System, Computer Network, Computing, Electronics, and Control **5**(4), 309–316 (2020)

26. Song, H., You, J., Chung, J.-W., Park, J.C.: Feature attention network: interpretable depression detection from social media. In: Proceedings of the 32nd Pacific Asia Conference on Language, Information and Computation (2018)

27. Ahmad, H., Zubair Asghar, M., Alotaibi, F.M., Hameed, I.A.: Applying deep learning technique for depression classification in social media text. J. Med. Imaging Health Inform. **10**(10), 2446–2451 (2020)

28. Husseini Orabi, A., Buddhitha, P., Husseini Orabi, M., Inkpen, D.: Deep learning for depression detection of twitter users. In: Proceedings of the Fifth Workshop on Computational Linguistics and Clinical Psychology: From Keyboard to Clinic, pp. 88–97 (2018)

29. Wang, X., Chen, S., Li, T., Li, W., Zhou, Y., Zheng, J., Zhang, Y., Tang, B.: Assessing depression risk in chinese microblogs: a corpus and machine learning methods. In: 2019 IEEE International Conference on Healthcare Informatics (ICHI), pp. 1–5 (2019)

30. Shetty, N.P., Muniyal, B., Anand, A., Kumar, S., Nagendra Prabhu, S.: Predicting depression using deep learning and ensemble algorithms on raw twitter data. In: International Journal of Electrical and Computer Engineering (IJECE) (2020)

31. Aswathy, K.S., Rafeeque, P.C., Murali, R.: Deep learning approach for the detection of depression in twitter

32. Verma, B., Gupta, S., Goel, L.: A neural network based hybrid model for depression detection in twitter. In: International Conference on Advances in Computing and Data Sciences, pp. 164–175. Springer (2020)
33. Ghosh, T., Hasan Al Banna, Md., Jaber Al Nahian, Md., Abu Taher, K., Shamim Kaiser, M., Mahmud, M.: A hybrid deep learning model to predict the impact of covid-19 on mental health form social media big data (2021)
34. Sadeque, F., Xu, D., Bethard, S.: Measuring the latency of depression detection in social media. In: Proceedings of the Eleventh ACM International Conference on Web Search and Data Mining, WSDM '18, pp. 495-503, New York, NY, USA. Association for Computing Machinery (2018)
35. Sekulic, I., Strube, M.: Adapting deep learning methods for mental health prediction on social media. CoRR, abs/2003.07634 (2020)
36. Zogan, H., Razzak, I., Jameel, S., Xu, G.: Depressionnet: a novel summarization boosted deep framework for depression detection on social media (2021). arXiv:2105.10878
37. Wongkoblap, A., Vadillo, M.A., Curcin, V.: Classifying depressed users with multiple instance learning from social network data. In: 2018 IEEE International Conference on Healthcare Informatics (ICHI), pp. 436–436. IEEE (2018)
38. Mann, P., Paes, A., Matsushima, E.H.: See and read: detecting depression symptoms in higher education students using multimodal social media data. In: Proceedings of the International AAAI Conference on Web and Social Media, vol. 14, pp. 440–451 (2020)
39. Wang, Xiaofeng, Chen, Shuai, Li, Tao, Li, Wanting, Zhou, Yejie, Zheng, Jie, Chen, Qingcai, Yan, Jun, Tang, Buzhou: Depression risk prediction for chinese microblogs via deep-learning methods: content analysis. JMIR Med. Inform. 8(7), e17958 (2020)
40. He, K., Zhang, X., Ren, S., Sun, J.: Deep residual learning for image recognition. In *Proceedings of the IEEE Conference on Computer Vision and Pattern Recognition*, pp. 770–778 (2016)
41. Xie, S., Girshick, R., Dollár, P., Tu, Z., He, K.: Aggregated residual transformations for deep neural networks. In: Proceedings of the IEEE Conference on Computer Vision and Pattern Recognition, pp. 1492–1500 (2017)
42. Bojanowski, Piotr, Grave, Edouard, Joulin, Armand, Mikolov, Tomas: Enriching word vectors with subword information. Trans. Assoc. Comput. Linguist. 5, 135–146 (2017)
43. Peters, M.E., Neumann, M., Iyyer, M., Gardner, M., Clark, C., Lee, K., Zettlemoyer, L.: Deep contextualized word representations. CoRR, abs/1802.05365 (2018)
44. Garcia Retamero, P.R., Segura-Bedmar, I.: Early risk prediction by means of deeplearning. In: CLEF (Working Notes) (2019)
45. Losada, D.E., Crestani, F., Parapar, J.: Overview of erisk 2019 early risk prediction on the internet. In: International Conference of the Cross-Language Evaluation Forum for European Languages, pp. 340–357. Springer (2019)
46. Ranganathan, A., Haritha, A., Thenmozhi, D., Aravindan, C.: Early detection of anorexia using rnn-lstm and svm classifiers. In: CLEF (Working Notes) (2019)
47. Amini, H., Kosseim, L.: Towards explainability in using deep learning for the detection of anorexia in social media. In: International Conference on Applications of Natural Language to Information Systems, pp. 225–235. Springer (2020)
48. Losada, D.E., Crestani, F., Parapar, J.: Overview of erisk: early risk prediction on the internet. In: International Conference of the Cross-language Evaluation Forum for European Languages, pp. 343–361. Springer (2018)
49. Marks, M.: Artificial intelligence-based suicide prediction. Yale JL & Tech. 21, 98 (2019)
50. Sawhney, R., Manchanda, P., Mathur, P., Shah, R., Singh, R.: Exploring and learning suicidal ideation connotations on social media with deep learning. In: Proceedings of the 9th Workshop on Computational Approaches to Subjectivity, Sentiment and Social Media Analysis, pp. 167–175 (2018)
51. Mesfin Tadesse, M., Lin, H., Xu, B., Yang, L.: Detection of suicide ideation in social media forums using deep learning. Algorithms 13(1), 7 (2020)
52. Mohammadi, E., Amini, H., Kosseim, L.: Clac at clpsych 2019: fusion of neural features and predicted class probabilities for suicide risk assessment based on online posts. In: Proceedings of the Sixth Workshop on Computational Linguistics and Clinical Psychology, pp. 34–38 (2019)

53. Shing, H.-C., Nair, S., Zirikly, A., Friedenberg, M., Daumé III, H., Resnik, P.: Expert, crowd-sourced, and machine assessment of suicide risk via online postings. In: Proceedings of the Fifth Workshop on Computational Linguistics and Clinical Psychology: From Keyboard to Clinic, pp. 25–36, New Orleans, LA. Association for Computational Linguistics (2018)

54. Mishra, R., Prakhar Sinha, P., Sawhney, R., Mahata, D., Mathur, P., Ratn Shah, R.: Snap-batnet: cascading author profiling and social network graphs for suicide ideation detection on social media. In: *Proceedings of the 2019 Conference of the North American Chapter of the Association for Computational Linguistics: Student Research Workshop*, pp. 147–156 (2019)

55. Grover, A., Leskovec, J.: node2vec: scalable feature learning for networks. In: Proceedings of the 22nd ACM SIGKDD International Conference on Knowledge Discovery and Data Mining, pp. 855–864 (2016)

56. Ophir, Y., Tikochinski, R., Asterhan, C.S.C., Sisso, I, Reichart, R.: Deep neural networks detect suicide risk from textual facebook posts. Sci. Rep. **10**(1), 1–10 (2020)

57. Sawhney, R., Joshi, H., Gandhi, S., Shah, R.: A time-aware transformer based model for suicide ideation detection on social media. In: Proceedings of the 2020 Conference on Empirical Methods in Natural Language Processing (EMNLP), pp. 7685–7697 (2020)

58. Wang, N., Luo, F., Shivtare, Y., Badal, V.D., Subbalakshmi, K.P., Chandramouli, R., Lee, E.: Learning models for suicide prediction from social media posts (2021). arXiv:2105.03315

59. Macavaney, S., Mittu, A., Coppersmith, G., Leintz, J., Resnik, P.: Community-level research on suicidality prediction in a secure environment: overview of the clpsych 2021 shared task. In: Proceedings of the Seventh Workshop on Computational Linguistics and Clinical Psychology: Improving Access, pp. 70–80 (2021)

60. Kshirsagar, R., Morris, R., Bowman, S.: Detecting and explaining crisis (2017)

61. Morris, R.R., Schueller, S.M., Picard, R.W.: Efficacy of a web-based, crowdsourced peer-to-peer cognitive reappraisal platform for depression: randomized controlled trial. J. Med. Internet Res. **17**(3). Publisher Copyright: Robert R Morris. Stephen M Schueller, Rosalind W Picard (2015)

62. Ribeiro, M.T., Singh, S., Guestrin, C.: "Why should I trust you?" explaining the predictions of any classifier. In *Proceedings of the 22nd ACM SIGKDD International Conference on Knowledge Discovery and Data Mining*, pp. 1135–1144 (2016)

63. Malhotra, Anshu, Jindal, Rajni: Multimodal deep learning based framework for detecting depression and suicidal behaviour by affective analysis of social media posts. EAI Endorsed Trans. Pervasive Health Technol. **6**(21), e1 (2020)

64. Benton, A., Mitchell, M., Hovy, D.: Multi-task learning for mental health using social media text (2017)

65. Ive, J., Gkotsis, G., Dutta, R., Stewart, R., Velupillai, S.: Hierarchical neural model with attention mechanisms for the classification of social media text related to mental health. In: Proceedings of the Fifth Workshop on Computational Linguistics and Clinical Psychology: From Keyboard to Clinic, pp. 69–77 (2018)

66. Shing, H.-C., Nair, S., Zirikly, A., Friedenberg, M., Daumé III, H., Resnik, P.: Expert, crowd-sourced, and machine assessment of suicide risk via online postings. In: Proceedings of the Fifth Workshop on Computational Linguistics and Clinical Psychology: From Keyboard to Clinic, pp. 25–36 (2018)

67. Coppersmith, G., Dredze, M., Harman, C., Hollingshead, K.: From adhd to sad: analyzing the language of mental health on twitter through self-reported diagnoses. In: Proceedings of the 2nd Workshop on Computational Linguistics and Clinical Psychology: From Linguistic Signal to Clinical Reality, pp. 1–10 (2015)

68. Losada, D.E., Crestani, F., Parapar, J.: erisk 2020: self-harm and depression challenges. In: European Conference on Information Retrieval, pp. 557–563. Springer (2020)

69. Losada, D.E., Crestani, F., Parapar, J.: Overview of erisk 2020: early risk prediction on the internet. In: Experimental IR Meets Multilinguality, Multimodality, and Interaction, pp. 272–287, Cham. Springer International Publishing (2020)

70. Kim, J.: https://jina-kim.github.io/dataset/20srep-mental. [Online; accessed 27-Mar-2022]

71. Coppersmith, G., Ngo, K., Leary, R., Wood, A.: Exploratory analysis of social media prior to a suicide attempt. In: Proceedings of the Third Workshop on Computational Linguistics and Clinical Psychology, pp. 106–117 (2016)
72. MacAvaney, S.: Clpsych2021-Shared-Task/Practice-Dataset. https://github.com/seanmacavaney/clpsych2021-shared-task/tree/main/practice-dataset. [Online; accessed 27-Mar-2022]
73. Prakhar Sinha, P., Mishra, R., Sawhney, R., Mahata, D., Ratn Shah, R., Liu, H.: # suicidal-a multipronged approach to identify and explore suicidal ideation in twitter. In: *Proceedings of the 28th ACM International Conference on Information and Knowledge Management*, pp. 941–950 (2019)
74. Coppersmith, G., Dredze, M., Harman, C.: Quantifying mental health signals in twitter. In: *Proceedings of the Workshop on Computational Linguistics and Clinical Psychology: From Linguistic Signal to Clinical Reality*, pp. 51–60 (2014)
75. https://github.com/eddieir. Depression_detection_using_Twitter_post . https://github.com/eddieir/Depression_detection_using_Twitter_post. [Online; accessed 20-Feb-2022]
76. Cohan, A., Desmet, B., Yates, A., Soldaini, L., MacAvaney, S., Goharian, N.: SMHD: a large-scale resource for exploring online language usage for multiple mental health conditions. In: Proceedings of the 27th International Conference on Computational Linguistics, pp. 1485–1497, Santa Fe, New Mexico, USA. Association for Computational Linguistics (2018)
77. Wongkoblap, A., Vadillo, M.A., Curcin, V.: A multilevel predictive model for detecting social network users with depression. In: 2018 IEEE International Conference on Healthcare Informatics (ICHI), pp. 130–135. IEEE (2018)

Predicting Mental Health and Nutritional Status from Social Media Profile Using Deep Learning

Zakir Hussain📵 and Malaya Dutta Borah📵

Abstract In recent times, discussion about mental health is getting importance alongside physical health. Due to the pandemic, people got stuck inside their home resulting reduction in physical activity. This has worked as a catalyst in boosting the mental health issues. Due to the pandemic, social media has been used as a medium of communication to a greater extend. Social media posts can tell a lot about the personality and states of mind of people. A section of people with mental health issues provide some hint through their social media handles. In this study, we collect a dataset containing Facebook confessions and apply Convolutional Neural Networks (CNNs) and Recurrent Neural Networks (RNNs) for predicting mental health status. We find a training accuracy of 71.67% and test accuracy of 73.94% on CNN. We implement SimpleRNN that yields results with 71.50% accuracy on training data and 70.10% accuracy on test data. We also implement Long Short-Term Memory (LSTM) network and get exactly the same accuracy as that of SimpleRNN for both training and test data. These results show that the implemented models predict the mental health status with comparably good accuracy.

Keywords Mental health · CNN · RNN · LSTM · Social media · Deep learning

1 Introduction

Most of the times, if we talk about the health, our thought directly strikes the concept of physical health. Generally, we understand that health means a disease free body. But according to World Health Organization (WHO), "Health is a state of complete physical, mental and social well-being and not merely the absence of disease or

Z. Hussain (✉) · M. D. Borah
Department of Computer Science and Engineering, National Institute of Technology Silchar, Cachar, Silchar 788010, Assam, India
e-mail: zakir_rs@cse.nits.ac.in

M. D. Borah
e-mail: malayaduttaborah@cse.nits.ac.in

T.-P. Hong et al. (eds.), *Deep Learning for Social Media Data Analytics*,
Studies in Big Data 113, https://doi.org/10.1007/978-3-031-10869-3_10

infirmity". So, mental health and social well-being is as important as the physical health. Now, physical health means, a body with adequate amount of nutrients and with immunity to win the fight against the germs that enter into the body. That means, the standard functioning of the body at all levels. Normal functioning of the biological processes and its balance with the environment [14]. According to the Department of Health & Human Services of U.S., mental health denotes a person's emotional, social as well as psychological well-being. A sound mental health is not only defined by the absence of anxiety, depression or other types of disorders. It is also defined based on a person's ability to enjoy life, tackling the situation after a difficult experience, adaptation to adversity, balancing family and finances, feeling safe and secure, achieving their full potential etc.

Physical health problems and mental health problems are closely related. Long term presence of physical health issues significantly increase the risk of developing mental health problems, and vice versa. It is often observed that nearly 34% of people with a long-term physical health problem suffers from a mental health issue, most often depression or anxiety [21]. If a chronic ailment disrupts a person's capability of completing their regular tasks, that can lead to stress and depression [13]. With improvement in testing, now-a-days, doctors are able to recognise some physical signs for some kind of mental ailment in CT scans and even in genetic tests.

Nutritional status has great impact on physical and mental health. Inadequate amount of nutrition in the body leads to development of malnutrition [7]. Malnutrition comes in the form of under-nutrition and over-nutrition [8]. Less amount of nutrition causes under-nutrition while excess amount of nutrition causes over-nutrition. Both the situations are against the stability of the health [2, 9]. Similar to the physical health, mental health is also impacted by nutritional status. According to the study in [16], the nutrients like omega-3 fatty acids, cholesterol, niacin, phospholipids, folate, vitamin B6, and vitamin B12 are advantageous for mental health. The study in [17] talks about the beneficial effects of macro-nutrients and micro-nutrients on mental health specifically based on the western diet. The study in [19] found that the patients with depression consume less amount of vitamin B6 and B12. The folic acid level was found to be very low. Also they found that abdominal obesity is more in depressed patients. So, it is clear that, nutritional status has significant impact on both physical and mental health.

Social network is a network for social interactions and individualized association [3]. And social media is the websites and applications that enable people to create and share content and to take part in social networking [1]. These are the exclusive websites or applications which allow users to communicate with each other by posting information, messages, comments, images, videos etc. These are some of the powerful mediums to reach wider audience at a time [11]. When we say social media, Facebook pops up in our mind. But actually Facebook is not the only social media platform. There are various types of social media platforms. Some of those can be named as- Community blogs, Social networking sites, Image sharing sites, Social review sites, Video hosting sites, Sharing economy networks, Discussion sites etc. [6, 20]. These are all social media platforms for different types of audiences.

Now-a-days almost every people with access to internet use social media of some type. They use social media for variety of purposes. They share a lot of information about their personal life as well. Each and every social media posts (of any kind) keep a thought in the back. So, these posts carry a lot of information about the mental state of the person [15]. These posts can be gathered and analysed to predict the mental status of that person [4, 12]. Traditionally, doctors consult the patient using the method of question-answering, body language, behaviour and some clinical tests. But for mental state it is less important than the former methods. In social media, we get the answers to a lot of questions without asking them. So, the posts can be used as a great source of information to predict the mental state of a person.

Social media is widely used by billions of people and hence it carries a big amount of data [5]. These data are mostly unorganized data. Though deep learning is a small part of machine learning, yet it works with large amount of data and also with unorganized data unlike traditional machine learning models that can work with small amount of data, and with organized data [22]. Also, if traditional machine learning models are applied with big amount of data, it takes more time to execute. Deep learning works with less effort and less amount of time with large amount of data.

1.1 Related Studies

To get an idea of the current status of this work, we search for some similar kind of works and find that only a few works have been carried out in this area. The study in [23] talks about detection of depressive symptoms from the tweets by developing a multi-modal framework and employing statistical techniques. This study utilized anonymized users in the datasets as per the approved Institutional Review Board (IRB). They have used two datasets- one is age enabled ground-truth dataset and another one is gender enabled ground-truth dataset. Their experiment resulted in 5% improvement in the average F1-score. The study in [10] applies unsupervised learning models on the posts and tweets from tweeter to classify the users based on the scale of their behavioural changes. The proposed model achieved an accuracy of 76.12%.

1.2 Scope for Contribution

From the above discussions it has been found that limited tasks have been accomplished in this field of study. There are a lot of research contributions in the prediction of mental health and nutritional status using traditional clinical approach. Contribution in the field of mental health and nutritional status using the information from social media platforms has left tremendous scope to explore. We have decided to contribute in the prediction of mental health and nutritional status using social media data with the application of deep learning model.

1.3 Research Objectives

The main objective of this study is to predict mental health status and nutritional status from the social media profile data using deep learning. To achieve this objective, we have completed the following intermediate works:

- Collection of dataset containing the required social media data.
- Application of deep learning model on the dataset to predict mental heath and nutritional status.
- Check performance of the model and tuning (if needed) to improve the performance.

2 Methodology

To perform the prediction task, we plan to use deep learning models. Deep learning is a subset of machine learning that is based on artificial neural networks. There are many deep learning models. Some are supervised and some are unsupervised. Some of the supervised models are- Convolutional Neural Networks (CNNs), Recurrent Neural Networks (RNNs), Multilayer Perceptrons (or Classic Neural Networks). Likewise some unsupervised models are- Self-Organizing Maps (SOMs), Boltzmann Machines, AutoEncoders etc.

2.1 Used Model

The collected dataset has predefined class labels. So, we use supervised deep learning models. In this experiment, we use Convolutional Neural Networks (CNNs), and Recurrent Neural Networks (RNNs). These models are briefly discussed below:
Convolutional Neural Networks (CNNs) A convolutional neural network is generally made up of several layers and these layers are mostly of three types:

- Convolutional layer: This layer consists of rectangular mesh of neurons and needs the previous layers to be rectangular as well. Each neuron accepts inputs from a rectangular section of the preceding layer and the weights for this rectangular section are also same. These wights designate the convolution filter.

- Max-Pooling: A pooling layer can be placed after each convolutional layer. This layer accepts small rectangular blocks from the convolutional layer and sub-samples it to generate a single output. There may have several ways to perform the pooling, like- taking average, taking maximum or linear combination etc.
- Fully-Connected: The fully-connected layers perform high-level reasoning in the network after several convolutional and pooling layers. It takes all neurons whether it is fully connected, pooling or convolutional in the preceding layer and connects with every single neuron it possess.

In the forward propagation phase, suppose, we have $N \times N$ neuron layers in the convolutional layers. If we use an $f \times f$ filter k, the output of our convolutional layer will be of size $(N - m + 1) \times (N - f + 1)$.

In the backward propagation phase, let us consider that, we have some error function E and we are aware of the error values at convolutional layer. The error we know and the error we need to compute for preceding layer is the partial of E with respect to each neuron output $\left(\frac{\delta E}{\delta y_{ij}}^{l} \right)$.

Recurrent Neural Networks (RNNs) Technically, recurrent networks are representation of feed forward networks with additional state variables and a recurrent loop. So, each element of the input sequences of RNNs is processed as:

$$s_t^{in} = I_t W_s + s_{t-1} U + B_s$$
$$s_i - f_s(s_t^{in})$$
$$o_t^{in} = s_t \times W_0 + B_0 \tag{1}$$
$$o_t = f_0(o_t^{in})$$

Where, I_t is the input at time-step t or t^{th} element of the input sequence.
s_t is the hidden state at time-step t.
o_t is the output at time-step t.
f_s is the activation function at input layer.
f_o is the activation function at output layer.

As in recurrent networks, it is possible to have a sequence as an input, there is a loss and it is represented as sum of losses at each output or sometimes average of them. Mathematically it is shown as:

$$L = \sum_t L_t$$
$$\frac{\delta L}{\delta \theta} = \sum_t \frac{\delta L_t}{\delta \theta} \tag{2}$$

Where, θ is basically represented by any trainable variable.

Most commonly, we use Cross-Entropy loss and the derivative of the loss with respect to o before the activation function can be presented as:

$$\frac{\delta L_t}{\delta o_t^{in}} = -y_t \times (1 - \sigma(o_t^{in})) = -y_t \times (1 - o_t) = y_t \times (o_t - 1) \tag{3}$$

2.2 Dataset Details and Pre-processing

To achieve the objective of this work, we search for datasets that contain confessions shared in different social media platforms. We find a dataset mentioned in the study [18], but that is not publicly available. We use a dataset called "Facebook Confessions Dataset" downloaded from kaggle.com. This comes under the "Mental Health Dataset" category. The details of the dataset are shown in Fig. 1 and Table 1. The dataset contains three columns. Out of those three columns, one records the serial number, next one records the class labels and the third one contains the confessions. Out of all these confessions, 72% confessions are labeled as "None". This means that those confessions are not categorized into any category. 8% confessions are adult texts while rest 20% confessions fall under other categories like- "Mental Health",

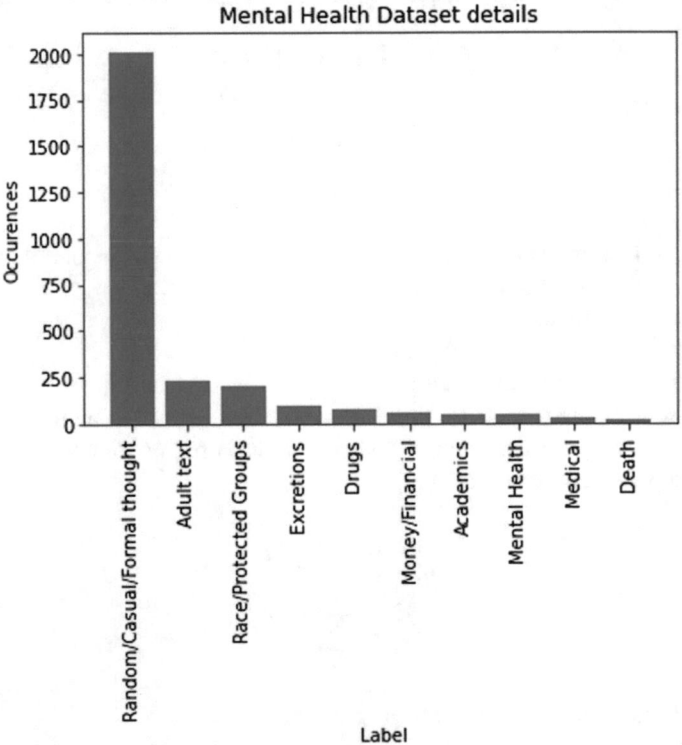

Fig. 1 Visual of the dataset details

Table 1 Details of the dataset

Name of the dataset	Facebook confessions dataset
Category	Mental health dataset
File name	labeled.csv
Number of columns	03
Number of rows	2802
Number of class labels	10
Number of attributes	02
Size	931.56 kB
Source	https://www.kaggle.com/apaul1218/mental-health-dataset

"Money/Financial", "Medical", "Drugs", "Race/Protected Groups", "Excretions", "Academics", and "Death". Out of the two columns of the dataset, first column contained the text and the second column contained the label of that text. As mentioned above that 72% of the text are labeled as "None". This means those are not assigned any class. We read the texts and find that those text express some random thoughts, some casual thoughts, and some formal thoughts as well. So, for better understanding we assign those texts a class called "Random/Casual/Formal thoughts". We know that, for working with the models, many times the data need to be converted into numbers. We convert the labels into numerical values and assign 1 to the "Random/Casual/Formal thoughts", 2 to the "Adult text", 3 to the "Mental Health", 4 to the "Money/Financial", 5 to the "Medical", 6 to the "Drugs", 7 to the "Race/Protected Groups", 8 to the "Excretions", 9 to the "Academics", and 10 to the "Death" class.

3 Results

We use the processed dataset on CNN and RNN models. The CNN is used for working with images in general. But it has another version for working with text. That version is called TextCNN. We use that version for working with our dataset as our dataset contains only textual content. We use three models namely- CNN, SimpleRNN and LSTM. The results obtained from the models are individually presented below.

3.1 Results Obtained from Convolutional Neural Network

We split the dataset into 75:25 ratio. That means, we use 75% of the data for training and 25% data for testing. We use tokenizer on both the training data and test data for vectorizing the texts. This converts the text input into vector and assigns individual

Table 2 CNN model summary

Model: "sequential"		
Layer (type)	Output Shape	Param #
embedding (Embedding)	(None, 100, 200)	1838000
conv1d (Conv1D)	(None, 96, 128)	128128
global_max_pooling1d (Global)	(None, 128)	0
dense (Dense)	(None, 10)	1290
dense_1 (Dense)	(None, 1)	11
Total params: 1,967,429		
Trainable params: 1,967,429		
Non-trainable params: 0		

coefficients. We use post padding method to ensure equal length of the sequences. We keep on adding 0s at the end up to a maximum length of 100. We add five layers in the CNN. First one is for word embedding, that provides dense vector representation of words and their relative meanings. This is the first layer of our CNN model and here we have used the vocabulary size as the value of the argument "input_dim", the value of "output_dim" is taken as 200 and "input_length" is the maximum length of the sequence that is 100. The output of this layer is 2D vector with one embedding for each word in the input sequence of words. Next we use a Conv1D layer with 128 numbers of filters, kernel size 5 with activation function "relu", i.e. rectified linear unit. The third layer is GlobalMaxPooling1D with the default values i.e. channel_last data_format and keeping the keepdims value as False. We use this layer to under-sample the input representation by taking maximum value. Next, we use a Dense layer with 10 dimensional output space and "relu" activation function. The last layer is again a Dense layer with one dimensional output space with "sigmoid" activation function. We compile the CNN with "adam" optimizer and "binary_crossentropy" loss function. We evaluate the model using accuracy. The details about the model is summarized in Table 2. We train the model considering 100 epochs and a batch size of 10. The accuracy and loss in each layer is plotted as shown in Fig. 2. We get the graphs by plotting values for each epoch. During the training of the model, it shows the accuracy and loss for each epoch. Also, the time required for that particular epoch. The plot for training accuracy and training loss use blue color and for test accuracy and test loss, we use red color. After completing training and testing for all the epochs, we find that, the overall training accuracy of the model is 71.67% and test accuracy is 73.94%.

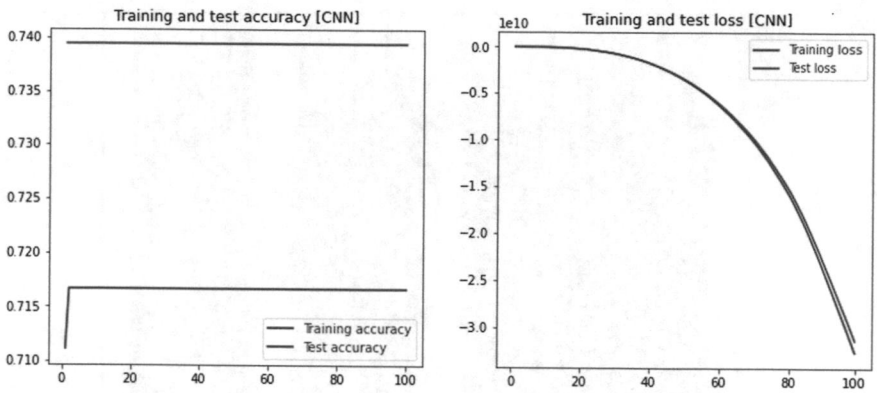

Fig. 2 Accuracy and loss during training and testing on CNN

3.2 Results Obtained from Simple Recurrent Neural Network

To work with the RNN models, initially, we segregate 25% of data as test data and create a separate dataset. The remaining 75% data remains as training data in the original dataset. We scale and translate the features of the datasets using MinMaxScaler in the range of (0, 1). The scaled and transformed training dataset is shown in Fig. 3. Then we reshape the data as per the format supported by the model. We use a total of 05 (Five) layers. The first layer is a SimpleRNN layer with 50 dimensional output space, "tanh" activation function and return_sequence = True, that means, we are expecting full sequence of the output. We provide input_shape of the type (shape[1], 1). The second layer is also a SimpleRNN layer with output space of 50 dimensions and "tanh" activation function. The third layer is same as the second layer, i.e. a SimpleRNN layer with 50 dimensional output space and an activation function of "tanh". The fourth layer is also a SimpleRNN layer of 50 dimensional output with default activation function. The fifth and the last layer of the model is a Dense layer with one dimensional output space. In the compilation phase, we use "adam" optimizer and "mean_squared_error" type of loss function. For evaluating the model, we use accuracy. The summary of the model is shown in Table 3. We train the model for 100 epochs with a batch_size of 32. We plot the accuracy and loss of every epochs as shown in the Fig. 4. Here also, we plot individual results obtained from each epoch. The individual results were the accuracy, the loss and required time for completing each epoch. Similar to the previous plot, we use blue color for training accuracy and training loss and red color for test accuracy and test loss. After completion of all epochs, we obtain the overall training accuracy of the model as 71.50% and test accuracy as 70.10%.

After training and testing of the model, we perform prediction on a test dataset. The results obtained on this predictive task is shown in Fig. 5. We use two different colors for the labels. We use grey color for real data labels and blue color for the predicted

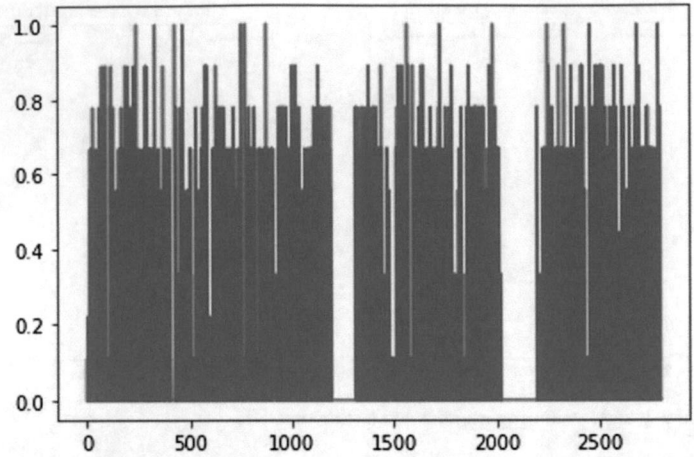

Fig. 3 Visual of the scaled training set

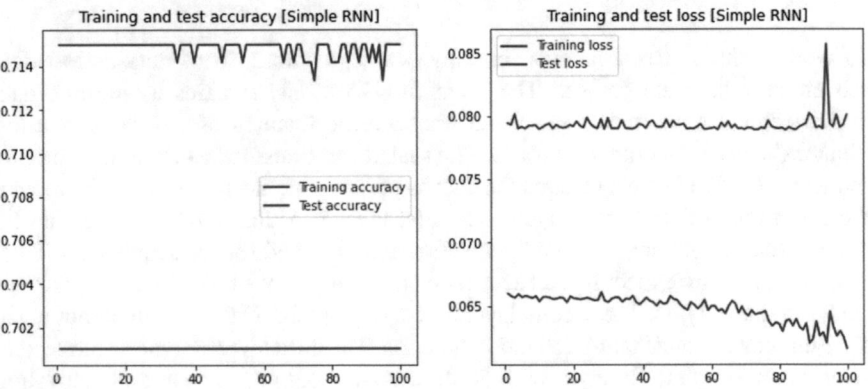

Fig. 4 Accuracy and loss during training and testing on Simple RNN

data label. We use a timestamp and the prediction of the text label is performed on the basis of the timestamps.

Table 3 Simple RNN model summary

Model: "sequential"		
Layer (type)	Output shape	Param #
simple_rnn (SimpleRNN)	(None, 50, 50)	2600
dropout (Dropout)	(None, 50, 50)	0
simple_rnn_1 (SimpleRNN)	(None, 50, 50)	5050
dropout_1 (Dropout)	(None, 50, 50)	0
simple_rnn_2 (SimpleRNN)	(None, 50, 50)	5050
dropout_2 (Dropout)	(None, 50, 50)	0
simple_rnn_3 (SimpleRNN)	(None, 50)	5050
dropout_3 (Dropout)	(None, 50)	0
dense (Dense)	(None, 1)	51
Total params: 17,801		
Trainable params: 17,801		
Non-trainable params: 0		

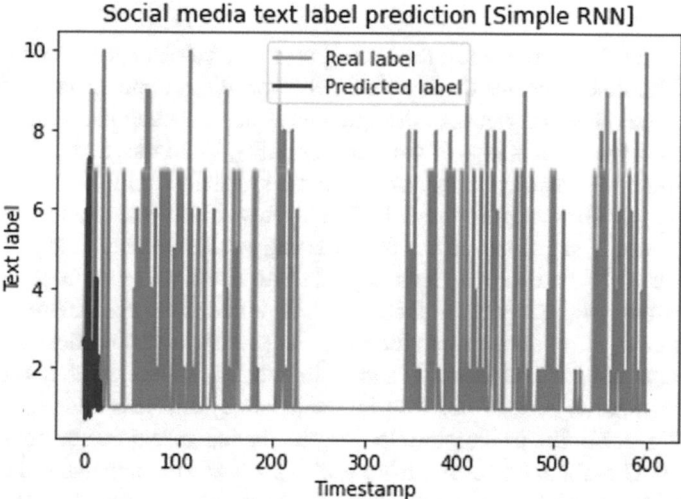

Fig. 5 Obtained results for Simple RNN

3.3 Results Obtained from Long Short-Term Memory Network

To work with LSTM, we divide the dataset into two parts. We segregate 25% of the data as test and remaining 75% of the data as training data. Similar to the SimpleRNN model, we transform and scale the data as shown in Fig. 3. After reshaping the data as per the format of the model, we proceed for building the model by deploying

Table 4 LSTM model summary

Model: "sequential_1"

Layer (type)	Output shape	Param #
lstm (LSTM)	(None, 50, 50)	10400
dropout_4 (Dropout)	(None, 50, 50)	0
lstm_1 (LSTM)	(None, 50, 50)	20200
dropout_5 (Dropout)	(None, 50, 50)	0
lstm_2 (LSTM)	(None, 50, 50)	20200
dropout_6 (Dropout)	(None, 50, 50)	0
lstm_3 (LSTM)	(None, 50)	20200
dropout_7 (Dropout)	(None, 50)	0
dense_1 (Dense)	(None, 1)	51

Total params: 71,051

Trainable params: 71,051

Non-trainable params: 0

different layers. We use exactly the same parameter values for different layers of LSTM. In the first layer, we deploy a LSTM layer with output space of 50 dimensions, we need all the sequences of output and hence we keep the return_sequence as "True", and the input_shape of the type (shape[1], 1). In the second layer, we use another LSTM layer with 50 dimensional output space and "True" return_sequence. In the third layer also we use another LSTM layer with 50 dimensional output space and "True" return_sequence. In the fourth layer, we use another LSTM layer with output space of 50 dimensions. In the last and fifth layer, we use a Dense layer with one dimensional output space. In the compilation phase, we use "adam" optimizer, "mean_squared_error" type of loss function. For evaluating the model, we use accuracy. The summary of the model is shown in Table 4. We train the model for 100 epochs with a batch_size of 32. We plot the accuracy and loss of every epochs as shown in Fig. 6. Similar to SimpleRNN, we plot the individual results obtained from each epoch. The obtained results from each epoch are the accuracy, the loss and time needed to complete that epoch. Here also, blue color has been used for plotting the training accuracy and training loss and red color has been used for plotting test accuracy and test loss. After completion of all the epochs, we find exactly the same accuracy as SimpleRNN, i.e. 71.50% for training and 70.10% for test.

After the training and testing of the model, we perform prediction on test dataset. The results of the prediction operation is shown in Fig. 7. Similar to the previous task of SimpleRNN, we use grey color for presenting original data labels and blue color for the predicted data labels. Here also, we use a timestamp and the prediction is performed based on the timestamp.

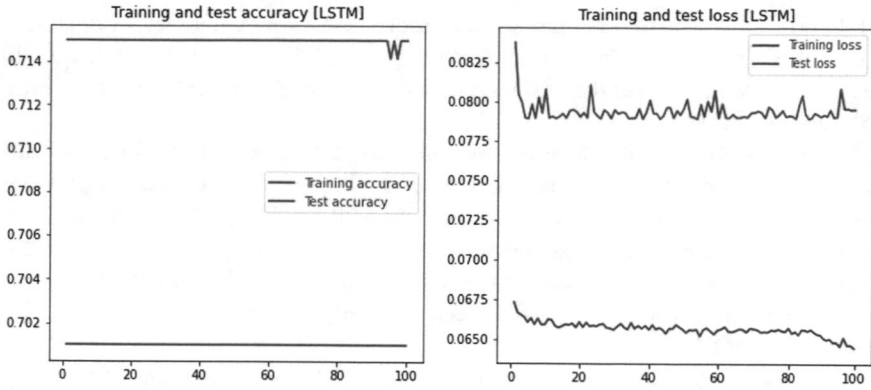

Fig. 6 Accuracy and loss during training and testing on LSTM

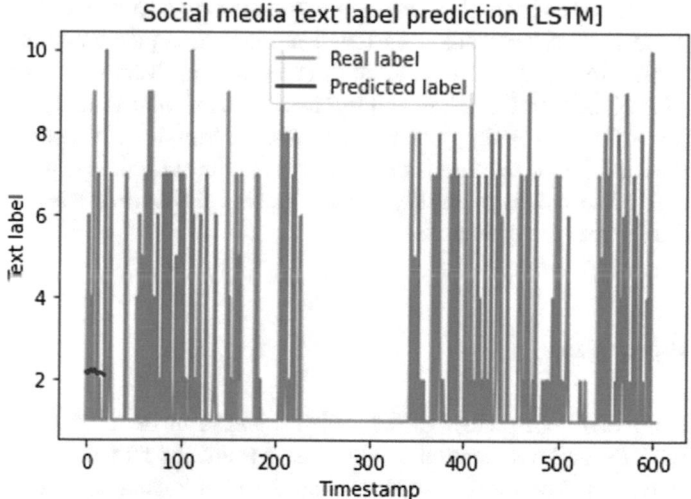

Fig. 7 Obtained results for LSTM

4 Discussions

From the obtained results we find that, for the initial a few epochs, the training accuracy of CNN increases from ≈71% to ≈71.50% and for rest of the epochs it remains same. For the training set, the accuracy remains constant around 74%. Similarly for the loss, we observe that, for both training and test data, the loss remains zero. Then it slowly comes down till the epochs around 40, and then it exponentially decays to > -3.0.

For SimpleRNN, the training accuracy remains constant around 71.5% till the epochs ≈27. After that, the accuracy fluctuates in between an approximated range of

71.1% to 71.5%. The training loss shows minor fluctuations and gradually decreases from ≈6.6% to ≈6.1%. The loss on the testing phase also shows minor volatility till the epochs around 96 and then shows a high fluctuation and then again comes down to a level of ≈7.9%.

For LSTM, the training accuracy remains constant around 71.5% till the epochs ≈96, after that it shows minor fluctuation and again remain constant at the previous level. The test accuracy remains constant around 70.1% for entire epochs. The loss in the testing phase shows huge volatility in the initial a few epochs and after that it keeps on showing minor fluctuations for rest of the epochs. The training loss shows minor fluctuations for entire epochs and it gradually decreases to a level of 6.45% from a level of 6.75%.

From the above discussion, we find that the results obtained from CNN is significantly different from SimpleRNN and LSTM. We can find a bit similarity in the test accuracy for all three models. But, the training accuracy for all three models are different. There is very high fluctuation in the training accuracy of SimpleRNN. However, little bit similarity could be found in the training accuracy of CNN and LSTM. Training loss and test loss for all three models are very different from one another. In CNN, both training loss and test loss are of similar kind. But these have no similarity with the training loss and test loss of SimpleRNN and LSTM. For SimpleRNN and LSTM, training loss are almost similar with little difference in the epochs near 98. The test loss volatility is seen more in LSTM while it is more volatile for SimpleRNN near the epochs 98.

4.1 Comparison

We could find two high quality closely related works in the studies [10, 23]. In the study [23], the authors mention that they use a multi-modal framework that uses feature selection and Random forest (RF). They also mention that, the model achieves 5% improvement in F1-score. The study in [10] uses their own approach called Unsupervised approach (UA) that modifies traditional machine learning algorithms and adds fuzziness. The highest accuracy obtained by one of their modified version is 76.12%. In this study, we apply CNN and RNN. For CNN, we achieve an accuracy of 71.67% for training data and 73.94% for test data. In case of SimpleRNN, it is 71.50% for training data and 70.10% for test data. In the case of LSTM, it is also 71.50% for training data and 70.10% for test data. Though we are comparing the results with the results in [10, 23] as shown in Table 5, the comparison is not even. Because, the datasets used in these studies are not same. The models that are implemented are also not same. The former study has used a multi-modal framework that implements RF with feature selection techniques. The latter one proposed their own unsupervised model that added fuzziness in traditional machine learning algorithms. Now, we have implemented deep learning models and comparison of obtained results from deep learning techniques with the obtained results from traditional machine learning

Table 5 Comparison of performances

Reference	Model	Training accuracy	Test accuracy	Remark
[23]	RF	NA	NA	5% improvement in F1-score
[10]	UA	76.12%	NA	NA
This	CNN	71.67%	73.94%	NA
This	Simple RNN	71.50%	70.10%	NA
This	LSTM	71.50%	70.10%	NA

models are uneven. Our implemented models have shown accuracy in a range of 71–74%, which is a considerable and comparable accuracy.

4.2   Prediction of Mental Health and Nutritional Status

Besides evaluating the models based on the accuracy, we perform prediction on a separate set of data. The results obtained for the prediction on SimpleRNN is shown in Fig. 5 and the prediction on LSTM is shown in Fig. 7. We find that, the prediction using SimpleRNN is almost similar to the real data. However, for the LSTM, the prediction is seen poor with respect to original data. This prediction activity is for all the ten labels on the dataset. Out of these labels, we need to consider the label for the mental health. Also, it is mention worthy that the major portion of the dataset is for Random/Casual/Formal thoughts. This portion of data contains a lot of information that indicate their physical health and nutritional status. Also, in the medical sciences, it is considered that the person with some sort of problem with mental health lacks some micro-nutrients that are meant for the stability of mental health.

5   Conclusion and Future Work

Now-a-days social media has become an inseparable part of daily life. People use this medium to communicate their feelings and emotions with near and dear ones, virtually known friends, colleagues etc. of each corner of the world. The posts from their profiles say a lot about their interest, taste, intention or even mental and physical health status. These posts are a great source of information to predict the mental and physical health status. This study covers the prediction of mental health and nutritional status using posts from Facebook only. We apply CNNs and RNNs and find that the models predict with impressive accuracy. As a future work, this study can be extended to other social media platforms. Also, further study can be carried out to predict other social issues from these posts.

Acknowledgements The required data to carry out the experiment has been downloaded from https://www.kaggle.com. The authors are thankful to the owner of the dataset for making it publicly available. Also, the authors will always be indebted to the website.

References

1. Aichner, T., Grünfelder, M., Maurer, O., Jegeni, D.: Twenty-five years of social media: a review of social media applications and definitions from 1994 to 2019. Cyberpsychology Behav. Soc. Netw. **24**(4), 215–222 (2021) https://doi.org/10.1089/cyber.2020.0134, pMID: 33847527
2. Asamane, E.A., Greig, C.A., Thompson, J.L.: The association between nutrient intake, nutritional status and physical function of community-dwelling ethnically diverse older adults. BMC Nutr. **6**(1), 36 (2020). https://doi.org/10.1186/s40795-020-00363-6
3. boyd, d.m., Ellison, N.B.: Social network sites: definition, history, and scholarship. J. Comput.-Mediat. Commun. **13**(1), 210–230 (2007). https://doi.org/10.1111/j.1083-6101.2007.00393.x
4. Chan, W.S., Leung, A.Y.: Use of social network sites for communication among health professionals: systematic review. J. Med. Internet Res. **20**(3), e117 (2018). https://doi.org/10.2196/jmir.8382, http://www.jmir.org/2018/3/e117/
5. Chancellor, S., De Choudhury, M.: Methods in predictive techniques for mental health status on social media: a critical review. npj Digit. Med. **3**(1), 43 (2020). https://doi.org/10.1038/s41746-020-0233-7
6. Fraidaki, K., Pramatari, K., Doukidis, G.: Living in the era of social media: How the different types of social media may affect information acquisition process. In: Meiselwitz, G. (ed.) Social Computing and Social Media, pp. 178–185. Springer International Publishing, Cham (2014)
7. Hussain, Z., Borah, M.D.: Birth weight prediction of new born baby with application of machine learning techniques on features of mother. J. Stat. Manag. Syst. **23**(6), 1079–1091 (2020) https://doi.org/10.1080/09720510.2020.1814499
8. Hussain, Z., Borah, M.D.: Nutritional status prediction in neonate using machine learning techniques: a comparative study. In: Bhattacharjee, A., Borgohain, S.K., Soni, B., Verma, G., Gao, X.Z. (eds.) Machine Learning, Image Processing, Network Security and Data Sciences, pp. 69–83. Springer Singapore, Singapore (2020). https://doi.org/10.1007/978-981-15-6318-8_7
9. Hussain, Z., Borah, M.D.: Nicov : a model to analyse impact of nutritional status and immunity on covid-19. Med. Biol. Eng. Comput. **60**(5), 1481–1496 (2022). https://doi.org/10.1007/s11517-022-02545-9
10. Joshi, D., Patwardhan, D.: An analysis of mental health of social media users using unsupervised approach. Comput. Hum. Behav. Rep. **2**, 100036 (2020) https://doi.org/10.1016/j.chbr.2020.100036, https://www.sciencedirect.com/science/article/pii/S2451958820300361
11. Kaplan, A.M.: Social Media, Definition, and History, pp. 2662–2665. Springer New York, New York, NY (2018). https://doi.org/10.1007/978-1-4939-7131-2_95
12. Karim, F., Oyewande, A.A., Abdalla, L.F., Chaudhry Ehsanullah, R., Khan, S.: Social media use and its connection to mental health: a systematic review. Cureus **12**(6), e8627–e8627 (2020). https://doi.org/10.7759/cureus.8627, https://pubmed.ncbi.nlm.nih.gov/32685296
13. Kim, J., Lee, D., Park, E.: Machine learning for mental health in social media: bibliometric study. J. Med. Internet Res. **23**(3), e24870 (2021) https://doi.org/10.2196/24870, https://www.jmir.org/2021/3/e24870
14. Koipysheva, E.: Physical health (definition, semantic content, study prospects), pp. 601–605 (2018). https://doi.org/10.15405/epsbs.2018.12.73
15. Kuss, D.J., Griffiths, M.D.: Online social networking and addiction-a review of the psychological literature. Int. J. Environ. Res. Public Health **8**(9), 3528–3552 (2011). https://doi.org/10.3390/ijerph8093528, https://www.mdpi.com/1660-4601/8/9/3528

16. Lim, S.Y., Kim, E.J., Kim, A., Lee, H.J., Choi, H.J., Yang, S.J.: Nutritional factors affecting mental health. Clin. Nutr. Res. **5**(3), 143–152 (2016). https://doi.org/10.7762/cnr.2016.5.3.143, https://pubmed.ncbi.nlm.nih.gov/27482518

17. Muscaritoli, M.: The impact of nutrients on mental health and well-being: insights from the literature. Front. Nutr. **8**, 97 (2021) https://doi.org/10.3389/fnut.2021.656290, https://www.frontiersin.org/article/10.3389/fnut.2021.656290

18. Ríssola, E.A., Bahrainian, S.A., Crestani, F.: A dataset for research on depression in social media. In: Proceedings of the 28th ACM Conference on User Modeling, Adaptation and Personalization, p. 338-342. UMAP '20, Association for Computing Machinery, New York, NY, USA (2020). https://doi.org/10.1145/3340631.3394879

19. Santos, A., Kaner, G., Soylu, M., Yüksel, N., Inanç, N., Ongan, D., Başmısırlı, E.: Evaluation of nutritional status of patients with depression. BioMed Res. Int. **2015**, 521481 (2015). https://doi.org/10.1155/2015/521481

20. Voorveld, H.A.M., van Noort, G., Muntinga, D.G., Bronner, F.: Engagement with social media and social media advertising: the differentiating role of platform type. J. Advert. **47**(1), 38–54 (2018). https://doi.org/10.1080/00913367.2017.1405754

21. Wongkoblap, A., Vadillo, M.A., Curcin, V.: Researching mental health disorders in the era of social media: Systematic review. J. Med. Internet Res. **19**(6), e228 (2017). https://doi.org/10.2196/jmir.7215, http://www.jmir.org/2017/6/e228/

22. Yazdavar, A., Mahdavinejad, M., Bajaj, G.K., Thirunarayan, K., Pathak, J., Sheth, A.: Mental health analysis via social media data, pp. 459–460 (2018). https://doi.org/10.1109/ICHI.2018.00102

23. Yazdavar, A.H., Mahdavinejad, M.S., Bajaj, G., Romine, W., Sheth, A., Monadjemi, A.H., Thirunarayan, K., Meddar, J.M., Myers, A., Pathak, J., Hitzler, P.: Multimodal mental health analysis in social media. PLOS ONE **15**(4), 1–27 (2020). https://doi.org/10.1371/journal.pone.0226248

Impact of Artificial Intelligence-Based Chatbots on Customer Engagement and Business Growth

Chitra Krishnan, Aditya Gupta, Astha Gupta, and Gurinder Singh

Abstract For a variety of reasons, artificial intelligent devices are becoming increasingly vital for enterprises. One of the major characteristics that makes AI enhanced services ideal for use in organisations is their ability to complete jobs faster and more accurately than humans. Almost all firms are utilising AI-based Chatbots on social media and messaging applications such as Whats App, Facebook, and others to engage with their enormous consumer base and provide real-time help. But AI is capable of much more. For many individuals, having a discussion with a bot that sounds like a human is a new disruption, and it serves as a tool to engage and entice customers to the point where they leave the website. But AI do more than just that. Having a conversation with a bot which feels like a human talking is a new disruption for a lot of people and it works as a tool to engage these customers and attract them in such a way that they leave the website after making the purchase. And as far as disruption is concerned from guiding the customer how to order pizza to describing through complex sales processes, AI Chatbots have been able to help both B2C and B2B interactions. This chapter provides an insight at how the AI Chatbots influences user interactions, how brands are using Chatbots for marketing and customer service, and why customers are attracted to interact with augmented agents such as Chatbots.

Keywords Chatbots · Marketing · AI · Customer service · Augmented reality · Conversational AI

1 Introduction

Over the years, we've seen people use the internet to find solutions and improve their daily lives by receiving quick and easy answers to their problems. This basic feature contributed to the moment when all businesses began to increase their online presence and sought to serve their clients through websites, blogs, and other means. We can't think of a single company today that isn't wanting to connect with its

C. Krishnan (✉) · A. Gupta · A. Gupta · G. Singh
Amity International Business School, Amity University, Noida, India
e-mail: ananya.chitra@gmail.com

© The Author(s), under exclusive license to Springer Nature Switzerland AG 2022
T.-P. Hong et al. (eds.), *Deep Learning for Social Media Data Analytics*,
Studies in Big Data 113, https://doi.org/10.1007/978-3-031-10869-3_11

customers over the internet and provide them with a hassle-free experience of what they have to offer, whether it's a product or a service [1]. Companies that have been able to build this bond with their customers through the internet are reaping the rewards now, as we are rapidly approaching a period in which the internet and online relationships will be all that matters. Chatbots, also known as conversational agents, are software applications that may be integrated into a variety of platforms to respond to incoming messages or create natural-sounding conversations. The idea isn't brand new. On the other hand, Chatbot adoption by various enterprises will continue to rise in 2021 and beyond. Chatbots have grown in popularity in recent years thanks to artificial intelligence and other underpinning technologies Natural language processing and machine learning are two examples [2]. The business world is shifting to a more customer-centric approach. A company needs cultivate client loyalty through good customer service in order to make money. Chatbots simplify giving excellent customer service, which leads to improved sales, which Chatbots can also assist with. Customer engagement is difficult since it necessitates an in-depth understanding of your consumers' preferences, dislikes, and loves, among other things.

Chatbots have, as we all know, taken over the job of a human customer support agent in many firms by engaging customers in productive conversations. The bulk of customers have become accustomed to Chatbot service and now prefer it to human customer support professionals. Companies have used the internet in a number of ways to communicate with customers and provide customer service. Emails, relay chats, customer care calls, and other techniques were employed, but they were out-dated in terms of providing real-time help and relevant information to customers who wanted to learn more about the firm and purchase something. As a result, these strategies failed to engage customers and deliver information about the brand's goods and services. Then there was the concept of integrating live chat into websites or messaging apps. This way of offering real-time client support has shown to be effective and valuable. Chatbots were the most popular tool for putting this theory into practise, and they've proven their worth as more firms turn to them for customer service.

Chatbots are proving to be an effective tool for learning more about your clients. It keeps track of how long the conversation lasted and what response irritated the consumer [3]. A five-minute delay in replying to a query can reduce the likelihood of gaining a lead, according to a Harvard Business School and Inside Sales study. A 10-min delay might reduce your chances of making a successful contact by 400%. Facebook even made some eye-opening data public, demonstrating the value of Chatbots:

- In the two-way communication process between enterprises and people, almost 2 billion messages are exchanged.
- Sixty-six percent of people would prefer to message rather than phone customer care.
- 53% of potential clients prefer to do business with companies with whom they can communicate.

A tool that handles problems as complicated as these has significant drawbacks, and AI Chatbots may fall short in a few areas. Because Chatbots are frequently perceived as an interruption for visitors who are browsing a company's website, the Chatbots must be placed in the appropriate location. Misunderstanding sentiments and offering appropriate responses are among the other issues, as are data breaches, security, and misunderstandings from the human user's end. There is room for a lot more research and development in this area to improve the use of AI Chatbots even further. Although clients prefer Chatbots for speedier service, a human touch can be the cherry on top. Bots have proven to be effective agents in answering basic inquiries. They are programmed to emulate human written speech using simple rules and AI logic. Almost all customers demand customer care agents available 24 hours a day, seven days a week, as their wants and preferences change frequently and redefine the corporate structure. If handled by a human agent, this is quite unlikely. Chatbots are proving to be effective in giving the proper support and service in this situation. A human-AI hybrid can redefine the customer experience, with humans focusing solely on channeling AI's speed, efficiency, and reach, while the Chatbots is intended to provide complete customer satisfaction by promptly resolving concerns. Technology has aided the Chatbot in comprehending the context of each discussion and responding to each customer's questions. It is important to recognize that as humans, we always want to speak with a real person; but, a real person cannot constantly be online [4]. Self-operating AI-driven technologies continue to fascinate humans. Chatbots are a relatively new trend that has piqued the interest of the majority of the tech industry. And, as a result of all the research and development in the industry, programming is getting more human-like while remaining mechanical. Because of the mix of fast response and permanent connectivity, they are an engaging variation on the web application trend.

How does a machine communicate naturally? When considering the use of chatbots, this is a question that many people ask themselves. The real method of how chatbots are coded and work is hard and involves extensive technical knowledge, but the idea of their conversing in a humanized manner is fascinating and simple to grasp. Chatbots use knowledge bases to process and respond to incoming requests as efficiently as possible. The incoming text is parsed, and the message's intent is determined using a combination of algorithms and patterns.

The scientific community concentrated on natural language processing (NLP), also known as computer linguistics, in 1980 to build an algorithm for data that was freely available in abundance. This linguistic approach was initially developed to cope with massive amounts of data using a statistical model. The term "big data" has since been renamed "machine learning" Natural language processing is used by chatbots to determine the message's intent (NLP). This method aids chatbots in extracting useful data from the text. NLP is the foundation that allows conversational assistants to grasp the context of a conversation and decode the intent (positive or negative), allowing the bot to choose the right auto-response or escalate the issue to a customer service agent. The original generation of chatbots was only designed to conduct text-based discussions, but thanks to technology advancements and excellent natural language processing, there are now bots that can conduct both text-to-text and speech-to-speech chats (e.g. Siri, Alexa, etc.).

The architecture of the conversation system, the first natural language understanding system, contains three important components that are responsible for decoding the message supplied by the user and passing it to the other system for further processing. Second, there's the conversation management system, which keeps track of the context from which the output is derived, and third, there's natural language generation, which displays the final result in natural language. Nowadays, machine learning approaches are utilized to deal with systems that comprehend the linguistic structure of human dialogue in terms of semantic and syntactic information.

NLP is a language processing engine that allows the chatbot to deduce the user's purpose in a message and select the best response from its database. NLP comprehends the syntax, semantics, discourse, and purpose of a communication in order to engage in a human-like conversation, regardless of which language a machine is learning. It focuses on how to construct computers that can efficiently and effectively analyse large amounts of natural language input, relieving humans from certain tasks and allowing machines to perform others. NLP solves the problem of computers interpreting human input and creating structured data from unstructured language. The chatbot's task, however, does not end there. After determining the query's intent, bots must search the knowledge base framework for relevant structured material and provide it as unstructured text responses.

The meaning of the user's input is deciphered by Natural Language Understanding (NLU). NLU gets the chatbot to understand what a body of text means, and it's mostly focused on machine reading comprehension. NLU is nothing more than comprehending the text and categorizing it into appropriate intents. LU can be used in a variety of processes. NLU can do everything from categorizing text, gathering news, and archiving particular pieces of text to analyzing content. Small tasks like delivering short commands based on text understanding to some degree like routing an email to the proper receiver based on basic syntax and a moderately big lexicon are real-world instances of NLU.

Natural Language Generation (NLG) is a piece of software that generates human-readable texts. NLG approaches can be used to create symbiotic systems that take advantage of both human and machine knowledge and capabilities. Any text contained within a speech stream as part of a document, report, explanation, or other help message can be used as the input, and any non-linguistic representation of information can be used as the output. The NLG can use any communication database as a knowledge source. The primary goal of chatbots is to assist customer care representatives in addressing commonly requested inquiries, allowing them to focus on more pressing issues while providing clients with their most valuable asset. Chatbots are a win-win situation in this regard since they give both the firm and its consumers significant time back.

Most chatbots will employ a series of predefined responses generated by the customer service team to answer FAQs, making it much easier to manage incoming queries. More powerful chatbots may even decode complicated structural data and provide plain text summaries for them with NLG in situ, freeing up human operators to work on other tasks. Overall, this technique reduces response times by allowing a significant number of enquiries to be answered directly by the chatbots, while more difficult issues are forwarded to a human customer service agent.

2 Chat Bot in Customers Engagement

The growing popularity of chatbots powered by artificial intelligence (AI) and recent advances in the area of natural language processing (NLP) are gradually transforming the way people interact with businesses [5, 6]. In this context, the usage of chatbots for a range of consumer interactions is becoming increasingly popular as a way to enhance both service quality and customer experience [7]. These agents help customers by connecting to them via text-based communication as a first contact point for minor client queries, but they're also increasingly being used in more complex support activities, such as customer advisory [6].

Customers, on the other hand, have been slow to adopt chatbots [8], contrary to industry projections. One possible explanation, according to experts, is that the development of chatbots was sparked by technological advancements rather than market need. As a result, the desires and needs of users were not properly met [9]. Consumers are generally thought to be distrustful about new technologies, therefore industry reports frequently emphasise a lack of confidence [9]. The fact that chatbots are becoming more human-like may stress the necessity of trust [6]. To summarise, chatbots are increasingly being utilised for a variety of customer service interactions, but getting clients to transition to these new service channels owing to a lack of confidence remains a difficulty.

Little is known about how these one-of-a-kind human-computer interfaces may affect users' perceptions of the service delivery process (i.e., trust), and also the ramifications for users. Recent research has looked at the impact of individual design features on a variety of user perceptions, including trust and the ability to act on social cues [10]. However, no research has been done to compare the perception of trust when utilising a chatbot with a static interface [11]. As a result, it's difficult to say whether the change to a conversational interaction logic or the greater usage of anthropomorphic design elements is creating perceptual shifts. Understanding these mechanisms, on the other hand, would allow for more accurate development of these agents for certain contexts.

Striking a balance between efficiency and quality is one of the most difficult tasks for customer service providers: According to studies and practitioners, client self-service has a number of potential benefits, including increased time efficiency, cheaper costs, and a better customer experience [8, 12]. Chatbots have the ability to increase service quality and provider-customer interactions as a self-service tool. Chatbots can reduce response times, free up employees for other tasks, and answer up to 80% of common questions, according to studies, resulting in a 30% reduction in global corporate expenditures connected with 265 billion customer support requests per year [13]. Chatbots are likely to save businesses more than 8 billion in customer-supporting expenses by 2022, up from a projected savings of 20 million in 2017. In the form of 24/7 electronic channels to serve clients, chatbots claim to be rapid, convenient, and cost-effective [8].

Individualised support, as well as easy-to-use and adaptable self-service channels, are frequently valued by customers. As a result, companies shouldn't rely solely on customer self-service channels, especially in the start of a client engagement [12], Online transactions may suffer as a result of the lack of a human social actor. Chatbots, on the other hand, have the potential to actively impact service interactions and operate as surrogates for service professionals by accomplishing tasks that would otherwise be completed by humans. Instead of calling a call centre or sending an email, customers can use Chatbots to ask inquiries or make complaints. Chatbots are available seven days a week, 24 hours a day. This self-service channel will become more essential (i.e., intelligent assistants working as a service interface) rather than human-driven as the relationship between organisations and consumers "gradually transforms to become technology dominant" [14].

A range of companies can benefit from artificial intelligence chat possibilities. Companies like Pizza Hut, Dominos, Nike, Spotify, and others are using chatbots to engage customers and boost their marketing campaigns. Because of its ability to learn from human interactions, AI Chatbots are a critical tool for businesses to discover what their customers want and how they react to their products or services. These Chatbots also gather and analyse data in order to assist the firm in improving its marketing and communication strategy. Chatbots can be used by product and service firms to increase the likelihood of a transaction. A bot can imitate human communication in a cost-effective manner, establishing confidence between the buyer and the store. This technology can be simply integrated into any website management strategy, encouraging customers to stay longer on the site. Chatbots have become more common in customer service as shops rely more on mobile solutions and tailored messaging. Calls and emails can be inconvenient, and long wait times may deter potential consumers. Because of chatbots, these wait times are no longer a concern.

3   Chatbot for Businesses Growth

Chatbots could be a very useful tool for developing strong client relationships. Your organisation may be able to create deep bonds with website users by engaging and talking with them. By incorporating chatbots, you can achieve marketing goals, increase sales, and improve customer service. In a variety of ways, a chatbot can help customers connect more efficiently and promote business growth [15].

3.1 Increased Customer Engagement

User data can be used to drive engagement, which can then be amplified through the usage of conversational AI chatbots. Furthermore, bots may deliver consistent responses, allowing you to avoid providing irrelevant information to your consumers. Customers will stay on your website longer and prolong the conversation if you respond to them in a timely and appropriate manner. Now that we know more about conversational AI and why it's so important for brands, let's talk about how to leverage it to boost consumer engagement:

- Conversational AI can assist you in better understanding the wants, requests, and preferences of your consumers. Conversational AI can assist you remember and retain knowledge that your clients have previously told you about their concerns. According to a Hubspot poll, customers don't want to repeat information they've already provided. Adding an artificial intelligence layer to your existing chatbot interface will help you improve your response rate, which will benefit your customer service team.
- According to Forbes, 71% of customers desire a consistent experience across all media. It's a reasonable request. It's critical to stay up to date and be where your customers are, especially as the world becomes more distant and the variety of channels to engage with customers grows. While WhatsApp and SMS conversation have the highest open rates, all businesses, large and small, must use other channels and make the most of them. Brands are swiftly catching up to Facebook's recently unveiled Messenger API for Instagram.
- Accessibility and receptivity 82% of your clients want a speedier service seven days a week, 24 hours a day, according to HubSpot research. Having rehearsed responses on hand is essential. Customers should be able to contact your company at any hour of the day or night, and prompt and accurate responses can make or break your customer service. Your agents and automated chatbots will benefit from pre-filled answers and comprehension of multiple intentions. Make a concerted effort to respond as soon as possible. If you can't, include a disclaimer and a time estimate for when you'll respond to the consumer.
- According to a Sales force study, 52% of customers want customisation. Remembering names, consumer preferences, important dates, and other facts will help brands a lot. Brands can engage with customers who are more aware of their requirements, likes, and dislikes thanks to conversational AI. Marketing departments, in particular, can benefit from this by focusing on offering engaging and relevant experiences to each customer/lead through social media, email marketing, and other channels.
- Customers are more engaged when self-service is rapid and intuitive. Organisations must understand and align their customer service department with their customers. Self-service is expected by 73% of respondents. Have a system in place to allow your customers to self-serve. Instead of expecting the consumer to fill out all of their questions, provide a resolution at the touch of a button. Create a conversational

AI platform that allows customers to be automatically directed to support agents, resulting in a more seamless customer experience.

Chatbots lack the ability to be human. They can, however, think and respond to queries in a human-like manner. Text is gradually displacing voice as the most preferred means of communication. Thanks to advances in artificial intelligence and natural language processing, machines can now understand and respond to questions in the same manner that humans can. More importantly, machines are catching on and developing their capacity to handle ever-more complicated interactions. It's exactly what we do as individuals, and it's elevating customer engagement to new heights.

Customer communication must be timely, whether for sales, marketing, or support. Customers will never return to your business if they had a negative experience. AI-powered chatbots might really shine in this situation, as they can automate customer communication and considerably improve support. Customers want more firms to use chatbots to improve their communication strategy and create a better customer experience, according to 35% of customers. The need to be available at all times is at the heart of chatbots' growing popularity in a variety of industries. You'll also need to use a bot at some point if you want to keep clients engaged around the clock and improve their experience. This will help with consumer demand management by giving quick responses and enhancing customer satisfaction.

3.2 Improve Lead Generation

Lead generation is a top priority for most businesses since without leads, there would be no sales and no revenue. Bots have progressed to the point where they can communicate with customers in a personalised way throughout their journey. In reality, chatbots can help your business create leads by supporting customers in making quick judgments. A pre-programmed quiz can be used by AI bots to attract users to complete a form and create leads, resulting in greater conversion rates.

4 Customer Service with Automation

Customer support solutions are gaining popularity as consumers become more aware of the wide range of applications for which they may be employed. Specifically, they want to know how they can improve customer service [16].

The following are some examples of how customer service automation makes things easier:

- Client data such as page visits, previous transactions, and other information can be used to provide personalisation throughout the customer experience. As a result, you can better understand the consumer and make better recommendations.

- Many customers believe that a rapid resolution time equates to a positive customer experience. As a result of automation, agents can obtain real-time information to serve clients as rapidly as possible.
- Innovative technology, like CSA, is available around the clock, 365 days a year. It can assist in providing high-quality services for a fraction of the cost. Because excess staffing and outsourcing support to contact centers are additional overhead costs that can be avoided, they should be avoided wherever possible.
- Provide proactive help by gaining a knowledge of the customer's needs and what they're searching for before they contact customer with a question or problem. This has the potential to drastically cut cart abandonment rates increasing customer happiness.
- It enables organizations to scale quickly by including many different functions into the support platform. Incorporating the Arabic language into your AI bots, for example, will assist you in expanding your consumer base and, consequently, in growing your business.

5 Usage of AI Chatbot by Brands

Emirates Vacations is a luxury travel company that debuted in 2018. The organisation has seen an 87% boost in user interaction since introducing a chatbot into its website [17]. Another example is Kia, a vehicle manufacturer. The chatbots on their site serves as an informative portal for users, and it has a conversion rate of 21% [17]. In addition, the bot created three times the amount of weekly user engagement when compared to comparable business websites. Nykaa, on the other hand, has a live chat option on its website, the company was able to engage with 99.7% of all its users in less than a minute, according to the company [18]. That is a significant sum of money! Live chat is used by AbhiBus, a bus ticket booking website, to automate the process of responding customer questions. The advanced feature resulted in a 33% growth in agent productivity and a managing time of under a minute.

5.1 *Emirates Vacations Introduces a Chatbot to Their Banner Advertisements*

Emirates Vacations, the airline's trip operator, is employing artificial intelligence to develop a chatbot that can be accessed within display ads, as shown in Fig. 1. People can utilise the ads to ask inquiries about travel and excursions, and they'll get quick answers from within the ad unit. Artificial intelligence, on the other hand, may have a significant impact on another application, according to the company: reducing additional friction points in search. Based on the context of users' questions, information on the website where the chat bot is placed, and Emirates Vacations' inventory, the

chat bot will recommend sites and vacation packages. If a hotel in Toronto is not available through Emirates Vacations, the chat bot will not recommend one. "Explore the world without ever having to leave the page," stated the advertisement.

The company's goal, as according Ailsa Pollard, senior vice president of Emirates Vacations, who collaborated on the project with Way Blazer, was to "look at the customer journey and remove as many friction places as possible." The commercials were sorely tested in a 30-day campaign by Emirates Vacations, and the outcomes were positive. Advertisements in major cities across the United States on sites like the New Yorker, Lonely Planet, Time, and Smithsonian.

When compared to standard click-through advertisements, Emirates Vacations saw an 87% boost in interaction. Emirates Vacations tested the new chat ad type with 550,000 impressions and compare results to the company's previous display ad format with the same number of impressions. Emirates Vacations provided no interaction statistics for its original display advertising, only stating that the improved version would be utilised in the future.

The Maldives, Bangkok, Milan, and the Seychelles are the four destinations presently advertised in chat rooms. For example, if someone types in a general question about where they might go on vacation, the chat advertisement will recommend vacation packages in these places. In addition, the ads are strategically positioned beside stories that mention these websites. FEmirates Vacations positioned one of their chat advertisements next to an item on People.com about singer Jordin Sparks visiting the Maldives, which grabbed the attention of potential customers. According to Pollard, "conversation advertising breathes new life into traditional display ads." Traditional display adverts have led to consumer annoyance with online advertising for years, and many companies are looking into ways to provide more value to customers. Travel brands, in particular, Pollard says, must pay great attention to the user experience.

According to Digiday, after implementing a chatbot on Facebook Messenger to aid automobile buyers in finding solutions to their questions, Kia experienced an increase in sales conversion rates.

5.2 Domino's

Domino's provides a robust online ordering experience to its guests through its website chatbot. Three tabs allow you to place orders, track them, and reorder them. The user-friendly UI makes ordering pizza a breeze. The chatbot can answer the majority of questions, relieving customer service representatives of their duties. They can be more productive and focus on the most critical things as a result.

5.3 KIAN Chatbot

Kia's newest AI-powered assistant provides a full purchase experience via Facebook Messenger, including product information, local inventories, current pricing offers, and even competitor vehicle comparisons. By developing a single digital specialist who can provide one-on-one support, Kia builds on NiroBot's success.

Kia built a holistic chatbot (shown in Fig. 2) experience for its whole vehicle model line-up in collaboration with CarLabs, a marketing-tech and conversational commerce platform supplier, and its digital agency Ansible. Kia developed a comprehensive chatbot experience for its whole automobile model lineup in collaboration with CarLabs, a marketing-tech and conversational commerce platform supplier, and its digital agency Ansible. Kian will also answer a few simple questions in order to assist buyers in finding the right partner. Customers can also use Kian's superior natural language processing capacity to ask specific questions on a variety of topics.

Fig. 2 Kia Motors America
Introduces AI-Powered
Virtual Assistant *Source*:
https://www.kiamedia.com/

All of this can be done without ever leaving Facebook Messenger, allowing shoppers to stay in touch with their friends and family.

Kian offers a mobile-first buying experience that resembles a discussion with a human automotive specialist, and it's smarter because it evaluates millions of vehicle data points in real time, merging and organising them in a data stack that serves as Kian's brain. Every auto shopper will now experience as if they are conversing with a virtual Kia assistance about Kia's world-class vehicles. Kian is the company's follow-up to the well-known NiroBot, which was created to make automobile shopping and purchasing more convenient and pleasant.

5.4   Kiehl's

Kiehl's, a 116-year-old cosmetics brand, just launched its first Facebook Messenger bot. The bot offers a suggestion engine, searches categorised by product category and skin disorders, customer assistance, and other features to its consumers. After making their selections, users can buy the items they've picked on the spot.

The bot's personalisation aspect, like that of other bots, is crucial, especially in the cosmetics business and skin care. For inexperienced customers, Kiehl's has a huge assortment of goods to choose from, which might be intimidating. The bot improves the entire experience by simplifying the choosing process. A unique feature of the bot is that it allows users to designate their gender as "other" instead of one of the usual gender identities. One of the very few brands that knows this is Kiehl's.

5.5   The Wall Street Journal

The Wall Street Journal chatbot for Facebook Messenger makes it simple to stay up to date on the latest big news and financial quotes. You can also make your own personalised alerts by entering a few simple commands. You'll get business information, critical financial information, current stock quotes, and the most recent news updates all in one location.

5.6   Hazel

Hazel is a well-known programme that helps people manage their work environments. It uses a website chatbot to collect data and give important workplace management insights. To increase efficiency, the website chatbot recognises pain points and provides customised growth strategies.

5.7 Spotify

Spotify is a digital music platform claims to have millions of songs available to subscribers. Customers can use their chatbots to walk them through the process of making a playlist direct from Facebook. After probing you for more information about your musical preferences, the smart robot proposes songs for your playlist depending on your activity, interests, preferred genre, and mood. Thanks to Spotify chatbots, you can now browse and search for your favourite music using Facebook's ubiquitous messaging app. Users may now search for and exchange music through a Facebook Messenger chatbot designed by the firm. The system will make use of Facebook chat extensions to provide music recommendations, search features, and sharing capabilities. Users can now send 30-second audio snippets to their Facebook friends using the chatbot. You may either listen to the audio excerpt within the application or go to Spotify's application to hear the full song.

You may also utilise the chatbot to search Spotify's music library. You can share albums, songs, and playlists from the library with your online chat buddies. Previously, users could send Spotify links to their Messenger conversations that featured an album or song title and a thumbnail. On the other side, the new 30-second clip features offer a totally different experience. The chatbot can be launched by hitting the blue '+' icon on the left side of the Messenger chat. A new window emerges when you press on Spotify, enabling you to browse for songs to share to your friends.

5.8 Staples

"The Staples Easy System delivers the 'on-demand' world to companies, allowing customers to order anytime, anywhere, via any device they like," said Ryan Bartley, director of mobile and applied innovation at Staples. "IBM Watson will power the app, which will use bots to answer simple questions like 'I need to return something.'" "IBM Watson's speech-to-text skills will also be used in the app, allowing users to speak their order or question to their phone." The Easy System also incorporates the 'Scan My List' feature, which allows consumers to scan their lists and produce a Staples.com order by taking a picture with their phone."

5.9 Uber

The Uber chatbot allows consumers to sign up for Uber with a single click. Instead of installing the Uber app, customers can book a ride using Uber chatbot on Messenger. The chatbot is powered by Uber's API. To get receipt and updates on journeys, you can watch a private message with Uber on Messenger. This tool makes keeping track of payments and ride history a breeze. The Uber Facebook Messenger chatbot includes the following features: The ability to share facts about the current Uber trip with your friends via Messenger. Send your friend a location on Messenger, and he or she can book a ride to that destination by just tapping on the address. In a Messenger group chat, you may also request an Uber. This notifies your friends when you'll be arriving at a particular spot [19].

The whole process of getting an Uber trip becomes as simple as sending a message when you use Messenger's features to browse, request, and pay for your transport. Slack and Telegram have recently been added to the chatbot service.

5.10 Starbucks

Starbucks is one of the world's most popular coffee brands. The popular coffee shop uses an intelligent chatbot to boost customer happiness and make buying and customising easier. Customers' preferences can be learned over time by the online chatbot, which can then assist them in rapidly ordering their favourite drink. It also informs customers when their order is ready, allowing them to enjoy their beverage without having to wait in line.

5.11 Nike

The Nike chatbot always have something for you on Facebook Messenger. We've got you covered whether you need new office equipment or new gym gear for your next summer getaway. Nike's innovative new chatbot, Stylebot, is designed to help customers style outfits or create magic. Users can either design their own sneakers or select from a variety of models. For each genre, there are shoes for the sporty girl, the working woman, and a range of other types. You may discover anything about any outfit you like on Nike's website. There are lots more to select from if you don't like one. The women's-specific style choice simplifies and expedites searching across Nike's collection.

The bot's ability to create personalised sneakers, on the other hand, remains the most exciting feature. It only requires you to submit a photo of your favourite pair of sneakers and choose a colour scheme from the NIKEiD 24-color palette. This feature is perceived by users to be more unisex than the style feature. The chatbot matches any image with Nike Air Max 90, whether that's a painting, scenery, or other image. This option in the Nike chatbot is an excellent example of product personalization based on the customer's preferences. Customers try out a variety of options on their own to see the ones they like most. After that, they'll be able to place orders for the shoes.

6 Conclusion

In today's corporate world, chatbots have become a need. Chatbots are a one-time investment that saves time and money while also enhancing customer connection. Chatbots, as we've seen in this post, can assist a company with a variety of tasks. From delivering information to making recommendations and processing orders, chatbots can handle practically every aspect of customer service and marketing for a business. Brands can provide seamless customer service by combining AI technology with a human touch. Customers can self-serve while also contacting a human agent when needed by combining chatbots and live chat technologies, for example. By using AI-enabled chatbots as the main interface, your company can deliver real-time support, reply to inquiries rapidly, and boost customer satisfaction. However, there may be situations when the bot is unable to discern the user's purpose and must rely on a human to perform the operation. These chatbots can also be incorporated into Facebook chats for easier access and conversion. Take, for instance, Siri and Google Assistant, which are both automated chatbots. On the other hand, these chatbots recently been more readily available (and affordable).

References

1. Mariappan, J., Edassery, J.R., Krishnan, C., Odamkulath, A.P.: Social media-as a pedagogical tool for universities. Amity Glob. Bus. Rev. **32**, 1 (2019)
2. Hill, J., Ford, W.R., Farreras, I.G.: Real conversations with artificial intelligence: a comparison between human-human online conversations and human-chatbot conversations. Comput. Hum. Behav. **49**, 245–250 (2015)
3. Mero, J.: The effects of two-way communication and chat service usage on consumer attitudes in the e-commerce retailing sector. Electron. Mark. **28**(2), 205–217 (2018)
4. Krishnan, C., Baba, M.M., Singh, G., Mariappan, J.: Viral marketing: a new horizon and emerging challenges. Princ. Soc. Netw. New Horiz. Emerg. Chall. **246**, 161 (2021)
5. Dale, R.: The return of the chatbots. Nat. Lang. Eng. **22**(5), 811–817 (2016)
6. Nordheim, C.B., Følstad, A., Bjørkli, C.A.: An initial model of trust in chatbots for customer service- findings from a questionnaire study. Interact. Comput. **31**, 317–335 (2019)
7. Adam, M., Wessel, M., Benlian, A.: AI-based chatbots in customer service and their effects on user compliance. Electron. Mark. **31**(2), 427–445 (2021)
8. Meuter, M.L., Bitner, M.J., Ostrom, A.L., Brown, S.W.: Choosing among alternative service delivery modes: an investigation of customer trial of self-service technologies. J. Mark. **69**(2), 61–83 (2005)
9. De Bellis, E., Johar, G.V.: Autonomous shopping systems: identifying and overcoming barriers to consumer adoption. J. Retail. **96**(1), 74–87 (2020)
10. Feine, J., Gnewuch, U., Morana, S., Maedche, A.: A taxonomy of social cues for conversational agents. Int. J. Hum. Comput. Stud. (2019)
11. Wambsganss, T., Winkler, R., Schmid, P., Söllner, M.: Unleashing the potential of conversational agents for course evaluations: empirical insights from a comparison with web surveys. Twenty-Eighth Eur. Conf. Inf. Syst. 1–18 (2020)
12. Scherer, A., Wünderlich, N.V., Von Wangenheim, F.: The value of self-service. MIS Q. **39**(1), 177–200 (2015)

13. Techlabs, M.: Can chatbots help reduce customer service costs by 30? (2017). https://chatbotsmagazine.com/how-with-thehelp-of-chatbots-customer-service-costs-could-be-reduced-up-to30-b9266a369945

14. Pelau, C., Dabija, D.C., Ene, I.: What makes an AI device human-like? The role of inter-action quality, empathy and perceived psychological anthropomorphic characteristics in the acceptance of artificial intelligence in the service industry. Comput. Hum. Behav. **122**, 106855 (2021)

15. Krishnan, C., Goel, R., Garg, V.: Role of artificial intelligence and its impact on the tourism industry of India. In: Garg, V., Goel, R. (eds.) Handbook of Research on Innovative Management Using AI in Industry 5.0, pp. 291–302. IGI Global (2022). https://doi.org/10.4018/978-1-7998-8497-2.ch021

16. Ayyagari, N.: Customer Support Automation and Its Importance (2020). https://verloop.io/blog/website-chatbot-examples/

17. https://www.marketingdive.com/news/emirates-vacations-digital-display-ads-integrate-chatbots-that-give-travel/518243/

18. https://verloop.io/case-studies/nykaa/

19. Kim, L.: 9 Great Examples of How Brands are Using Chatbots (2018). https://www.socialmediatoday.com/news/9-great-examples-of-how-brands-are-using-chatbots/524138/

Social Media Security Analysis

Do Not 'Fake It Till You Make It'! Synopsis of Trending Fake News Detection Methodologies Using Deep Learning

Rishabh Misra and Jigyasa Grover

Abstract The modern bloom of social media has propelled a new pattern of information propagation termed *push journalism*, where a certain piece of news is shoved in the faces of as many people as possible with a sliver of hope that it will reach the people who need that information the most. This form of news reporting, especially via social media campaigns has boosted the access and fabrication of bogus reporting, or what is referred to as *fake news*. Fake news, in the form of clickbait, hoax, satire, propaganda, hyperpartisan, deepfakes, or simply unreliable news has the power of influencing its readers to a dangerous extent, predominantly causing political, socio-economic, or psychological harm. In this chapter, we analyze the meaning of fake news in the world of social media, the various forms it can take, what causes its spread, and what are the rudimentary signs of such fake news. We will walk through a comparative study of the state-of-the-art deep learning models to approach the tasks of identifying phony information, verifying the validity of various claims and facts, catching fake content, and so on. The exposition will especially elucidate the adversarial approaches in deep learning to detect counterfeit content that could come in any form like text, images, videos, or audio. In doing so, we establish the importance of generating plausible and understandable explanations for model predictions with a special emphasis on algorithm fairness. With the fact that deep learning methods rely on comparatively larger datasets of top-notch quality, this chapter will also highlight the availability of relevant datasets in this space, as well as share pointers to curate one if needed. Even with sufficient data, however, detection problems in this domain are especially challenging since spammers and fake content generators are working tirelessly to evolve their strategies in parallel to the advancement in detection mechanisms. We will further shed some light on some recent and upcoming trends from the aspect of fake news contributors, and critically evaluate how our current state-of-the-art deep learning techniques fare against those.

Both the authors contributed equally.

R. Misra (✉) · J. Grover
Twitter, Inc., San Francisco, CA, USA
e-mail: rmisra@twitter.com

J. Grover
e-mail: jgrover@twitter.com

© The Author(s), under exclusive license to Springer Nature Switzerland AG 2022
T.-P. Hong et al. (eds.), *Deep Learning for Social Media Data Analytics*,
Studies in Big Data 113, https://doi.org/10.1007/978-3-031-10869-3_12

In closing, we will leave readers with some thoughts on future directions for the development of better and smarter fake news detectors.

1 Introduction

The emergence and accelerating growth of social media has altered the human interaction style in tremendous ways. Boosted with the advent of high-speed internet and smart devices, real-time communication has reached a new pedestal involving more and more people every second. Along with its impact on breaking physical barriers for speedy communication, amplifying commerce via promotions and advertisements, enhancing entertainment options, augmenting educational and professional growth, and transforming many more vital applications it has had a crucial effect on the journalism domain. In partnership with social media, a new pattern termed *push journalism* has been garnering interest wherein information tidbits are deliberately propagated to the entirety of the population with an anticipation that it would ultimately converge on the targeted audience.

Oftentimes, in this process, bogus reporting takes over and misinformation or false information is fabricated and disseminated to the masses. Owing to social media, this proliferation is as effortless and rapid as it can get. The inaccurate information that is specifically concocted with an intent to manipulate or deceive people online is called *fake news*. Fake news is a misleading piece of information that is falsely constructed with no relevance to reality, containing no verifiable facts and no reference to credible sources or quotes. It has the potential to influence the readers, and cause reputational damage, thus causing political, socio-economic, and psychological harm. Though fake news is a very subjective and sensitive topic, it should not be confused with pieces of information that do not align with our views.

Fake news can be manifested in various forms, be it a rumor, clickbait, hoax, satire, propaganda, hyperpartisan, or the modern-day deepfakes. Any piece of unreliable or uncertain news published with no verified sources or fact-checking is called a *rumor*. A *clickbait* is an eye-catching, sensationalized piece of content, like an image preview, or a quirky title, created with an aim to garner attention and lure people to a particular web page with little relevance or no meaningful content. *Hoax* is a falsely curated piece of information purposefully made to pass it off as the truth and is different from jokes due to their evil intention. *Satire* or *parody* is exaggerated, ironical content mostly created with an aim to humor people or take a dig at realistic situations, however, some uninformed readers might take it as gospel. Biased or misleading news fundamentally generated to manipulate people and further an agenda, for instance, to promote political ideologies, is termed as *propaganda*. On similar lines, content be it political or socio-economic claiming to be unbiased yet reeking of partial vibes is called *hyperpartisan*. And in contemporary times, we have synthetically generated media, images, or videos, which might or might not bear resemblance to an existing human being called *deepfakes*. This type of fake manipulative audio-visual content is created using deep learning, hence the name,

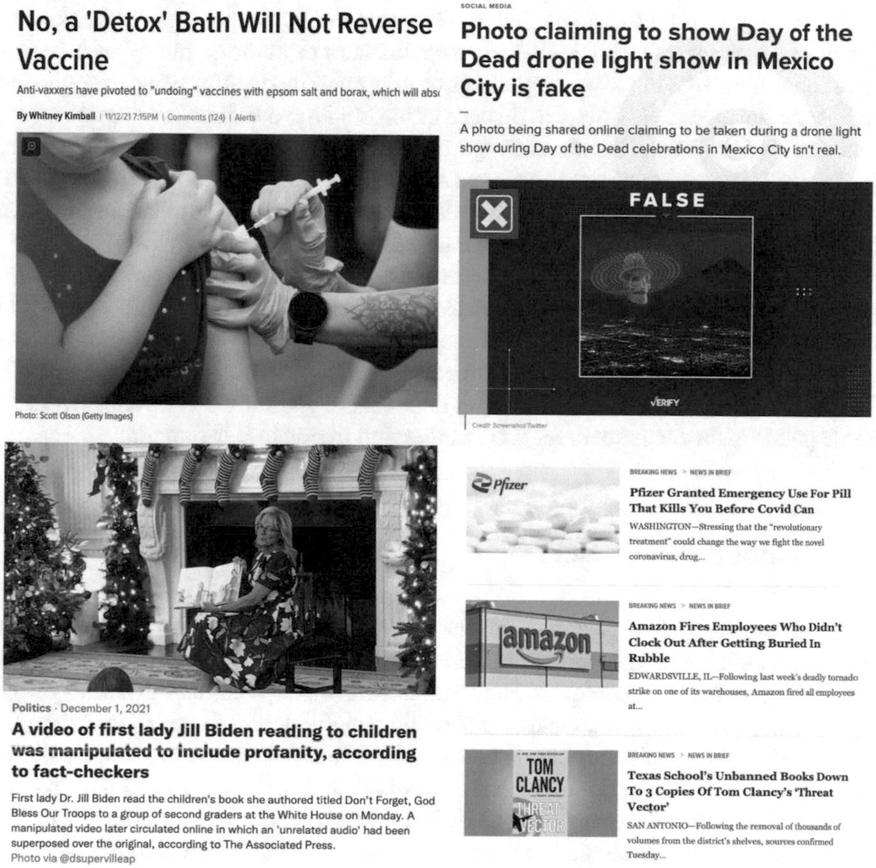

Fig. 1 Snapshots from social media exhibiting misinformation, deepfake, and satire

and has the potential to create major havoc. These are just a few examples of the most common forms of fake news and are generally created by unprofessional journalists, people wanting to make money regardless of the content they push out, satirists who want to entertain, or partisans wanting to influence people (Fig. 1).

With social media becoming ubiquitous, it has become one of the major dispersing platforms for fake news. Since the news on the platform is scarcely targeted and is very subjective in nature, it is ingested by readers that appeals to their emotions. This leads to a manipulated reality for many people, especially the ones who have a hard time distinguishing it from the real news, thereby intensifying societal conflicts. It not only causes mistrust amongst the people, instigates violence but also distracts from the important issues which often are left unresolved due to this. Hence, it is very important to create a robust mechanism to detect and filter these misleading pieces of content from the real news.

Telltale signs of fake content are lack or misquoting of original sources, unknown (or in some cases imitating well-known) publishers or authors, phony websites or apps, poor language in terms of spellings or grammar, incoherent story, exaggerated image previews, etc. However, in this age of information overload, it is cumbersome and almost impossible to manually go over each piece of content and scrutinize it. This is where the power of deep learning comes into play, to swiftly and accurately identify the counterfeit from factual content with minimal human intervention thereby mitigating the spread of inaccurate information.

In this chapter, we first formally define the problem and explore various contemporary deep learning techniques to handle fake news with due categorization into misinformation, clickbait, satire, and deepfakes. We further discuss the limitations of the deep learning methods along with a few insights into the fairness, interpretability, and accountability of the fake news detection algorithms. In the end, we leave some pointers for the readers regarding emerging trends in this domain.

2 Formal Problem Definition

Now that we are well-versed with the characterization and types of fake news doing rounds on social media, let us explore some ways to detect this as early as possible by plying deep learning techniques. However, before we explore the intricacies of the multiple approaches one can take, we should formally define the problem at hand.

Consider a piece of information, or a news article on social media, denoted by N. It is composed of two major elements, the *publisher details*, say P, and the *content*, say C_N. The *publisher details* comprise the domain where the information or the news article is published, author's name, profile, etc., and other publisher-related attributes. The *content* component is the actual piece of news involving the headline, text, images, videos, tags, and so on. Furthermore, we can signify the *engagements* over N on social media via E_N where E is a set of tuples $\{e_{it}\}$ denoting interaction of a user u_i with the given news article at a given time t and further propagating it via their post p_i. In that case, e_{it} can be represented as a unique combination of (u_i, p_i, t).

Hence, the problem can be formally defined as, given a news article N and all the related attributes like P_N, C_N, and E_N, the task is to predict whether the news article N is a fake news piece or not, i.e. $F : (P, C, E) \rightarrow \{0, 1\}$ such that, $F(n) = 1$, if n is a fake news article and $F(n) = 0$, if it is a genuine news article, where F is the prediction function we aim to learn.

That being, fake news detection is a straightforward case of binary classification problem in the machine learning world where the outcome of the prediction function is either 0, or 1. Or to say, the given piece of news can either be genuine or fake. Shu et al. [1] define fake news detection as a *distortion bias* based on previous research media bias theory which is usually modeled as a binary classification problem since fake news is nothing but a distortion bias introduced by the publisher in a world of genuinity.

3 Deep Learning Techniques for Fake News Detection

Being an emerging area of research, fake news detection is garnering a lot of traction and there is a continuous development of new tools and technologies to combat this social evil of bogus reporting. Past research and pragmatic studies have shown that deep learning techniques have been quite successful in creating a predictive model to identify fake news with high rates of accuracy. Let us walk through a few different approaches, with an attempt to categorize them according to the types of fake news they help tackle.

3.1 Misinformation

Social media is empowering novel forms of communication for global reach and along the way unleashing innovative journalism strategies. This in turn has accelerated the dispersion of false and inaccurate content, also referred to as *misinformation*. These fallacious pieces of news might be shared inadvertently due to lack of awareness or simply if the fabricated content is too convincing. However, if the same phony news is created and shared with malicious intent, it gets termed as *disinformation*. Misinformation online is a pressing public issue in political, socio-economical, and many other domains and is known to cause real-world consequences. Personalized ranking algorithms are further aggravating the issue by promoting this sort of misleading, sensational, and conspiratorial content. With tremendous amounts of data being churned every second, it is important to move on from human fact-checking mechanisms to automated artificially intelligent mechanisms to detect and adjourn the spread of misinformation online.

Advancing from knowledge-based detection mechanisms, style-based detection which analyzes the content of the news article is making headway. Zhou et al. [2] describe knowledge-based detection methods like the ones that flag fake news by cross-checking the knowledge dissipated in the given news article with facts. Whereas, style-based detection methods are the ones that focus on how the content is actually written. Here, we emphasize style-based detection methods since they assess the intention to spread misinformation.

Fusion of Neural Networks A benchmark study by Khan et al. [3] points to the superior performance of deep learning models over traditional machine learning models in detecting fake news, albeit requiring large-sized datasets. In that direction, Wang et al. [4] contributed the largest public dataset at the time, *LIAR*, which has politicians' statements along with the label of whether they were genuine or misinformation. In their work, they also propose a hybrid neural network framework to integrate both text and speaker metadata. The architecture has a convolutional layer to capture speaker metadata with standard max pooling on the latent space followed by a bidirectional LSTM layer. These embeddings are then concatenated with the max pooled textual representation from another CNN and then fed to a fully con-

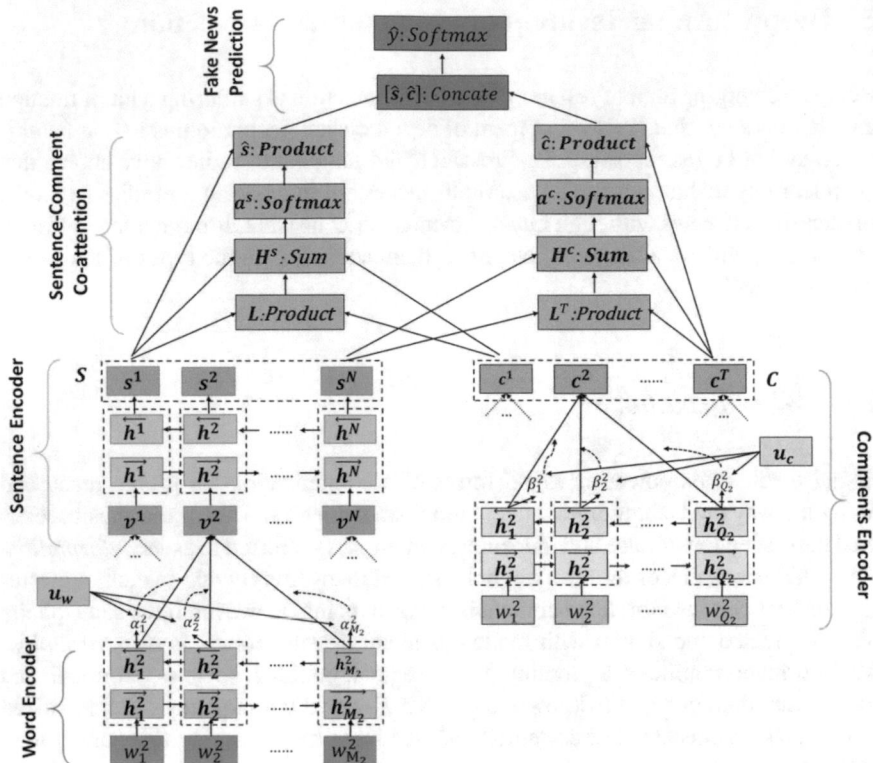

Fig. 2 dEFEND framework: news content encoder, user comment encoder, sentence-comment co-attention component, and fake news prediction component

nected layer with a softmax activation function to generate the final prediction. In another instance, Singhania et al. [5] propose a *Three-level Hierarchical Attention Network (3HAN)*, a level each for words, sentences, and the headline of a news article, thereby effectively representing the input news article, by processing the article in a hierarchical bottom-up manner. The experiment yields 96.77% accuracy when evaluated on a real-world dataset. The visualization of the attention layer helps with qualitative analysis and provides an insight that fake news articles use an inverted pyramid writing style (i.e. distributing information in decreasing importance).

Expanding Data Signals In an extensive linguistic analysis of fake news titled *'Truth of Varying Shades'*, Rashkin et al. [6] found that misinformation uses more first-person and second-person pronouns, more subjectives, superlatives, and modal adverbs, and less assertive words. Their work uses a simple LSTM model with pre-trained embeddings, however, the experiment demonstrates that crafted linguistic features provide a lot of value (Fig. 2).

Based on the learnings from the past works, Shu et al. [7] in their study *'dEFEND'* reason that the explainability of a fake news detector is critical, and how a user

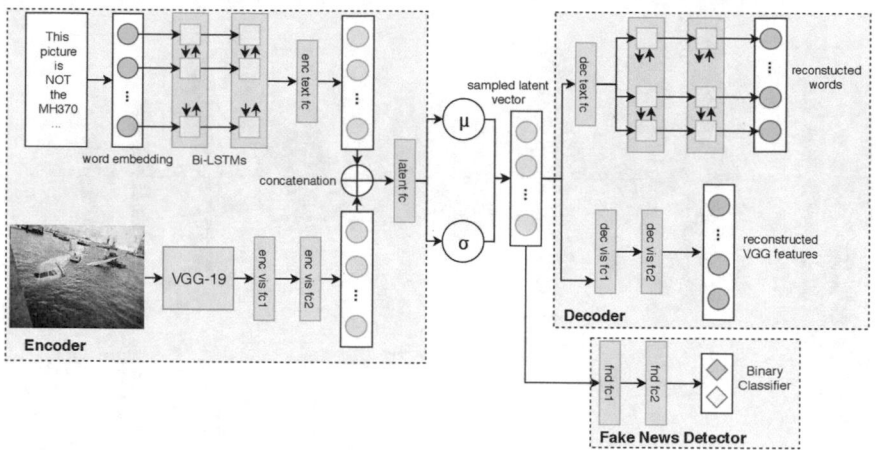

Fig. 3 Components of MVAE framework: encoder, decoder and fake news detector

comments on news articles on social media can be utilized to facilitate that. The authors propose having a news content encoder (at the sentence level) as well as a user comments encoder, which are then fed to a co-attention sub-network to exploit both news content and user comments (Fig. 2). The experiment setup indicates that this technique not only outperforms state-of-the-art fake news detection methods by at least 5.33% in F1 score but also (concurrently) identifies top-k user comments that explain why a news piece is fake (Fig. 3).

Additionally, Khattar et al. [8] further augment the data signals by using visual information along with textual information to detect misinformation. Their proposed framework, termed *Multimodal Variational Autoencoder (MVAE)*, uses a bi-modal variational auto-encoder to learn the representations from textual and visual data, which is then fed to a fully-connected feed-forward neural network for the task of fake news detection (Fig. 3). This setup is evaluated on datasets from *Weibo* and *Twitter* and the results show that across the two datasets, on average this model outperforms state-of-the-art methods by ∼6% in accuracy and ∼5% in F1 scores (Fig. 4).

Advancements with Language Models Presently, advanced pre-trained language models like BERT, ELECTRA, ELMo are receiving sizable attention for several natural language tasks including text classification, and rightly so, the misinformation detection domain has seen some work in this direction as well. Jwa et al. [9] propose using a BERT-based model to detect fake news that analyzes the relationship between the headline and the body of the news article. Since BERT is pre-trained on generic data, the authors incorporate news domain data for further fine-tuning and see a considerable improvement in the performance. The deep-contextualizing nature of BERT improves the F1 score by 0.14 over the previous state-of-the-art models. The *FakeBERT* model proposed by Kaliyar et al. [10] further suggests improvement by feeding BERT representations to three parallel blocks of 1d-CNN having different

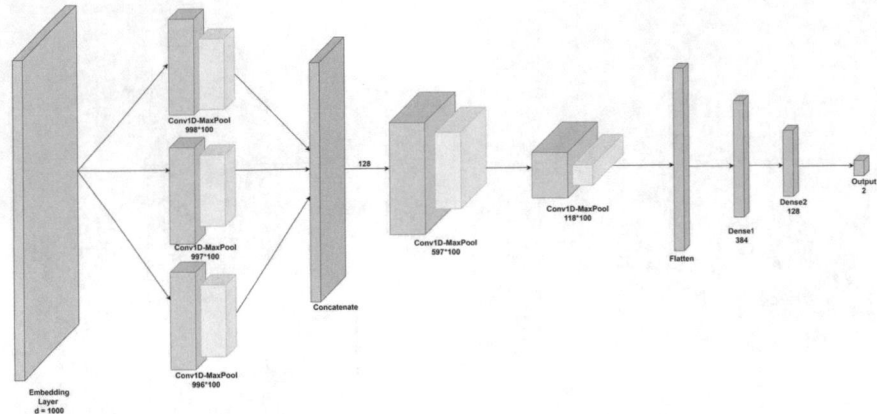

Fig. 4 FakeBERT architecture

kernel-sized convolutional layers with different filters for better feature extraction (Fig. 4). The authors illustrate an accuracy of 98.90% which is a 4% improvement over the baseline approaches and is a promising direction for the fake news detectors' development.

3.2 Clickbait

In social media, *clickbait* is a text or a link with exaggerated or eye-catching headlines that lure a reader to *'click'* on it. Clickbait is a nuisance in the online user experience since it exploits a reader's curiosity and lures them to poor quality or inaccurate information. It is one of the most common types of fake news and is often used for advertisements, where mass content generators make money on phony content using a click-based model to optimize it. Clickbait takes advantage of some vulnerable users and abuses the purpose of *user-generated content*. This form of deception is considered a fraudulent activity on social media and is frowned upon in the news reporting circles since it leads to obstruction of real news propagation. Hence it is important to combat this category of fake news by detecting and flagging clickbaits early on in the network.

Post 2016 US presidential elections, this domain is getting a lot of attention, as experts debated that clickbait headlines on social media and other fake news might have influenced the decision making. Some of the earlier works [12] characterize clickbait using certain linguistic cues like *Unresolved Pronouns, Forward Referencing, Backward Referencing*, and *Action Words*. However, since headlines contain limited information only, it becomes challenging to encode these features in the model. Forging ahead, we dive into a few deep learning techniques that take advantage of the knowledge from past research and have proven to effectively detect clickbait.

Utilizing Textual Data In the past, CNNs have demonstrated effectiveness in various sentence classification tasks hence it is instinctive to try them out for the clickbait detection task as well. Contrary to the traditional machine learning approaches, CNN obviates the need to curate meaningful features, which might or might not be as helpful. One of the early works in this domain uses CNN with pre-trained *word2vec* embeddings to extract meaningful information from the headlines of the news articles which is further used for prediction [13]. This model, termed *TextCNN*, performs remarkably well compared to its antecedent techniques, however, there is a huge room for improvement with respect to clickbait's domain.

In their attempt to optimize the TextCNN model for clickbait detection, Zheng et al. [14] recognize that different types of clickbait articles tend to use different ways to draw users' attention. For their approach, coined as *ClickbaitCNN*, the authors collect headlines from four famous Chinese news websites that fall into four article types: news, blogs, *BBSs*, and *WeChats*. They propose using a new word-embedding layer that takes both overall and the article type-related word meanings into consideration. They also propose a new loss function to regulate the influence of article type-related meaning of the word. It is found that employing these techniques improves the performance over TextCNN by 2–3% in terms of precision and recall.

Encoding Sequential Information Though CNN-based techniques have proven helpful for language-related tasks, such as this, they are limited by their nature since they can not leverage sequential information. The clickbait detection problem could benefit a lot from RNN especially if posed as a multi-class classification problem since they can encode sequential/contextual information well.

On these lines, Zhou et al. [15] propose a self-attentive RNN based model to infer the levels of importance of the text tokens in predicting clickbait. The data is sourced from *Twitter* and manually annotated as either *'not clickbaiting'*, *'slightly click-baiting'*, *'considerably clickbaiting'*, or *'heavily clickbaiting'*. *GloVe* embeddings are used to effectively learn from the representations and dropout regularization is applied to the outputs of the word embedding layer, on the outputs of the *bidirectional Gated Recurrent Unit (biGRU)* encoding layer, as well as on the outputs of the self-attentive layer. Experimental runs indicate an improvement over the ClickbaitCNN by 4% in terms of F1 score with very low computational cost (Fig. 5).

Dong et al. [16] push the state-of-the-art further by exploiting the relationship between the misleading titles and the content, which is found to give important clues for solving this problem. The authors propose a deep attentive similarity model to capture both global and local similarities of the pair of inputs (i.e. title and content) (Fig. 5). The representations of the textual input are obtained by using an attention-based biGRU model, similar to Zhou et al. [15]. The global similarity is learned via cosine similarity between the title and the content by minimizing it for mismatched pairs. The local similarity on the other hand is computed on pieces of text selected using block size and strides, again by minimizing the value for mismatched pairs. These similarities are concatenated with the latent representations and fed into a fully connected layer to produce the output. This approach leads to an improvement over Zhou et al.'s self-attentive RNN model by ~4% in terms of F1 score.

Linguistic Analysis of Headlines Furthermore, Naeem et al. [11] exhibit the success of modeling the intrinsic characteristics of clickbait for knowledge discovery and using it for decision making. The said knowledge discovery is done by performing a linguistic analysis of the news headlines using the *Part of Speech Analysis Module (POSAM)*. The idea is to understand the underlying structure and syntax and accordingly adjust the structure of the LSTM module for the conclusive classification (Fig. 6). POSAM is a variation of the original n-gram classifier based on *Part of Speech (POS)* tagger, which reveals a stark difference in the occurrences of *WH-Determiners* and *Personal Pronouns* amongst others in clickbait text. Moreover, the observation that POS tokens that create an information gap exist in the latter half of the sentence motivates the decision to have a loopback of five words that effectively double the weights of the second half. This framework is evaluated on a newly collected dataset from *Reddit* and produces an F1 score of 0.973 (Fig. 6).

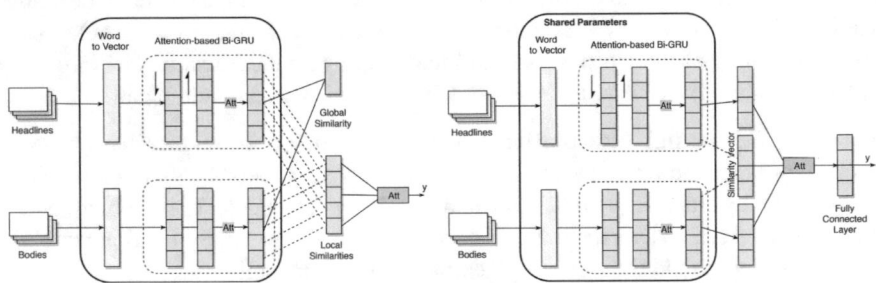

Fig. 5 L: Learning of the global and local similarities. R: Combined method for final prediction

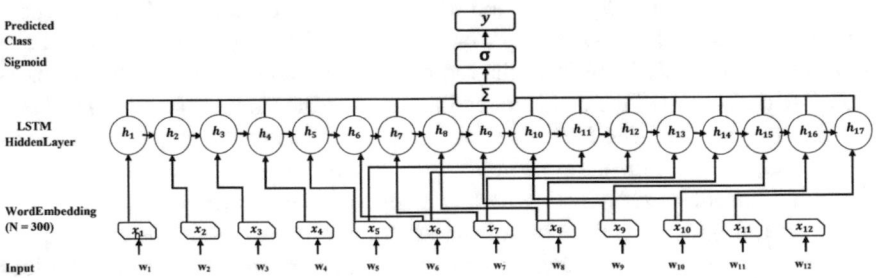

Fig. 6 LSTM using POSAM uses a single layer with a time stamp of 10 units and a loopback of half the words in an average headline to focus on the words that occur its last half

3.3 Satire

Satire is a very interesting genre that involves ridicule, humor, irony, or exaggeration and is primarily considered a form of entertainment. However, oftentimes readers might not be able to recognize it if the cues are too subtle to be identified or they lack relevant context. In cases where people perceive it as the truth, satire becomes a type of fake news thereby spreading inaccurate information and causing mistrust in society. As reported in the *Guardian*, it was found that regardless of the ridiculous content, people not only believe satirical content but also share it in their networks frequently.

The task of detecting satire in an article, which mimics genuine news, is understandably a binary classification problem. Past research in the domain of fake news might not directly apply to satire detection since it is a distinct domain with its unique traits. For instance, the works by Xiao et al. [17], Dong et al. [18], and Ge et al. [19] focus on tackling misinformation and fake news by tending to discover the truth through the knowledge base and truthfulness estimation, however since there is no ground truth for satire generally, these techniques might not work for this category of fake news. Owing to the similarity to the problem of deception detection, we can utilize its solutions involving analysis of psycholinguistic features [20], writing analysis [21], and cultural differences [22] to tackle the satire detection problem. Though it should be noted that these techniques consider features at the document-level however it is observed that satirical cues are found only in certain paragraphs thus indicating that document-level features might even be superfluous for this use case. In this section particularly, we will discuss a few deep learning techniques that leverage past research in devising clever frameworks to effectively address the problem at hand

Exploiting Lexical and User Signals One of the early works by Amir et al. [23] exploits the fact that the way satirical content is delivered depends heavily on the writer. Therefore, their CNN-based deep learning framework learns and uses the user embeddings in addition to the lexical signals without the need for manual feature engineering to tackle the problem effectively. The textual embeddings which take advantage of the lexical signals are produced by employing multiple CNNs with different filter sizes (to generate multiple feature maps) on pre-trained word embeddings. Whereas, the user embeddings encode latent aspects of users and capture homophily. Both these embeddings are concatenated to provide holistic contextual information and fed to a fully connected network, which ultimately produces the output. This approach was evaluated on a dataset derived from *Twitter*, collected using hashtag-based supervision, and demonstrated an edge over the state-of-the-art approaches (with a 2% accuracy increase) which leveraged an extensive set of carefully crafted features.

Improving Data and Model Quality Datasets collected from *Twitter* tend to have noisy labels, and language along with the fact that the tweets might not be self-contained, that is there is a trend of conversations in replies or the content being broken into threads. To address these shortcomings, Misra et al. [24] collect a high-quality dataset by leveraging two popular web sources: *HuffPost* (for the real news) and *The Onion* (for the satirical news). Furthermore, the authors improve the model quality by recognizing that RNNs work better in encoding sequential information, hence they propose a hybrid framework where textual embeddings are learned via both CNN and self-attentive bidirectional LSTM. Then embeddings from both methods are concatenated and ultimately fed to fully connected layers for final prediction. Apart from increasing classification accuracy by 5%, the work also visualizes the attention layer on various headlines for improved model interpretability. Qualitative results show that the network emphasizes the co-occurrence of incongruent word phrases within each sentence such as *'oppressing other people'* & *'insane k-pop sh*t during opening ceremony'*, which are important cues to detect satire for us, humans, too.

Utilizing Paragraph-level Linguistic Features Work by Yang et al. [25] showcases that satirical cues are often reflected in certain parts of the document, hence a hierarchical neural network with an attention mechanism to extract paragraph-level linguistic features is successful in tackling the problem of detection satire. They propose using four levels of features: *character-level features* that are extracted using CNN as they recognize morphological information and name entities well, *word-level features* are generated by applying bi-GRU on top of character-level representations, *paragraph-level features* are generated by applying bi-GRU on word representations concatenated with engineered linguistic features, and *document-level features* are generated using attention on paragraph-level representations concatenated with engineered linguistic features. The linguistic features include psycholinguistic, writing stylistic, readability, and structural features from the given text. The qualitative evaluation suggests readability features support the final classification while psycholinguistic, writing stylistic, and structural features are beneficial at the paragraph level. In addition, this exposition reveals that the writing of satirical news tends to be emotional and imaginative.

Source-agnostic Adversarial Training The approaches discussed so far are built upon corpora labeled automatically based on the source of the article. McHardy et al. [26] hypothesize that this encourages the models to learn characteristics from the publication sources, rather than characteristics of satire to some degree, e.g. it is satirical if from *The Onion* and it is real if from *HuffPost*, thus leading to poor generalization performance on unseen publication sources. The authors propose having an adversarial component to control for the confounding variable of the publication source. The framework consists of three parts: feature extractor, satire detector, and publication identifier (Fig. 7). The *feature extractor* takes pre-trained word embeddings and feeds them into bidirectional LSTM with attention. *Satire detector* and *publication identifier* have a softmax layer to output the prediction of respective tasks. Since the goal is to control for the confounding variable of publication sources, the training is done considering the publication identifier as an adversary, i.e. classifier's parameters are updated to optimize the publication identification while the parameters of the shared feature extractor are updated to fool the publication identifier. This technique was evaluated on a dataset collected from 4 genuine news and 11 satirical German web

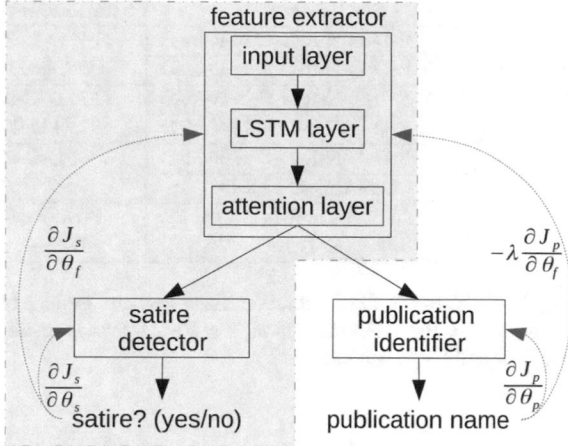

Fig. 7 Source-agnostic adversarial training model

sources, and the qualitative analysis shows that the adversarial component enables the model to pay attention to linguistic characteristics of satire (Fig. 7).

3.4 Deepfake

With the advancement in image processing and machine learning technology, it has become so much easier to generate images, audio, and video of situations that did not happen in reality, which is quite daunting. These synthetic audio-visual content, called *deepfakes* are leading to the propagation of misinformation on social media, for instance—celebrity pornography, tweaked videos of political leaders to induce conflict, hoax calls, and so on. It should be noted that the idea of faking content is not novel, it has existed for a while now via *'Photoshop'* and regular video editing, however, deepfakes are such a powerful convincing set of media that it is almost impossible to tell the fact from the fiction. The creation of these phony images and videos is done via deep learning technology using the sophisticated *Generative Adversarial Networks (GANs)*. *CycleGAN* by Zhu et al. [27] is a popular technique that uses GANs to generate a new image with the same characteristics as that of the input image. The key characteristic of CycleGAN is that it applies a cycle loss function that enables it to learn the latent features and performs an image-to-image translation without the need for a paired example, thus falling under the unsupervised learning paradigm. The success of this technique can be owed to a humongous amount of data available for training on the web these days in the form of images and videos, which are now a popular medium of content, thanks to social media networks like *TikTok* and *Instagram*.

Evidently, deepfakes have a lot of potential to create havoc in our society by seeding and boosting social conflict, fraud, and revenge in society. It is thus of utmost

ShallowNetV1	ShallowNetV2	ShallowNetV3
C-N-R-D-C-N-R-D-C-N-R-M-D	C-R-D-C-R	C-R-D-C-R
C-N-R-D-C-N-R-D-C-N-R-M-D	-D-C-R-D	-D-C-R-M-D
C-N-R-D-C-N-R-D-C-N-R-M-D	C-R-D-C-R	C-R-D-C-R
C-N-R-D-C-N-R-D-C-N-R-M-D	-D-C-R-D	-D-C-R-M-D
C-N-R-D-C-N-R-D-C-N-R-M-D	C-R-D-C-R	C-R-D-C-R
C-N-R-D-C-N-R-D	-D	-D
F-De-R-N	F-De-R-N	F-De-R-N
-D-De-S	-D-De-S	-D-De-S

Fig. 8 Variations of the ShallowNet. Each row represents a block in the architecture. (*Note* C = Conv2D, N = Batch Normalization, R = ReLU, D = Dropout, M = MaxPooling, F = Flatten, De = Dense and S = Sigmoid)

importance to construct ways to accurately detect and flag this set of counterfeit images and videos as early on in the propagation journey as possible. Ironically, deep learning has proven to be one of the more precise methodologies in performing this act of distinction between genuine and fake content online.

Deepfake Image Detection To distinguish phony images from real ones, it is no surprise that one of the successful techniques is the use of GANs.

Plying Deep CNN In one of the simpler approaches based on face recognition techniques, Nhu-Tai et al. [28] use a deep CNN to tackle the deepfake image detection problem. First, they normalize the faces to a frontal view followed by the deep feature extraction to derive and normalize facial features. Then in the face matching process to distinguish between the real and fake images, they create a suitable model to calculate the distance between the facial feature vectors for face identification and verification. This is topped with a fine-tuning mechanism by adjusting the weights of the classifier layer to calibrate the extracted features as per the forensics data provided by the *National Research Foundation of Korea (NRF)*. The proposed method performs decently well on the given forensic data with an 80% accuracy.

Boosting Performance with Ensemble Going the extra mile, Tariq et al. [29] design an ensemble model in *'GAN is a Friend or Foe?'* with three different shallow CNNs complete with L2 kernel regularizer of 0.0001, batch normalization, max pooling, and dropout which they refer to as the *ShallowNet*. Different variations of the ShallowNet are created with different layer settings mentioned in (Fig. 8). V2 and V3 variations are similar in depth, however, shallower than V1, which leads to lower training durations. However, the introduction of the max pooling layer in V3 yields better performance on lower resolution images, on which V1 was noticeably poor. The ShallowNet achieves ∼99.99% accuracy and outperforms the well-known GAN-generated image detection neural networks like the VGG16, XceptionNet, and NASNet in terms of accuracy and AU-ROC. The results also indicate that as the resolution of the deepfakes goes down, it becomes harder to detect them as the model performance dips a bit.

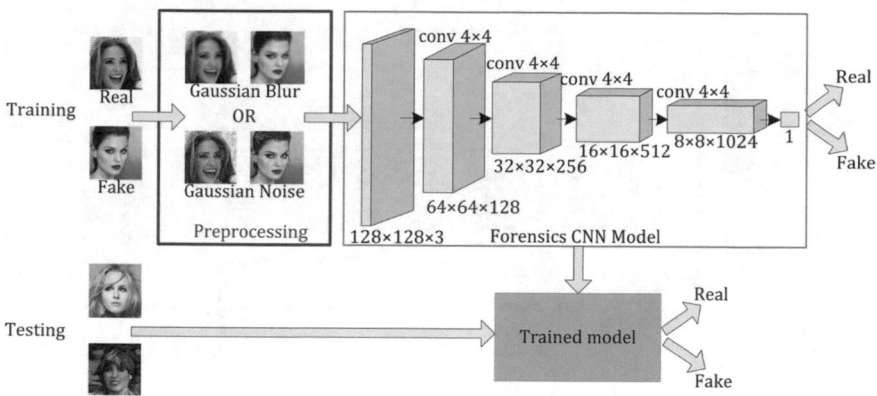

Fig. 9 Architecture of Deep Convolutional Generative Adversarial Network

Enhancing Model Generalization Ability using Image Preprocessing One of the major drawbacks of the above-stated techniques is their generalizability or to put it simply, the use of the same dataset for training and evaluating the model which is not well-suited in the real world scenario. To battle against the detection of an unseen variety of deepfakes and amplify the generalization ability of the deepfake detection models, Xuan et al. [30] propose a novel method of image preprocessing, namely *Gaussian Blur* and *Gaussian Noise*, in the training phase. Contrary to the generic methods, the idea is to destroy the unstable low-level high-frequency noise cues by adding Gaussian Blur and Gaussian Noise to the training images only to improve pixel-level statistical similarity between real images and fake images. This propels the model to learn more intrinsic and meaningful features, instead of simply learning the style of the fake image generating model. Instead of using a complex model, they use a fairly simple *Deep Convolutional Generative Adversarial Network (DCGAN)* as the discriminator (Fig. 9). The experimental results demonstrate the effectiveness of the proposed technique in enhancing the generalization ability although not by a lot owing to the inherent difficulty of this problem (Fig. 9).

Real-Fake Pairwise Learning So far, deepfake detection has been considered a binary classification problem, with an aim to compartmentalize a given image either as fake or genuine. Along with, previous studies have been focused on fake images which were partially reconstructed from a real image, i.e. those models can not be used to detect a full-fledged synthetic image that bears no resemblance to any being living or dead. Understandably, the latter is a hard problem since gathering a training set composed of only GAN-generated images is cumbersome. In other words, the supervised learning technique of learning from past GAN-generated images might not lead to generalizable models, since the detector will not be able to recognize an image that it might not have seen in the past during the training process. To overcome this problem, Hsu et al. [31] proposed a pairwise learning approach over a modified CNN, which they refer to as the *Common Fake Feature Network (CFFN)* (Fig. 10).

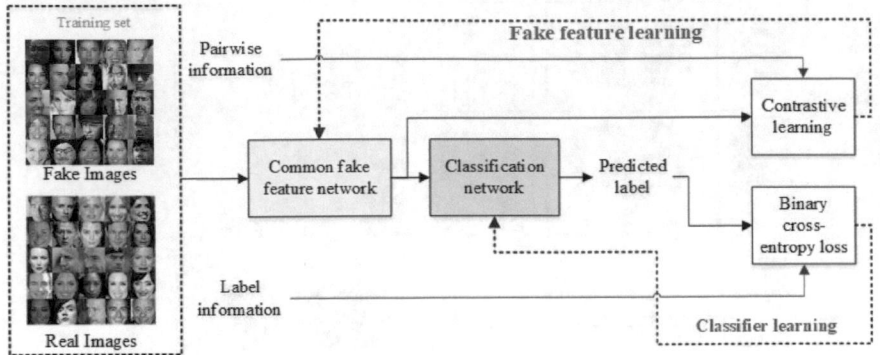

Fig. 10 CFFN based two-step learning approach

In their approach, the authors pair the fake and real images and use the pairwise information to construct the contrastive loss which helps the proposed CFFN to learn discriminative *common fake features (CFFs)*. CFFs can then be used further to discriminate between a fake and a genuine image. The CFFN is a Siamese network integrated with the DenseNet. Contrary to regular CNNs which are usually single-streamed, CFFN is a dual-streamed network having the ability to ingest pairwise input for CFF learning. Since CNNs use only high-level feature representation to detect fake from the real, the cross-layer feature concatenation in CFFN further helps in capturing fine-grained feature representation, which is where the CFFs of fake face images exist. This technique involves a two-step learning policy since it uses contrastive loss to learn the CFFs and then the classifier is optimized by minimizing the cross-entropy loss. Experimental results indicate that CFFNs had a higher generalization ability and effectiveness than the other methods.

Deepfake Video Detection At present, there has been a lot of major buildout in artificially generated videos with such sophisticated lip movements and eye syncs that it is hard to tell whether it is fake or not. And due to the loss of frame information after video compression, it is not viable to apply deepfake image detection mechanisms on each frame to unravel the fakeness.

Leveraging the Physiological Signal of Eye-Blink Li et al. [32] make a very valid point by utilizing the physiological signal of eye blinking to detect fake videos in their research *'In Ictu Oculi'*. Their novel approach employs *Long-term Recurrent Convolutional Neural Networks (LRCN)* which is a combination of CNN and RNN to capture the phenomenological and temporal regularities in the process of eye blinking. The idea is that on average resting blinking rate is 17 blinks/min which can go up to 26 blinks/min or as low as 4.5 blinks/min depending on the intensity of the conversation, however, it was found that synthetically generated face videos lacked eye blinking function, since training datasets seldom contain faces with closed eyes. The proposed LRCN model comprises three chief components. The feature extraction part converts the input eye region into discriminative features, followed by sequence

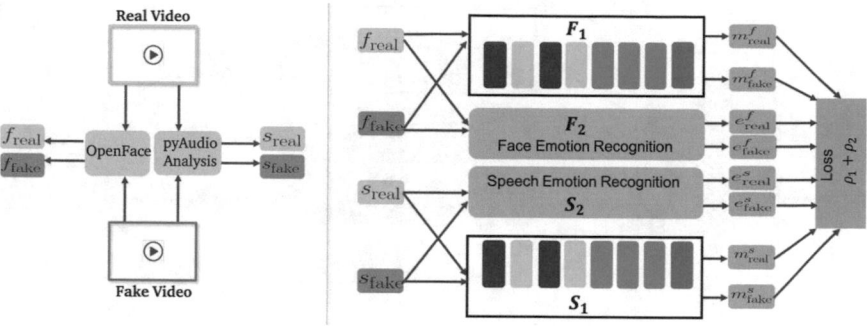

Fig. 11 Multimodal DeepFake Detection: Training Routine

learning which is implemented with RNN with LSTM cells to increase the memory capacity of the RNN model and avoid gradient vanishing while backpropagating. Ultimately, the state prediction component takes the output of the LSTM to generate the probability of the eye being open or closed using which we can plot an eye blinking time series. This method was evaluated using the *Closed Eyes (CEW)* dataset and the *DeepFake* generated videos. The experiments indicate that even though regular CNN was able to predict the state of the eye exceptionally well for an image, it lacks temporal knowledge which is where LRCN comes into play by taking advantage of the long-term dynamics to effectively predict eye state, which is more smooth and accurate.

Outside the Limitations of Visible Signals Building upon the use of biological signals, Ciftci et al. [33] approach this problem in depth by researching the effectiveness of signals like heart rate, blood flow which might not be visible to the human eye. Their research *'How Do the Hearts of Deep Fakes Beat?'* proposes extraction of *photoplethysmography (PPG)* signals from the image to detect the change in skin reflectance over time when the blood flows through the veins which would ultimately help us flag the fake video. The idea is to find a face in each frame of the video using a face detector and extract regions on the face that have as many stable PPG signals as possible. The power spectral density of raw value in the PPG cells in the different time windows are then computed which helps the classifier, which is a shallow CNN, flag it as fake or not. This setup is experimented with on public datasets like *CelebDF*, *Face2Face*, *FaceSwap*, etc. which contain fake videos, and is able to achieve about 97.29% accuracy which is considerable (Fig. 11).

Employing Emotive Cues In addition to the biological signals like the blinking of the eye and blood flow, there are other signals like the sync between the audio and video which can help detect if a video is fake or not. In their study *'Emotions Don't Lie'*, Mittal et al. [34] suggest the idea to fish for facial cues, speech cues, background context, hand gestures, and body posture, and orientation from a video and further analyze these modalities to identify a fake video from the real one. The experiment uses a Siamese network-based architecture that takes the input of a real video and its counterpart deepfake video and obtains the modality along with the per-

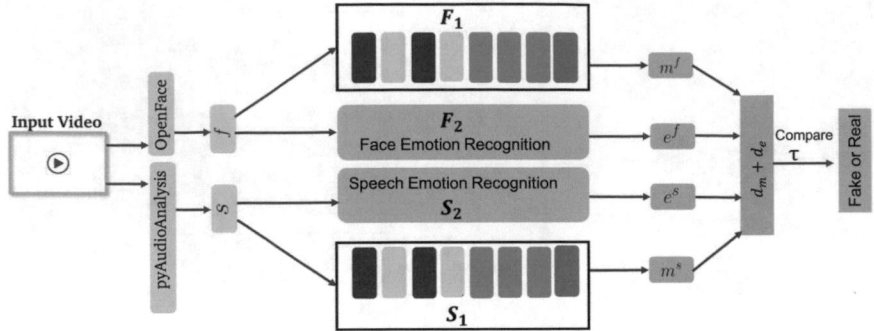

Fig. 12 Multimodal DeepFake Detection: Testing Routine

ceived emotion embedding vectors for the face and speech of the subject (Fig. 11). The embeddings are then used to compute the triplet loss function to maximize the similarity between modalities for the real video and minimize it for the fake video. The said approach is tested on two deepfake identification benchmark datasets, *DeepfakeTIMIT* and *DFDC*, and yielded an accuracy of 96.6% and 84.4% respectively (Fig. 12).

Going Beyond a Single Frame So far, the trend has been to do a frame-by-frame analysis to dig for clues that could indicate if a given video is artificially generated or not. Lima et al. [35] propose incorporating temporal aspects into various action recognition methods to apply for deepfake detection, which demonstrates an edge over the contemporary frame-based mechanisms. The intuition is that these synthetic videos lack temporal coherence due to various tempering. *CelebDF v2* dataset is used for this setup and the data is preprocessed by cropping beyond the face (considering the tampering is done only on the face), thus eliminating noise from the data. Experimental results exhibit the advantageous nature of incorporating temporal information in various architectures over the frame-based baselines. R3D network outperforms other methods and it consists of a sequence of residual networks which introduce shortcut connections bypassing signals between layers.

Undoubtedly, there is a massive surge in the fabrication of phony images and video, and with such perfection. The tools for the creation of deepfakes are becoming more and more accessible, and social media is providing a prominent pedestal for their propagation. The above-mentioned research in deep learning for deepfake detection are just a few of the many work streams going on in parallel to fight this social evil.

4 Limitations of Deep Learning Approaches

Though deep learning models have proven to be successful in flagging fake news by utilizing textual and temporal attributes to an extent, fake news detection remains a tough challenge to tackle. With sophisticated feature extraction methods and state-of-the-art architecture, it is hard to distinguish genuine information from counterfeit without cross-referencing, fact-checking, or additional information. This is because the fake content is being generated with such poise to deceive even the well-informed readers.

Most of the techniques we have discussed in this chapter adopt *Natural Language Processing (NLP)* to analyze the writing styles and lexical patterns of news articles to flag as fake news or not. These methodologies are quite shallow in their analysis since they check whether the news articles adhere to the standard styles generally used by professional journalists or not. Zhou et al. [36] in their study suggest using crowdsourced fact-based knowledge checks on top of these NLP-based fake news detection techniques to achieve a more robust mechanism. With their experiments on *Fakebox*, a fake news detector, they highlight the vulnerabilities of the NLP-based fake news detection methods since without deeper semantic knowledge they can lead to inaccurate results. They are also highly likely to be fooled by malicious users with adversarial attacks via fact distortion, content exaggeration, linking of two independent events, exchange of subject-object, and so on. These kinds of attacks are much more subtle since they don't change the overall writing style of news articles and thus have the potential to evade similarity detection.

5 Fairness and Interpretability

There is a growing focus on deep learning-based approaches to research for more novel, accurate, and robust fake news detection techniques. In this attempt, one territory that is lagging behind the rapid development is the concentration on rational and ethical considerations. Since the societal repercussions of unfair fake news detection models are far more severe than comparative pieces of work, say spam filters or speech recognition, the ethics and fairness aspects of fake news detection also warrant some attention.

Deepak et al. [37] analyze the ethical considerations of the fake news detection methodologies across the key dimensions: mismatch of values, data-driven nature, and domain attributes. The context of the content matters a lot while approaching this problem, hence it is but obvious that there is a tautness between the accuracy of the news and the fairness aspect of the model. The rules of conduct learned by a model might be different from that of the society since politics, legality, and emotions is a spectrum and oftentimes there is no black and white view. Since the deep learning techniques to detect fake news, be it supervised or unsupervised, are nothing but a statistical model built from the *past data*, they end up encoding assumptions based on the bygone events which might or might not be relevant in the present age. There can also be instances where the decision timelines and reversals of laws or any quotes are not factored in based on the architecture of the model. Hence this implicit assumption of the static nature of the context of the fake news is another dimension that can hinder a wholly fair and ethical insight. Another aspect that makes this problem unique is that fake news is a *universal* attribute, there is hardly any instance where it might be genuine for a certain set of people and phony for others. Therefore, any kind of inconsistencies in the results are not acceptable especially if viewed from an ethical point of view.

Despite significant advancements in fake news detection methodologies, there remain corners that are yet to be fully discerned. Interpretability of fake news detection techniques is one such domain since fake news detection is a wide spectrum problem depending on the type of fake news we are dealing with. Interpretable models can help with debugging and model validation along with assisting users in identifying bias in algorithms by putting forward an explanation as to why and how a decision was made. Moheseni et al. [38] propose the *algorithmic interpretability*—ability to visualize model parameters to inspect model behavior, *human interpretability*—transparency for end-users with understandable explanations of hows and whys, and *supporting evidence*—verified claims related to the news, as the three dimensions which can help improve fake news detection research.

6   Emerging Trends

Investigation in fake news detection is gaining attention and remains a growing area of research and development, especially amongst ML practitioners and scientists.

Given the sensitive and high-stakes nature of the fake news domain, accountability of fake news detection algorithms is another emerging area of exploration. This has become essential due to possible adversarial attacks via fact distortion, content exaggeration, linking of two independent events, or exchange of subject-object, which can deviate even the more rugged techniques from their normal mode of operation due to lower quality data. Bogaert et al. [39] study the robustness of the fake news detection against fabricated adversarial examples in *'Can Fake News Detection be Accountable?'*. To tackle the issues, one suggestion is to have an adversarial component built into the detection frameworks, automate fake news generation by morphing genuine news, and allow detection models to better discern the syntactic and semantic patterns. For accountability of such methods in the long term, Bogaert et al. suggest developing ways to evaluate such countermeasures.

Recently, the popular social network, *Twitter*, introduced a *'Read before you retweet!'* prompt in an attempt to promote informed discussion amongst its users. Since the headlines can be often misleading, this prompt encourages users to read the whole article before sharing it, in case it turns out to be yet another fake news article. These types of prompts are a smart way to encourage media literacy and control the viral spread of fake news on social media. In another effort to combat fake news, *Twitter* introduced a community-driven effort called *Birdwatch* that allows people to identify information in tweets they believe is misleading and write notes that provide informative context. Such data could be really valuable for algorithmic accountability and help in improving the fake news detection methods.

Ahead of the elections, a law in Singapore called *Protection from Online False-hoods and Manipulation Act (POFMA)*, now pushes social media companies to adhere to their repressive law that aims to combat fake news by putting warnings next to posts the authorities deem to be false, and in extreme cases get them taken down. Though these laws were considered a tool for potential censorship, the authorities claim that since social media companies put profit above principle by attracting eyeballs via fake news propagation, this is a necessary action to curb this nuisance. In compliance with POFMA, Twitter now has an additional *'Legally Required Notice'* flag to indicate misinformation.

An interesting take on this topic is to analyze the propagation of fake news on social media, and one of the major elements in the network is the users. Lopez et al. [40] provide a compelling argument of how certain users can be flagged as malicious based on their history and thus indicating the intention of spreading fake news. Since *Twitter* is one of the popular social media networks where people actively share updates, the authors worked on profiling the users as fake news spreaders or not, for now focusing on English and Spanish tweeters only. This is an area having scope to research further, for instance, based on multiple languages, other demographics, temporal or psychology-related features, and so on.

References

1. Shu, K., Sliva, A., Wang, S., Tang, J., Liu, H.: Fake news detection on social media: a data mining perspective. ACM SIGKDD Explorations Newsl **19**(1), 22–36 (2017)
2. Zhou, X., Zafarani, R.: A survey of fake news: fundamental theories, detection methods, and opportunities. ACM Comput. Surv (2020)
3. Khan, J.Y., Khondaker, Md.T.I., Afroz, S., Uddin, G., Iqbal, A.: A benchmark study of machine learning models for online fake news detection. Mach. Learn. Appl. (2021)
4. Wang, W.Y.: "liar, liar pants on fire": a new benchmark dataset for fake news detection (2017). arXiv:1705.00648
5. Singhania, S., Fernandez, N., Rao, S.: 3HAN: A Deep Neural Network for Fake News Detection (2017). https://doi.org/10.1007/978-3-319-70096-0_59
6. Rashkin, H., Choi, E., Jang, J.Y., Volkova, S., Choi, Y.: Truth of varying shades: analyzing language in fake news and political fact-checking. In: Proceedings of the 2017 Conference on Empirical Methods in Natural Language Processing, pp. 2931–2937 (2017)
7. Shu, K., Cui, L., Wang, S., Lee, D., Liu, H.: defend: explainable fake news detection. In: Proceedings of the 25th ACM SIGKDD International Conference on Knowledge Discovery and Data Mining, pp. 395–405 (2019)
8. Khattar, D., Goud, J.S., Gupta, M., Varma, V.: MVAE: multimodal variational autoencoder for fake news detection. In: The World Wide Web Conference (WWW '19), pp. 2915–2921. Association for Computing Machinery, New York, NY, USA (2019)
9. Jwa, H., Oh, D., Park, K., Kang, J.M., Lim, H.: exBAKE: automatic fake news detection model based on bidirectional encoder representations from transformers (BERT). Appl. Sci. **9**(19), 4062 (2019)
10. Kaliyar, R.K., Goswami, A., Narang, P.: FakeBERT: fake news detection in social media with a BERT-based deep learning approach. Multimed. Tools Appl. **80**, 11765–11788 (2021)
11. Naeem, B., Khan, A., Beg, M.O., Mujtaba, H.: A deep learning framework for clickbait detection on social area network using natural language cues. J. Comput. Soc. Sci. 1–13 (2020)

12. Chen, Y., Conroy, N.J., Rubin, V.L.: Misleading online content: recognizing clickbait as "False News". In: Proceedings of the 2015 ACM on Workshop on Multimodal Deception Detection, pp. 15–19 (2015)
13. Kim, Y.: Convolutional neural networks for sentence classification. EMNLP (2014)
14. Zheng, H.-T., Chen, J.-Y., Yao, X., Sangaiah, A.K., Jiang, Y., Zhao, C.-Z.: Clickbait convolutional neural network. Symmetry 10(5), 138 (2018)
15. Zhou, Y.: Clickbait detection in tweets using self-attentive network (2017). arXiv:1710.05364
16. Dong, M., Yao, L., Wang, X., Benatallah, B., Huang, C.: Similarity-aware deep attentive model for clickbait detection. Lecture Notes in Computer Science (including subseries Lecture Notes in Artificial Intelligence and Lecture Notes in Bioinformatics) (2019)
17. Xiao, H., Gao, J., Li, Q., Ma, F., Su, L., Feng, Y., Zhang, A.: Towards confidence in the truth: a bootstrapping based truth discovery approach. In: Proceedings of the 22nd ACM SIGKDD International Conference on Knowledge Discovery and Data Mining, 1935–1944 (2016)
18. Dong, X.L., Gabrilovich, E., Murphy, K., Dang, V., Horn, W., Lugaresi, C., Sun, S., Zhang, W.: Knowledge-based trust: estimating the trustworthiness of web sources (2015). arXiv:1502.03519
19. Ge, L., Gao, J., Li, X., Zhang, A.: Multi-source deep learning for information trustworthiness estimation. In: Proceedings of the 19th ACM SIGKDD International Conference on Knowledge Discovery and Data Mining, pp. 766–774 (2013)
20. Li, J., Ott, M., Cardie, C., Hovy, E.: Towards a general rule for identifying deceptive opinion spam. In: Proceedings of the 52nd Annual Meeting of the Association for Computational Linguistics (Vol 1: Long Papers), pp. 1566–1576 (2014)
21. Feng, S., Banerjee, R., Yejin, C.: Syntactic stylometry for deception detection. In: 50th Annual Meeting of the Association for Computational Linguistics, ACL 2012 - Proceedings of the Conference, vol. 2, pp. 171–175 (2012)
22. Taylor, P., Larner, S., Conchie, S., Zee, S.: Cross-cultural deception detection. Detecting Deception: Current Challenges and Cognitive Approaches, pp. 175 201 (2015). https://doi.org/10.1002/9781118510001.ch8
23. Amir, S., Wallace, B.C., Lyu, H., Carvalho Mário, P., Silva, J.: Modelling context with user embeddings for sarcasm detection in social media (2016). arXiv:1607.00976
24. Misra, R., Arora, P.: Sarcasm detection using hybrid neural network (2019). arXiv:1908.07414
25. Yang, F., Mukherjee, A., Dragut, E.: Satirical news detection and analysis using attention mechanism and linguistic features (2017). arXiv:1709.01189
26. McHardy, R., Adel, H., Klinger, R.: Adversarial training for satire detection: controlling for confounding variables (2019). arXiv:1902.11145
27. Zhu, J.-Y., Park, T., Isola, P., Efros, A.A.: Unpaired image-to-image translation using cycle-consistent adversarial networks. In: Proceedings of the IEEE International Conference on Computer Vision, pp. 2223–2232 (2017)
28. Do, N.T., Na, I.S., Kim, S.H.: Forensics Face Detection From GANs Using Convolutional Neural Network (2018)
29. Tariq, S., Lee, S., Kim, H., Shin, Y., Woo, S.: Detecting both machine and human created fake face images. In: the Wild, pp. 81–87 (2018). https://doi.org/10.1145/3267357.3267367
30. Xuan, X., Peng, B., Wang, W., Dong, J.: On the generalization of GAN image forensics. In: Chinese Conference on Biometric Recognition, pp. 134–141 (2019)
31. Hsu, C.-C., Zhuang, Y.-X., Lee, C.-Y.: Deep fake image detection based on pairwise learning. Appl. Sci. 10(1), 370 (2020). https://doi.org/10.3390/app10010370
32. Li, Y., Chang, M.-C., Lyu, S.: In ictu oculi: exposing ai created fake videos by detecting eye blinking. In: IEEE International Workshop on Information Forensics and Security (WIFS), pp. 1–7. IEEE (2018)
33. Ciftci, U.A., Demir, I., Yin, L.: How do the hearts of deep fakes beat? Deep fake source detection via interpreting residuals with biological signals. In: 2020 IEEE International Joint Conference on Biometrics (IJCB), pp. 1–10. (2020)
34. Mittal, T., Bhattacharya, U., Chandra, R., Bera, A., Manocha, D.: Emotions don't lie: an audio-visual Deepfake detection method using affective cues. In: Proceedings of the 28th ACM International Conference on Multimedia, pp. 2823–2832 (2020)

35. de Lima, O., Franklin, S., Basu, S., Karwoski, B., George, A.: Deepfake detection using spatiotemporal convolutional networks (2020). arXiv:2006.14749
36. Zhou, Z., Guan, H., Bhat, M.M., Hsu, J.: Fake news detection via NLP is vulnerable to adversarial attacks (2019). arXiv:1901.09657
37. Deepak, P., Chakraborty, T., Long, C., Santhosh, G.: Ethical Considerations in Data-Driven Fake News Detection (2021). https://doi.org/10.1007/978-3-030-62696-9_10
38. Mohseni, S., Ragan, E., Hu, X.: Open issues in combating fake news: interpretability as an opportunity (2019). arXiv:1904.03016
39. Bogaert, J., Carbonnelle, Q., Descampe, A., Standaert, F.-X.: Can fake news detection be accountable? The adversarial examples challenge. In: 41st WIC Symposium on Information Theory in the Benelux (2021)
40. López, Á., Martí, P.: Profiling fake news spreaders on Twitter. In: CLEF (Working Notes) (2020)

Towards Detecting Fake Spammers Groups in Social Media: An Unsupervised Deep Learning Approach

Jayesh Soni, Nagarajan Prabakar, and Himanshu Upadhyay

Abstract In recent years, the acceptance of social media has increased, where multiple users use various platforms to share different kinds of information. The current estimated number of online users is almost a billion. Today's web traffic has increased by more than a half within a year because of the increase in online activities. More data can be mined with more users performing online activities. Therefore, advertising and marketing administrations need to know how people express their views and their response can improve their business efficiency. Since social media are rooted at diverse scales in various data sources, they are often quite large. Nowadays, people review products online, which plays a vital role in making buying choices by most consumers in digital consumer markets. Spammers often write fake reviews to increase/decrease the value of specific products by taking advantage of online reviews. Past studies have focused on spotting distinct fake reviewer-ids of fake reviews. However, to target and control a specific product's sentiment, fake reviewers create multiple fake ids and work together in groups to write reviews. In this chapter, we address the problem of detecting such groups of fake reviewers. Explicitly, we propose a deep learning-based framework for fake reviewer groups detection data of reviewers using various unsupervised deep learning algorithms such as Self Organizing Maps and Restricted Boltzmann Machine. We perform the hyperparameter optimization of these algorithms on this reviewer graph data using advanced deep learning frameworks such as TensorFlow and popular machine learning frameworks such as scikit-learn. In the end, we discussed the practical implementation of grouping the fake reviewers on the real-world Google Play Store app review dataset, which has fractional ground-truth data on about 2207 fake reviewer-ids out of all 38,123 reviewer-ids in the original dataset. We validate with the experimental results that the

J. Soni (✉) · N. Prabakar
Knight Foundation School of Computing and Information Sciences, Florida International University, Miami, FL, USA
e-mail: jsoni@fiu.edu

N. Prabakar
e-mail: prabakar@cis.fiu.edu

H. Upadhyay
Applied Research Center, Florida International University, Miami, FL, USA
e-mail: upadhyay@fiu.edu

© The Author(s), under exclusive license to Springer Nature Switzerland AG 2022
T.-P. Hong et al. (eds.), *Deep Learning for Social Media Data Analytics*,
Studies in Big Data 113, https://doi.org/10.1007/978-3-031-10869-3_13

projected method can detect the group of fake reviewers with reasonable accuracy. It can also be extended to perceive opinion spammers groups in social media with semantic features, sequential affinity, and emotion analysis.

Keywords Fake spammer group · Social media · Deep learning · Unsupervised learning · Google play store

1   Introduction

Online reviews are becoming increasingly prevalent and exert a more significant influence on purchase decisions [1]. Recent consumer reports show that 78% of users look for reviews before making online purchases. Online reviews are now seven times more powerful than advertisements in influencing consumer decisions [2]. The rising economic influence of these reviews creates strong incentives for businesses to post fraudulent reviews to promote themselves and discredit their competitors [3]. Self-promotion in online reviews was first detected in 2004 when Amazon unmasked considerable self-reviewing by book authors [4]. The issue of fraudulent reviews has not only attracted the attention of researchers and affected platforms; regulatory authorities have also recognized the gravity of the problem and have begun to step in. The methods proposed in the existing literature on fraudulent review detection can be classified into behavioral and linguistic approaches. Behavioral approaches include finding distributional anomalies of ratings and observing unusually correlated patterns in the review metadata. For example, such patterns could reveal collaboration among a group of reviewers by checking whether these reviewers are systematically writing positive reviews for a group of products or negative reviews for a group of competing products [5]. Alternatively, linguistic approaches detect fraudulent reviews by focusing on the differences in the language patterns of genuine and fraudulent reviews. Research has shown that humans unconsciously use different language patterns when describing real versus imaginary events, a tendency that machines can utilize to detect deception automatically [6]. There are millions of apps on the Google Play Store and iOS where fraudsters write fraud reviews and become more refined in this task. Herefore, finding false reviews with fake reviewers becomes even more challenging. There is an upsurge in such groups that execute deceitful actions in clusters. Initially, fraudsters form many accounts, use various crowdsourcing methods to connect with reviewers, and finally engage in fraudulent activities. In addition, to prevent detection, some team members review the products indirectly related to the target products and review them as regular users to trick the spam detection tools. Such a team of reviewers works together to inscribe false reviews. With team reviewers, we indicate a set of reviewer-ids where the real reviewers can be one person with manifold identities, multiple people, or both. They are more dangerous than individual frauds. It is therefore essential to detect such groups as fraudsters. This chapter aims to study the application of multiple deep learning and machine learning-based unsupervised algorithms with their advantages and disadvantages to detect such a group of fraudsters.

The rest of the chapter is summarized as follows. Section 2 discusses the literature review. Section 3 gives an in-depth overview of learning-based algorithms. Section 4 explains various popular unsupervised deep learning and machine learning algorithms with the evaluation metrics. Section 5 discusses the open-source framework and library. Section 6 describes these algorithms' practical real-time use cases to solve fraud detection problems using the real Google Play App Store Dataset. Finally, we conclude in Sect. 7 with a future work discussion.

2 Literature Review

In recent years, the study of detecting fake reviews has been explored broadly by scholars. Mukherjee et al. [7] use transactions of product review to recognize candidate review spammer groups. Xu et al. [8] proposed a kNN-based method to categorize each reviewer as the spam/nonspam label for particular spammer groups. Allahbakhsh et al. [9] detect biclique by training two separate clustering algorithms to validate the spamicity. Wang et al. [10] utilized the topological arrangement of the reviewer graph and proposed a divide-and-conquer algorithm. The textual content of the review is a bot used by their algorithm as a feature. Akoglu et al. [11] used subgraphs of the top-ranked spammers and designed a technique based on graph clustering. They exploit the effect of networks on the reviewers and products. Choo et al. [12] use the structure of community and reviews' interaction data to perform sentiment analysis. They further use that analysis to discriminate spam groups from nonspam ones on the Amazon dataset. Their experimental results recommended robust optimistic communities in opinion group spammers. Rayana and Akoglu [13] proposed SpEagle to distinguish the spammer groups. They examine meta information and relational data. Ye and Akoglu [14] developed GroupStrainer, a revised two-step algorithm. First, they use a network footprint score to calculate the product's likelihood of being spam and then cluster the spammers. Dhawan et al. [15] use spam score to rank each reviewer group by proposing the DeFrauder model, which considers underlying graph structure. Bitarafan and Dadkhah [16] utilize the biconnected modules in the reviewer graph to recognize candidate spammer groups which are further used for classification. Ramnath et al. [17] use three different features of the application, namely rating, review, and ranking, for classifying a particular app to be fraudulent. Umer et al. [18] use natural language processing based Term Frequency-Inverse Document Frequency technique on the text review data for predicting the numerical rating for an application. Lin et al. [19] studied the applications of the iOS app store and developed a model to predict whether an application will be removed from the store in advance or not. The majority of the prevailing approaches tag fake reviews manually, which is a significant drawback. Thus, we propose a deep learning-based framework to detect the spammer groups for non-ground truth data.

3   Learning-Based Algorithms

AI is a wider area that shields everything connected to creating machines intelligent. Machine Learning (ML) is a subdivision of AI, and Deep Learning (DL) is a machine learning (ML) smeared to large and complex data sets (Fig. 1).

1.   *Artificial Intelligence (AI)*

People are fascinated by computerization since the advent of machinery. AI allows machines to learn it deprived of human interference. It is a wide-ranging arena of computer science. It comprises two words: "Artificial" meaning something which is not natural, and "Intelligence" pointing to the capability to comprehend or reason consequently.

2.   *Machine Learning (ML)*

ML systems can automatically learn the patterns in the data on their own without explicit programming. There are three types of ML Algorithms: Supervised, unsupervised learning, and Semi-Supervised.

3.   *Deep Learning (DL)*

DL is a method encouraged by the information processing mechanism of the neurons in the human brain. It makes use of Neural Networks (mathematical models of human brain neurons) to mimic such behavior. It helps to study the incoming data through layers for further prediction and classification purposes. Its network architectures can be classified into Recurrent Neural Networks [20], Convolutional Neural Networks [21], and Generative Neural Networks [22].

(3.1)   *Machine Learning Algorithms*

Machine learning algorithms are primarily categorized as supervised, unsupervised, and semi-supervised. We will limit our focus to supervised and unsupervised.

Fig. 1 Hierarchy

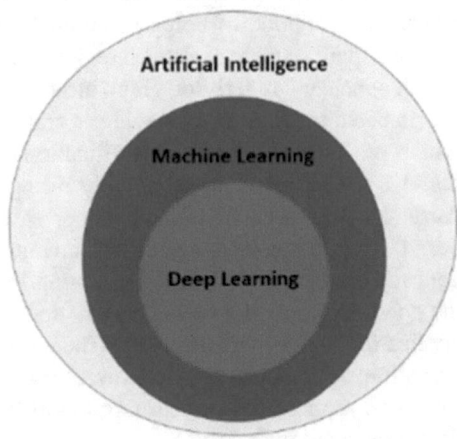

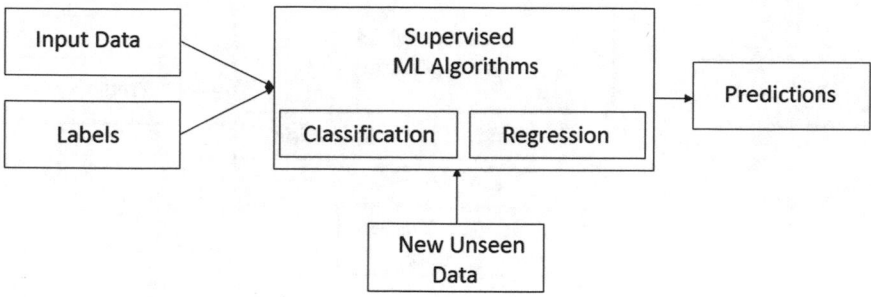

Fig. 2 Supervised ML algorithms

(3.1.1)   *Supervised ML Algorithms*

In supervised-based ML algorithms, we need an input dataset with labels while training.

For example, the input observations can be different raw images of flowers, and the labels can be the name of the flower type (Fig. 2).

These algorithms learn the patterns from the raw dataset and predict the label. While training, the input data is called the training data, with the labels attached for each data point. The algorithm learns the function that maps the input to its output. Next, it predicts the outcome and compares it with the actual output. Finally, it calculates the error based on the predicted and actual output difference and retrains accordingly.

Supervised algorithms can be divided into two categories:

(1)   Classification
(2)   Regression

Classification: The output label is categorical, e.g., "Cancer or No Cancer."

Regression: Output is actual value, e.g., House Price prediction.

Classification-based algorithms: Logistic Regression, Decision Trees, Random Forest, K Nearest Neighbor, Support Vector Machine, etc.

Regression-based algorithms: Decision Tree Regressor, Linear Regression, Support Vector Regressor, etc.

(3.1.2)   *Unsupervised ML Algorithms*

In unsupervised-based ML algorithms, we only have input datasets without labels while training. The algorithms learn the hidden structure from the unlabeled dataset (Fig. 3).

Unsupervised algorithms can be divided into two categories:

(1)   Clustering
(2)   Association

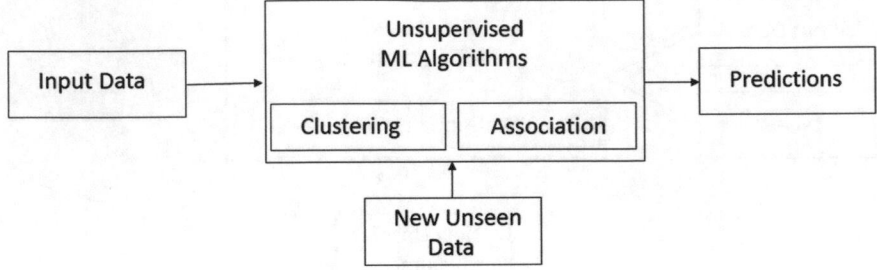

Fig. 3 Unsupervised ML algorithms

Clustering: It unveils the inherent structure of your data and clusters it into different groups. E.g., Clustering the test data into groups like politics, weather, etc.

Association: It discovers the association rules. E.g., If people purchase Y, they also tend to buy Z.

Unsupervised machine learning based algorithms: KMeans + + [23], Principal Component Analysis (PCA) [24], T-SNE [25], DBSCAN [26], mixture models [27] etc.

Unsupervised deep learning-based algorithms: Self Organizing Maps (SOM) [28], Restricted Boltzmann Machine (RBM) [29].

Let us go deep dive into these algorithms.

4 Unsupervised Algorithms

Among several unsupervised algorithms, the following three algorithms are chosen based on wide acceptance in academia and industry.

4.1 Self Organizing Maps

Teuvo Kohonen creates a Self-organizing Map (SOM). Geometric relationships between points on the lower-dimensional space developed by SOM from multi-dimensional space stipulate their similarity. It is a deep learning-based unsupervised algorithm that uses competitive learning concepts to generate the low dimensional subspace. Winning neurons are very similar to the sample input, and they use winning neurons to adjust weights. Furthermore, it goes through several iterations for training before final convergence.

4.1.1 Architecture of SOM

SOM has a single computational layer of neurons and has a feed-forward architecture. All neurons are densely connected to all the initial units' neurons (Fig. 4).

It has five steps:

1 *Choose Weights*: Each weight vector is selected randomly.
2 *Race*: Entice a random input from the training data vector.

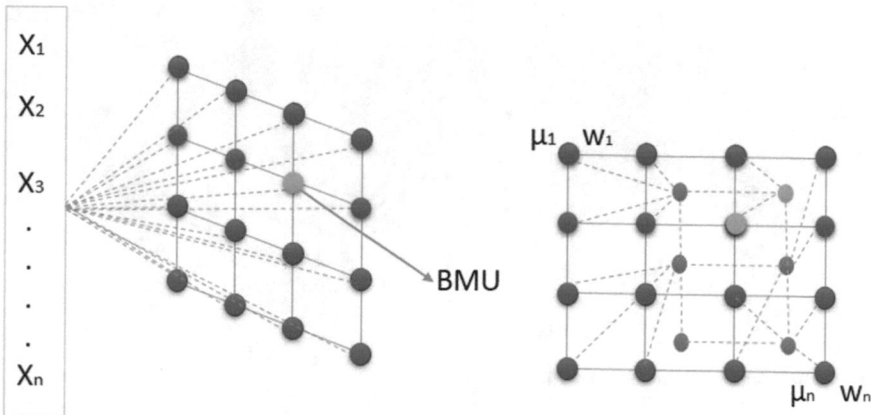

Fig. 4 Self-organizing maps

3   *Similarity*: Find the weight vector nearest to the input node and declare it as a winning neuron.
4   *Apply*: Update the weights.
5   *Maintenance*: Return to step 2 till the feature map converges.

It uses the following weight update formula:

$$Wv(i+1) = Wv(i) + \Theta(b, v, i).\alpha(i).(K(h) - Wv(i)) \tag{1}$$

where i is the step value, h is a sample training example, b is the best matching unit (BMU) for vector K(h), $\alpha(i)$ is a learning coefficient; $\Theta(b, v, i)$ is the function that gives the distance between the neuron u and the neuron v in a particular step.

Advantages

- The final feature map can be interpreted easily.
- Can organize multidimensional composite data sets.

Disadvantages

- Weights selection is difficult to regulate.
- It has the constraint requirement of having the nearby points behave similarly.

Applications

- Organization of a Massive Document Collection.
- Phonetic Typewriter.
- Classifying World Poverty.

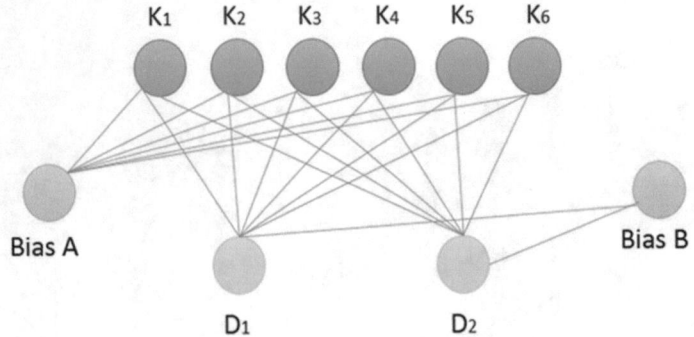

Fig. 5 Restricted boltzmann machine

4.2 *Restricted Boltzmann Machine*

RBM comprises one input layer and one hidden layer. In Fig. 5, (K1... K6) are the nodes for the input layers, whereas (D1, D2) are the nodes in the hidden layers, and finally, the matching biases are present. RBM does not need an output layer since its learning mechanism is different from the regular feed-forward networks.

They are probabilistic, where the model consigns probability values rather than discrete values. RBM is in a particular state at each point. Here, the state denotes the input layer (k) and hidden layer (d) neuron values. The joint probability distribution shown in the below equation gives the value of k and d at a certain state.

$$P(k, d) = \frac{1}{Z} e^{-E(k,d)} \tag{2}$$

$$M = \sum_{k,d} e^{-E(k,d)} \tag{3}$$

where M is the function, the total sum of all the possible values of k and d.

The below equation gives the conditional probabilities of state d given k and vice versa.

$$p(d|k) = \prod_i (k_i|d) \tag{4.1}$$

$$p(k|d) = \prod_i (k|d) \tag{4.2}$$

Finally, using the Bayes rule, a neuron t is activated using the below equation for a given input k:

$$p(d_{t=1}|k) = \sigma(b_t + \sum_i k_i w_{it}) \tag{5}$$

Advantages

- The imbalanced problem can be solved by SMOTE [30] method.
- Using Gibb's sampling, RBM finds the missing values.
- To solve the problem of noisy labels, RBM uses its reconstruction errors.
- It transforms the raw data into hidden units to overcome the problem of unstructured data.

Disadvantages

- The main disadvantage is that it is suggestively slower than backpropagation.

Applications

- Pattern Recognition.
- Recommendation Engines.
- Radar Target Recognition.

4.3   *KMeans*

Kmeans algorithm cluster the dataset into K different groups where K has to be pre-defined, and each point of the dataset will belong to only one cluster out of K clusters.

The following are the steps of the Kmeans algorithm:

1   Randomly select any point in the subspace as the initial K centroids.
2   Calculate the distance between all the points to all the K centroids
3   Assign points to the cluster based on the minimum distance criteria.
4   For each cluster Ki, calculate the mean value of the distance between its points and reassign the new cluster centroids.
5   Repeat steps 2 to 4 until the distance between the previous cluster centroids and the current concierges.

The objective function is:

$$G = \sum_{v=1}^{m} \sum_{k=1}^{K} w_{vk} ||x^v - \mu_k||^2 \tag{6}$$

where the centroid of v_i's cluster is μ_k.

Advantages

- Easy to code.
- It can quickly scale to big complex datasets.
- Convergence to saddle cluster centroids is guaranteed.
- It can generalize well with clusters of various forms and dimensions.

Disadvantages

- The number of clusters has to be pre-defined.
- The assignment of the data points to the clusters depends on the initial location of the cluster centroid.

Applications

- Market Segmentation.
- Document Clustering.
- Image Segmentation.

KMeans++ is a version of KMeans where centroid initialization is not random, and it depends on the first centroid initialization. The further assignment of the next centroid depends on the distance from this first centroid. Each centroid will be taken from the actual data point.

4.4   Evaluation Metrics

4.4.1   The Elbow Method

Intercluster Distance: The distance between the cluster and the points belonging to its cluster.

Intracluster Distance: The distance between the clusters.

The elbow method is one measure that can be used to determine the value of clusters [31]. In this method, on the x-axis, we have the number of clusters, and on the y-axis, we have the intercluster distance. As we increase the K value, the intercluster distance decreases and creates the shape of an elbow. Moreover, after that point, the value will move parallel to the x-axis. The elbow point is considered as the optimal cluster value.

4.4.2   The Silhouette Score

The silhouette score measures how well the point is towards its cluster and how far away a particular cluster is compared to the nearest cluster. It is a combination of intercluster distance and intracluster distance. Its value ranges from -1 to $+1$, where $+1$ indicates a good score and vice versa [32].

For a given data point d, silhouette coefficient sc(d) is given as:

$$sc(d) = \frac{m(d) - n(d)}{\max\{n(d), m(d)\}} \tag{7}$$

where n(d): the average distance between d and all the points in its cluster.

m(d): the least mean distance of point d to all other clusters to which d does not belong.

5 Framework and Libraries

The experimentation of the proposed research framework is performed with the following three libraries.

Tensorflow: Tensorflow is an open-source library from google for deep learning and machine learning [33].

Keras: Keras is a wrapper on Tensorflow and is used by many societies around the world. The code written in Keras is internally converted to TensorFlow for further execution. It has functional API (Application Programming Interface) and Sequential API [34].

Scikit-learn: Scikit-learn deals with a wide variety of learning-based algorithms (both supervised and unsupervised) [35].

6 Practical Implementation

This section provides a practical implementation of the unsupervised algorithms to detect the groups of fake google play store app reviewers.

Google Play store [36]: Google Play store provides various exciting applications that can run on the Android operating system (one of the popular mobile operating systems). It is developed and managed by Google, and applications can be downloaded for free or with charge (depending on the applications). Different applications from the Google Play store were downloaded over 82 billion times in 2016, and the total number of applications available was almost over 3.5 million apps in 2017.

Problem Definition: Developers create an application and upload it to the google play store to download. Users go through the reviews and ratings of an application to download it. Therefore, developers need to have good reviews and ratings for their applications to generate revenue. Nowadays, some developers create fraud by writing fake good reviews for their applications and fake bad reviews for their competent applications. Furthermore, they hire reviewers to commit such fraud in a group.

Dataset: We have a real Google Play Store app dataset with 640 applications reviewed. We have fractional ground truth data about 2207 fake reviewer-ids belonging to 23 unique clusters out of 38,123 reviewer-ids in the original dataset.

The problem is to use the ground truth data with the whole dataset to cluster the unknown reviewer-ids. We have the co-review graph where each node represents the individual reviewer, and the link between two reviewers (R1 and R2) indicates the total number of apps reviewed together (Fig. 6).

6.1   Framework Description

We have the co-review graph of the reviewers who review the application together, and we do not have any textual review dataset. So we use the deep walk approach to extract the feature from the graph. Let us discuss the deep walk [37] in brief.

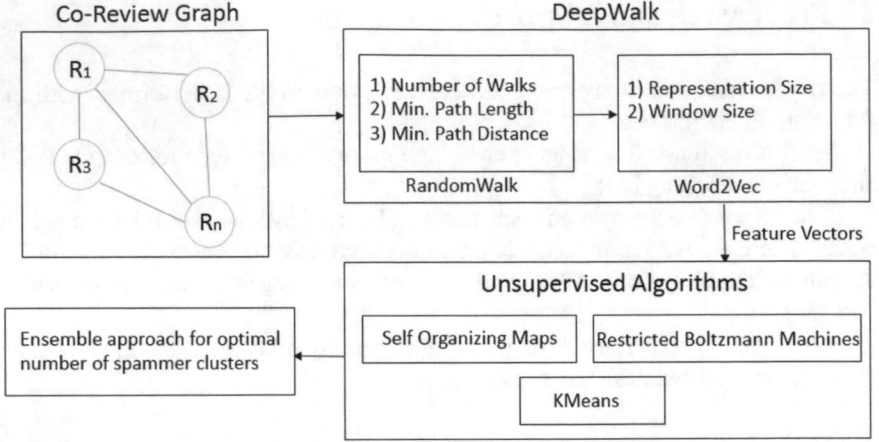

Fig. 6 Framework

DeepWalk

DeepWalk is a neural network architecture that gives the high-level feature embedding (representation) of a given graph with nodes and vertices. It has two sub-parts:

- RandomWalk
- Word2Vec

RandomWalk

Through Randomwalk, it generates the sequence of reviewer-ids by using random sampling. It has three hyperparameters that need to be tuned.

- Number of Walks: Total number of walks that need to be generated.
- Minimum path length: Total number of reviewer nodes that need to be used for a complete walk.
- Minimum distance: Total distance (the edge weight) to be covered.

Word2vec

Word2vec is a neural network that takes a particular walk from random-walk as an input and outputs the meaningful feature vectors.

It has two different architectures:

A. CBOW
B. SkipGram

Both the architecture learns the high-level representations from each reviewer node.

Fig. 7 CBOW

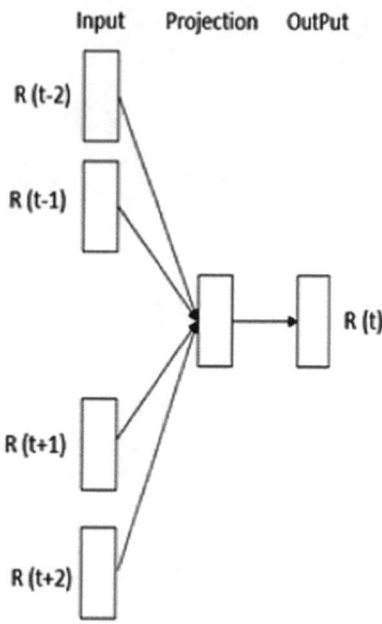

A.   *CBOW (Continous Bag-of-Words)*

CBOW [38] takes the context words (the surrounding reviewer ids) and predicts the word in the middle. (The middle reviewer id) (Fig. 7).

B.   *SkipGram*

SkipGram [39] is the opposite of the CBOW, where it takes a single reviewer id as an input and outputs the context reviewer. This means that given a reviewer id (context reviewer), it outputs the set of reviewer ids related to the context reviewer (Fig. 8).

Word2Vec has the following two hyperparameters that need to be tuned.

(1)   *Representation Size*: Number of a feature vector to be generated
(2)   *WindowSize*: Number of reviewer-ids to be used as a context vector.

6.2   *Experimental Results*

For any machine learning and deep learning algorithms, multiple hyperparameters need to be tuned. This is called hyperparameter optimization. There are various techniques such as:

(1)   *GridSearchCV*: This technique performs an exhaustive search on all the parameters of the algorithms. It takes a lot of time but gives the optimal set of hyperparameters.
(2)   *RandomSearchCV*: This technique search for the random value of parameters rather than an exhaustive search. It takes less time than the GridSearchCV and gives a near-optimal set of hyperparameters.

Fig. 8 SkipGram

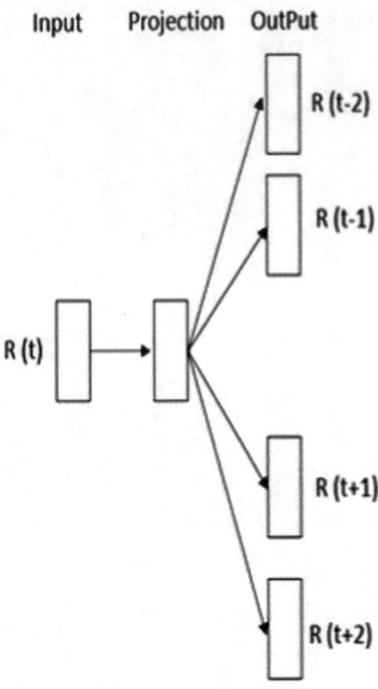

Here CV is K-Fold CrossValidation with K being 10.

Figure 9 shows the elbow curve from KMeans algorithms. The number of clusters shown on the X-axis and the Y-axis represents Within Cluster Sum of Square (WCSS) distance. The WCSS distance decreases with an increase in the number of clusters. The elbow point is between 300 and 400. Similarly, RBM created the clusters in the

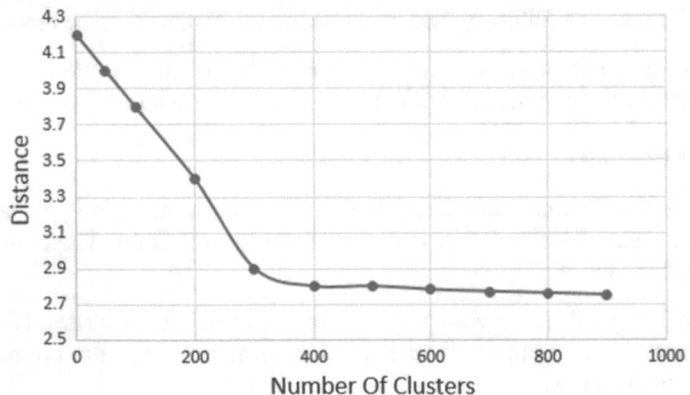

Fig. 9 Distance w.r.t number of clusters

range of 350–400, whereas SOM clustered the points in the cluster range of 370–420. The number of clusters represents the number of unique spammer groups in the data. Based on the results from these algorithms, we can say that the total number of optimal clusters is in the range of 300–420. Each cluster has its behavior. Thus the behavior of the new undetected spammer can be easily detected based on these clusters in an optimized way.

7 Conclusion

Spam reviews have been a significant concern due to fierce competition in the online market. Developers want their applications to be downloaded and used by millions of users to generate revenue, making them commit fraud by writing false reviews. Therefore, identifying such a spammer group is essential in preventing online customers from being influenced by fake reviews. This chapter discussed various machine learning and deep learning-based unsupervised algorithms such as KMeans, Restricted Boltzmann Machine, and Self Organizing Maps with their advantages and disadvantages. Furthermore, we gave a practical implementation of applying these algorithms to solving the fraud review problem using a real-world google play store app dataset.

We employed the DeepWalk based embedding approach for feature extraction. The deep walk method has to be retrained with the addition of a new node. Thus, for dynamic networks, we strategies to use the streaming graph-embedding method. This will dynamically rearticulate the spammer groups in real-time. We use fractional ground-truth information to optimize the hyperparameters for DeepWalk. We intend to extend this by exploring the node2vec technique. Additionally, appropriate natural language processing (NLP) methods will improve the precision of semantic association.

References

1. Luca, M., Zervas, G.: Fake it till you make it: reputation, competition, and Yelp review fraud. Manage. Sci. **62**(12), 3412–3427 (2016)
2. Zhang, K.Z.K., et al.: Examining the influence of online reviews on consumers' decision-making: a heuristic–systematic model. Decis. Support. Syst. **67**, 78–89 (2014)
3. Chen, P.-Y., Dhanasobhon, S., Smith, M.D.: All reviews are not created equal: the disaggregate impact of reviews and reviewers at Amazon.com. Com (May 2008) (2008)
4. Norris, G., Brookes, A., Dowell, D.: The psychology of internet fraud victimisation: A systematic review. J. Police Crim. Psychol. **34**(3), 231–245 (2019)
5. Mukherjee, A., Liu, B., Glance, N.:Spotting fake reviewer groups in consumer reviews. In: Proceedings of the 21st International Conference on World Wide Web (2012)
6. Mihalcea, R., Strapparava, C.:The lie detector: explorations in the automatic recognition of deceptive language. In: Proceedings of the ACL-IJCNLP 2009 Conference Short Papers (2009)

7. Mukherjee, A., Liu, B., Wang, J., Glance, N., Jindal, N.: Detecting group review spam. In: Proceedings pf the 20th International Conference on Companion World Wide Web, pp. 93–94 (2011)

8. Xu, C., Zhang, J., Chang, K., Long, C.: Uncovering collusive spammers in Chinese review websites. In: Proceedings of the 22nd ACM International Conference on Information and Knowledge Management pp. 979–988 (2013)

9. Allahbakhsh, M., et al.: Collusion detection in online rating systems. In: Proceedings of the Asia–Pacific Web Conference Berlin, pp. 196–207. Germany, Springer (2013)

10. Wang, Z., Hou, T., Song, D., Li, Z., Kong, T.: Detecting review spammer groups via bipartite graph projection. Comput. J. **59**(6), 861–874 (2016)

11. Akoglu, L., Chandy, R., Faloutsos, C.: Opinion fraud detection in online reviews by network effects. In: Proceedings of the ICWSM, vol. 13, nos. 2–11, p. 29 (2013).

12. Choo, E., Yu, T., Chi, M.: Detecting opinion spammer groups through community discovery and sentiment analysis. In: Proc. IFIP Annual Conference on Data and Applications Security and Privacy, pp. 170–187. Cham, Switzerland: Springer (2015)

13. Rayana, S., Akoglu, L.: Collective opinion spam detection: bridging review networks and metadata. In: Proceedings of the 21th ACM SIGKDD International Conference Knowledge Discovery Data Mining, pp. 985–994 (2015)

14. Ye, J., Akoglu, L.: Discovering opinion spammer groups by network footprints. In: Proceedings of the Joint European Conference on Machine Learning Knowledge Discovery Databases, pp. 267–282. Cham, Switzerland, Springer (2015)

15. Dhawan, S., Gangireddy, S.C.R., Kumar, S., Chakraborty, T.: Spotting collective behaviour of online frauds in customer reviews (2019). arXiv:1905.13649

16. Bitarafan, A., Dadkhah, C.: SPGD_HIN: spammer group detection based on heterogeneous information network. In: International Conference on Web Resources (ICWR), pp. 228–233 (2019)

17. Ramnath, M., Rubavathi, C.Y.: App assessment with three phase evidence system using sentiment analysis. In: 2021 Third International Conference on Intelligent Communication Technologies and Virtual Mobile Networks (ICICV), pp. 1180–1183 (2021).https://doi.org/10.1109/ICICV50876.2021.9388544

18. Umer, M., Ashraf, I., Mehmood, A., Ullah, S., Choi, G.S.: Predicting numeric ratings for google apps using text features and ensemble learning. ETRI J. **43**(1), 95–108 (2021)

19. Lin, F., Wang, H., Wang, L., Liu, X.: A longitudinal study of removed apps in iOS app store. In: Proceedings of the Web Conference 2021 (WWW '21). Association for Computing Machinery, pp. 1435–1446. New York, NY, USA. https://doi.org/10.1145/3442381.3449990

20. Soni, J., Prabakar, N., Upadhyay, H.:Behavioral analysis of system call sequences using LSTM seq-seq, cosine similarity and Jaccard similarity for real-time anomaly detection. In: 2019 International Conference on Computational Science and Computational Intelligence (CSCI). IEEE (2019)

21. Albawi, S., Mohammed, T.A., Al-Zawi, S.:Understanding of a convolutional neural network. In: 2017 International Conference on Engineering and Technology (ICET). IEEE (2017)

22. Serban, I., et al.: Building end-to-end dialogue systems using generative hierarchical neural network models. In: Proceedings of the AAAI Conference on Artificial Intelligence, vol. 30, no. 1 (2016)

23. Likas, A., Vlassis, N., Verbeek, J.J.: The global k-means clustering algorithm. Pattern Recogn. **36**(2), 451–461 (2003)

24. Wold, S., Esbensen, K., Geladi, P.: Principal component analysis. Chemom. Intell. Lab. Syst. **2**(1–3), 37–52 (1987)

25. Soni, J., Prabakar, N., Upadhyay, H.: Visualizing high-dimensional data using t-distributed stochastic neighbor embedding algorithm. In: Principles of Data Science, pp. 189–206. Springer, Cham (2020)

26. Schubert, E., et al.: DBSCAN revisited, revisited: why and how you should (still) use DBSCAN. ACM Trans. Database Syst. (TODS) **42**(3), 1–21 (2017)

27. Topchy, A., Jain, A.K., Punch, W.:A mixture model for clustering ensembles. In: Proceedings of the 2004 SIAM International Conference on Data Mining. Society for Industrial and Applied Mathematics (2004)
28. Kohonen, T.: The self-organizing map. Proc. IEEE **78**(9), 1464–1480 (1990)
29. Sutskever, I., Hinton, G.E., Taylor, G.W.:The recurrent temporal restricted boltzmann machine. In: Advances in Neural Information Processing Systems (2009)
30. Chawla, N.V., et al.: SMOTE: synthetic minority over-sampling technique. J. Artif. Intell. Res. **16**, 321–357 (2002)
31. Soni, J., Prabakar, N.: Effective machine learning approach to detect groups of fake reviewers. In: Proceedings of the 14th International Conference on Data Science (ICDATA'18). Las Vegas, NV (2018)
32. Ogbuabor, G., Ugwoke, F.N.: Clustering algorithm for a healthcare dataset using silhouette score value. Int. J. Comput. Sci. Inf. Technol. (IJCSIT) **10**(2), 27–37 (2018)
33. Dillon, J.V., et al.: Tensorflow distributions (2017). arXiv:1711.10604
34. Ketkar, N.: Introduction to Keras, pp. 97–111. Deep learning with Python, Apress Berkeley, CA (2017)
35. Pedregosa, F., et al.: Scikit-learn: machine learning in Python. J. Mach. Learn. Res. **12**, 2825–2830 (2011)
36. Viennot, N., Garcia, E., Nieh, J.:.A measurement study of google play. In: The 2014 ACM International Conference on Measurement and Modeling of Computer Systems (2014)
37. Soni, J., Prabakar, N., Upadhyay, H.:Feature extraction through deepwalk on weighted graph. In: Proceedings of the 15th International Conference on Data Science (ICDATA'19). Las Vegas, NV (2019)
38. Luo, Q., Xu, W., Guo, J.:A Study on the CBOW model's overfitting and stability. In: Proceedings of the 5th International Workshop on Web-Scale Knowledge Representation Retrieval & Reasoning (2014)
39. Guthrie, D., et al.: A closer look at skip-gram modelling. LREC **6** (2006)

A Deep Learning Approach
for Anomalous User-Intrusion Detection
in Social Media Network System

Nidhi Lal, Shishupal Kumar, and Garima Kaidan

Abstract Nowadays, artificial neural networks have the potential to learn and improve the performance of machines in terms of various aspects. It provides a platform for humans which make life simpler and more flexible. In ANN, machines are trained and learn to work efficiently with high accuracy. There is a wide variety of application fields like healthcare, military, finance, home appliances, etc. By detecting intrusion using an artificial neural network, the security of the system can be enhanced. Motivated by this, we proposed a novel approach that uses recurrent neural networks to classify the type of social media network intrusion according to the given input data. The proposed model is trained and tested on a large dataset resulting in high accuracy. Afterward, the proposed deep learning model is used for detecting intrusion in social media networks, and accordingly, classification is done with the new set of input data.

Keywords Recurrent neural networks · Social media network intrusion · Classification · Backpropagation

1 Introduction

In recent times, it has been noticed that there is a rise in computer social media network-related attacks. The number of internet users has grown rapidly over the years, providing more threats and vulnerabilities resulting in compromise of the social media network system via the internet. Any activity which is unauthorized on a computer network is known as social media network intrusion. The defenders play

N. Lal (✉) · S. Kumar · G. Kaidan
Department of Computer science & Engineering, Indian Institute of Information Technology, Nagpur (IIIT Nagpur), Maharashtra 440003, India
e-mail: skumar@iiitn.ac.in

G. Kaidan
e-mail: garima@thapar.edu

Thapar University, Patiala, Punjab, India
e-mail: nidhi@iiitn.ac.in

T.-P. Hong et al. (eds.), *Deep Learning for Social Media Data Analytics*,
Studies in Big Data 113, https://doi.org/10.1007/978-3-031-10869-3_14

255

an important role in detecting intrusion as having a clear and better understanding of how social media network attacks work is an advantage. In most situations, such unwanted social media network activity or intrusion causes social media network resources intended for other uses to be absorbed and also threatens the network's and its data's security. By properly designing and deploying Network Intrusion Detection Systems (NIDS) we can block the intruders.

An intrusion Detection System (IDS) is a software application or device that monitors the traffic and detects malicious activity or suspicious policy violations. Any activity is called suspected malicious if it is either reported by an administrator or is centrally collected by the use of a security information and event management system. IDS is a system that combines the outputs from various sources, and by using the alarm filtering technique this system differentiates whether it is a malicious activity or it is a false alarm. The types of IDS can vary depending on the scope ranging from a single computer system to a large number of computer social media network systems. The conventional classification of IDS is done in two types: NIDS and Host-based Intrusion Detection Systems (HIDS). The processes in the operating system are known as HIDS which observe and analyze vital files. However, NIDS belongs to the processes that monitor the incoming social media network traffic & detect the presence of an intrusion. Depending on the detection approach IDS can also be classified as signature-based and anomaly-based intrusion detection. The signature-based IDS includes malware detection in a system that uses pattern recognition. However, anomaly-based IDS involves the detection of deviations from a generated model based on "good" traffic.

1.1 Intrusion Detection Methods

Signature-based IDS It refers to intrusion attacks that are identified by specific searching patterns. These patterns can be previously known instruction sequences used by malware or malicious byte sequences in social media network traffic. These patterns are detected as signatures. Therefore, signature-based IDS can effortlessly identify the previously known attacks. But it can not identifies the new attacks in the system.

Anomaly-based IDS This IDS is used to detect unknown attacks because of the swift malware development. Machine learning is a primary technique used to create a model of trustworthy activity. Anomaly-based IDS is used as a benchmark model to recognize the new attacks. However, it suffers from false positives like an activity that is legitimate and new but is identified wrongly as malicious. In addition, it is a time-consuming intrusion detection process and leads to performance degradation. In this paper, we considered only Anomaly-based Intrusion Detection.

1.2 Motivation

Anomaly-based IDS is an effective method to detect previously known and unknown attacks in the systems. By continuous monitoring and modeling of behavior in networks threats are detected. There are three ways to categorize the anomaly: contextual, collective, or point. These three categories of anomalies are related to four types of attacks in social media network security named as, User to Root (U2R) attack, Denial of Service (DoS) attack, Remote to Local (R2L) attack, and Probe attack. The functionality of the collective anomaly matches the characteristics of the DoS attack. In a DoS attack, the computational resources are made full to serve valid networking requests. As a result, the services are denied to users. The functionality of contextual anomaly can be understood as a probing attack. In this attack, the weakness or vulnerabilities of the networking devices are scanned by the hacker. Afterwhich, the targeted system gets exploited. A probe attack is a common technique that is used oftenly during data mining. Illegal access to the account of the administrator, thereby leading to exploitation of one or more vulnerabilities is characteristic of U2R attack and termed a point anomaly. In this attack, the main focus is to exploit the system to abuse the vulnerabilities present in the system, and afterward, the user privileges are breached. In an R2L attack, attackers use the hit and trial method to determine the password and derive local access with privileges to transmit the packets on the network.

This paper aims to propose an efficient and reliable method to detect and classify anomaly-based IDS by using a deep learning approach. For achieving this, we used deep neural social media network architectures, and hence an approach is proposed which is robust and efficient. The main objective of this paper is to design a deep neural social media network that must be able to predict with a good degree of accuracy. On the presence of intrusion or attack in the system, the proposed approach can classify the type of intrusion among the four types of attacks namely: R2L, U2R, DoS, and probe.

2 Literature Review

During the survey, it was concluded that despite the advances in computation technology as well as research in the domain of NIDS, a large part of these detection techniques operate on less-capable signature-based methods instead of using anomaly-based detection methods [4]. There are various reasons for the unwillingness to change over anomaly-based detection techniques few of them are difficulty in obtaining reliable training data, high false error rates, behavioral dynamics of the system, and the training data longevity. Shortly, this condition may reach such a point wherein the dependence on signature-based intrusion detection techniques leads to ineffective, imprecise, and unreliable detection.

The task in hand is to propose a widely acceptable anomaly-based detection method that can overcome the limitations produced by the incessant changes prevailing in modern networks. This challenge in social media network security has two primary limitations. The first limitation is the continuous increase in the volume of social media network data. The increase in the level of connectivity among users, the Internet of Things (IoT) obtaining popularity, and the substantial increase in the use of cloud-based services are chiefly the cause of this growth in social media network data. To manage such large volumes of data the techniques should be able to analyze the data effectively and efficiently.

In-depth monitoring and granularity are required to aid in improving the accuracy & effectiveness is the second limitation. More detailing and contextual awareness is also required in NIDS analysis rather than abstract high-level observations. In recent years, research on NIDS mainly focuses on the application of Decision Trees (DT), Support Vector Machines (SVM), Naive Bayes, and other similar machine learning algorithm for intrusion detection. So far, solutions to these techniques have provided enhancements for accurate detections. But, there are certain constraints in these techniques, like knowledge required to preprocess the data, for example, identification of data patterns that are useful [4]. Thus this is an expensive process that turns out to be labor-intensive as well as prone to error. The training data required can be vast similarly, becoming a challenge in a dynamic and heterogeneous environment.

Prior research has focused on various traditional approaches in machine learning some of which include support vector machines (SVM) in [1, 5], random forests (RF) in [11, 12], artificial neural networks (ANN) [9], k-nearest neighbour (kNN) [7] and other classification methods as mentioned in the survey papers [10, 13]. These proposed methods have achieved commendable success for an IDS. Through reviewing literature in [2, 6, 8] and conducting the literature survey, we can conclude that deep learning for social media network intrusion detection has not been utilized and can apply for the same.

In [15], the authors have proposed a deep learning technique for flow-based anomaly detection using deep neural networks. From the results it can be deduced that in software-defined networks, anomaly detection can be done using deep learning. In [14], the authors have proposed a deep learning technique for a NIDS, based on self-taught learning (STL) the dataset used was the benchmark NSL-KDD dataset. However, the above class of references has focused on a deep learning approach to detect and classify intrusion attacks. The study done by the authors is an incremental approach where they use LSTM and achieve better accuracy.

These references have used deep learning techniques for pre-training while the task of classification has been performed by the traditional supervision model. It hasn't been common for the application of deep learning techniques for directly performing the task of classification, whereas in multiclass classification there is a need for the study of the model's performance. As claimed by [16], a Recurrent Neural social media network(RNN) is considered a reduced-size neural network. The author of this paper has proposed a three-layered architecture of the RNN with inputs having 41 features and the outputs having four intrusion categories for the misuse-based intrusion detection system. However, the neurons of layers have partial connections and the reduced RNNs don't display the capability of deep learning for the modeling of high-dimensional features. Also for the case of binary classification, the authors did not study the performance of the model.

In anomaly-based social media network intrusion detection, the system undergoes training with social media network traffic which is labeled as "normal", using which we can classify new traffic as "anomalous" or "normal". The characteristic which is expected to be present in the generated model is the capacity to adapt and learn so that its behavior can be generalized and should be able to cope and perform well in the dynamic social media network environment. That summarizes that a self-learning system is required. The availability of labeled data for training and validation of models for anomaly-based NIDS is a significant problem as the labels for "normal" data are available while those of intrusion or "anomalous" data are not. Hence in this case unsupervised or semi-supervised anomaly detection methods are preferred Like SVM [19–22].

The limitations mentioned above have been tackled recently, by the current research in the deep learning field that has recently received considerable interest across many domains. Deep learning is an improved subset of machine learning and it has helped in overcoming some drawbacks of shallow learning. The approach is to put forth a model for anomaly-based NIDS using deep learning. By using deep learning, issues such as the non-availability of training data, adaptability to social media network environments that are dynamic, high rate of intrusion detection that anomaly-based NIDS face is expected to be tackled. In the field of IDS, there are a few research proposals that have utilized deep learning but have not effectively exploited the power of deep learning methods.

In Deep learning, multiple layers in architectures that are of hierarchical type are often exploited for classification, pattern analysis and unsupervised feature learning. This method for the display the power of generative models resulting in high classification accuracy and sometimes even the capacity for extracting parts of useful data from incomplete training data. Deep learning is based on the principle of computing hierarchical representations or features of observational data. Here the lower-level factors or features specify the higher-level ones. These techniques aim to learn a valid feature representation from a huge unlabeled dataset to pre-train the model in an unsupervised learning approach.

Labeled data which is limited is then used for fine-tuning the model for a particular task in supervised classification. This approach of unsupervised training is then followed by supervised fine-tuning of the deep neural social media network model which is efficient and provides favorable results on many difficult learning problems. The initialization of weights near local minima can increase the chance of the model to improve the data representations and offer better generalization. Many deep learning techniques are reliable and are chosen based on the problem domain. The techniques in deep learning used in this paper are recurrent neural networks and long short-term memory networks.

3 Proposed Work

In this section, a detailed description of the steps performed in the proposed methodology is given.

3.1 Methodology

In this subsection, a deep learning architecture has been proposed which is used as an RNN. A variation of the recursive artificial neural social media network is an RNN, which is the one that links between the neurons to form a directed cycle unlike a Feedforward Neural social media network (FNN). This implies the current inputs as well as the previous step's neuron state effect. According to it, the output is determined. The RNN consists of input units in the input layer, output units that constitute the output layer, and a hidden layer. In this model, the flow of information is in a one-way manner from the units in the input layer to the units in the hidden layer. Figure 1 shows the one-way flow of information from a previous temporal concealment unit to a current time hidden unit. The units in the hidden layer is considered as the memory for the entire social media network and is used to remember the end-to-end information. Upon unfolding of RNN, it embodies deep learning. The recurrent neural networks technique can be used for the task of classification by supervised learning. The essential difference between RNNs from traditional FFNs is the introduction of a directional loop that is capable of memorizing the previous information to present the output. In RNN, the current output of a sequence is related to the previous output, and the neurons between the hidden layers have connections instead of being connectionless like in FNNs. In addition, the output of the input layer and previously hidden layers acts as an input for the next hidden layer. RNNs were created around the 1980s and are relatively old but have shown their real potential in recent years. The reason is, that an increase in the accessibility and availability of

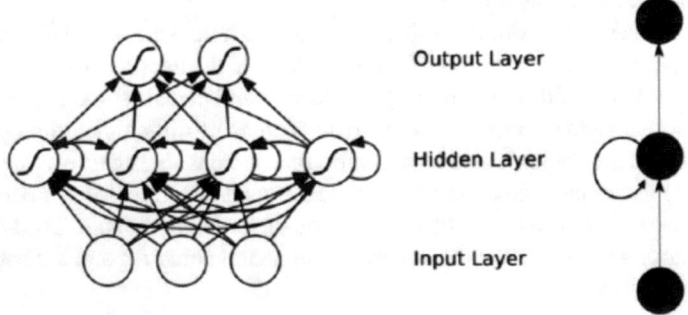

Fig. 1 Representation of a RNN

computational power leads to a huge amount of data available for the development of Long Short-Term Memory (LSTM) in the 1900s. The presence of memory in RNNs provides a feature to remember the characteristics of the received input data for making predictions. In the case of FNNs, the flow of information is from the input, through the hidden layers, and finally to the output layer in a single straight direction only.

There are two main issues faced by RNN: the problem of vanishing and exploding gradients. LSTM networks resolves these problems. The problem of vanishing and exploding gradient occur in deep neural networks during the training phase due to the backpropagation of the gradient values in the initial layer itself. The gradient values which are passed back from the deeper layers to the initial layers are continuously subjected to matrix multiplications due to the chain rule. In the case of having values lesser than 1, it tends to exponentially shrink up to a point of vanishing. Upon approaching earlier layers, it makes the task of learning impossible for the model and is defined as the vanishing gradient problem. The exploding gradient problem occurs when the value is greater than 1 and it leads to the crash of the network. The tendency of an ANN to abruptly and completely erase learned information and proceed to fresh ones is called catastrophic forgetting. It is caused by interference in the social media network that is termed catastrophic, hence it is called catastrophic forgetting.

LSTMs are a specific RNN architecture that is designed for modeling temporal sequences. Its long-range dependencies give more accuracy rather than usual RNNs. In LSTMs, the activation function is not used in its recurrent components. Therefore, the stored values are not modified and vanished during the training of the gradient. LSTM has an explicit memory unit called a cell which is inherently present in an LSTM network. Hence, the problem of 'forgetting' is solved be to 'adding' memory. Since LSTMs can incorporate memory and overcome the second drawback of RNNs. Generally, the implementation of LSTM units is done in 'blocks' with several units. There are three to four 'gates' in the blocks that decide the flow of information based on logistic function. Upon association of RNNs with LSTM units, they are known as long short-term memory networks or LSTM networks.

The basic LSTM unit consists of a cell, an input, an output, and forget gate. These three gates provide help in the regulation of information flow in and out of the cells. The cell also remembers the values over arbitrary intervals of time. The development of LSTMs tackles the problem of vanishing and exploding gradients that are faced during the training of RNNs. The vanishing gradient problem is solved using LSTMs by keeping the gradients sufficiently steep. Hence, the training would be shorter by obtaining a higher accuracy level. The RNN for intrusion detection was originally proposed by Yin et al. in [3]. However, the proposed RNN model differs from the original [3] in terms of using wide social media networks and deep social media networks of the RNN. In this paper, we discussed the implementation of both an RNN social media network and an LSTM social media network for the task of intrusion detection. Besides these two methods, other standard machine learning classification algorithms such as SVM, and Multi-Layer Perceptron (MLP) are used for comparison study against the proposed approach.

The implementation of the proposed model is done in Python. We utilized a Keras neural social media network framework that is powered by a CPU or GPU backend. Fine-tuning is done to improve the accuracy by adjusting the learning rate, activation function, and the number of iterations. The loss function to be used is the cross-entropy between output value, predicted, and the actual label value. Both the neural networks are evaluated for binary as well as multiclass classification of the intrusion in the network. As an evaluation metric, we considered the accuracy of the detection because it is a good performance indicator.

3.2   Dataset Description

The training and testing process of the deep neural social media network architecture was done using the NSL-KDD dataset. A detailed study on the NSL—KDD dataset is done in [17]. It was generated in 2009 which is an improvement over KDD Cup '99 data set, whose inherent problems are mentioned in [18], by the removal of redundant inherent records. This data set is known to be used as the benchmark dataset in recent research work about IDS. This dataset consists of the training set which is called the KDDTrain$^+$(includes attack-type labels—for binary as well as five-class classification) and the testing sets namely KDD Test$^+$ (full test set) and KDDTest^{-21} (a subset of file 'KDDTest$^+$' and doesn't contain records with the difficulty level of 21 out of 21).

The data set consists of 41 features with one class label for each traffic record. The basic features (from columns 1 to 10), content features (from columns 11 to 22) & traffic features (from columns 23 to 41) are the features included among the set of 41 features. The dataset has the attacks classified into four classes namely: R2L, U2R, DoS, and Probing attacks. The testing set provides a realistic theoretical basis for the detection of intrusion. it consists of certain attack types that are not present in the dataset provided for training.

3.3   Pre-processing the Dataset

The first step involves preprocessing the dataset. Preprocessing is necessary before solving any problem in machine learning is data preprocessing. The datasets used in machine learning generally require the data to be processed or transformed before the machine learning algorithm is trained on it. It means converting raw data to a clean dataset. This is done to achieve better results from the model designed in machine learning. The techniques used for preprocessing include normalization, scaling of the data, binarizing the data, and standardizing the data. The preprocessing done for the NSL - KDD dataset is americanization and normalization.

The dataset of 41 features consists of 38 numeric and 3 non-numeric features, it is required to convert to numeric format the non-numeric type features. This is since the RNN requires that the input values be provided in numeric format. The three non-numeric features are *'service'*, *'flag'* and *'protocol_type'* features, which are required to be converted into numeric format. For instance, the *'protocol_type'* feature has attributes of three types: *'icmp'*, *'udp'* and *'tcp'* that need to be numerically encoded as binary vectors using one-hot encoding method as (0, 0, 1), (0, 1, 0) and (1, 0, 0) respectively. The *'service'* feature has attributes of 70 types, and the *'flag'* feature has attributes of 11 types. As a result of one-hot encoding and numerical sizing, these features lead to the mapping of 41-D features into 122-D features.

There are another set of three features, *'dst_bytes'*, *'src_bytes'* and *'duration'*, where the variation between minimum & maximum value is huge and hence it is required to apply logarithmic scaling; following which every feature is mapped linearly, to a value in the range [0,1] using the formula of normalization as in Eq. 1 where *'min'* and *'max'* denote the minimum and maximum value of feature *'x_i'* respectively.

$$x_i = (x_i - min)/(max - min) \tag{1}$$

4   Experimental Details

In this section, we discussed in detail about the performance on the RNN. Afterward, the description of the implementation of the LSTM social media network is given. The parameters and hyper-parameters used for training and testing the proposed model are also discussed.

In the model of RNN, the input layer contains 122 nodes after preprocessing. The output layer contains two nodes in the case of binary classification where the class is Normal or Anomaly; while it contains five nodes in the case of multi-class classification and the class labels namely, are Normal, R2L, DoS, U2R, and Probe. The wide and deep models of the RNN generated for experimentation are as follows:

1. Wide social media network with a single (one) hidden layer with nodes varying between 30 and 300.
2. Deep social media network with the number of layers varying between 2 and 20.

In the case of a wide RNN network, the number of nodes in the single layer is initially set to 30. The social media network is trained on 75% of the data and validated on the remaining 25%. The model weights are saved for testing the data set. The second iteration of the wide social media network is done by setting the number of nodes in the hidden layer to 40 and the remaining parameters are the same. The used pattern takes the number of nodes is increased in increments of 10 until the social media network contains 300 nodes in one hidden layer. The saved models are tested and the accuracies are stored for comparison. Hence, the comparison plot for the wide model of the RNN social media network indicates the variation of test data accuracy concerning the increasing number of neurons in one hidden layer. The experiment is done for both binaries as well as multi-class classification.

A similar approach is followed for the implementation of a deep RNN network. Here, the number of nodes in every layer is kept fixed, while varying the number of hidden layers. The number of nodes is set to 40 in this implementation and it gave a better result. Therefore, the first model of the deep social media network contains 2 hidden layers with 40 neurons/nodes. The second iteration contains 3 hidden layers with each layer having the same number of nodes following up to 20 hidden layers keeping the remaining parameters the same. The social media network is trained on 75% of the data and validated on the remaining 25%. The model weights are saved for testing the model using the test data set. Here, the comparison plot for the deep model of the RNN social media network is generated by considering the variation of test data accuracy concerning the increasing number of hidden layers. The experiment is performed for both binaries as well as multi-class classification.

Hence, we have four models of the RNN social media network namely:

1. Wide social media network for binary classification
2. Deep social media network for binary classification
3. Wide social media network for multi—class classification
4. Deep social media network for multi—class classification

The wide and deep models of generated LSTM for experimentation are as follows:

1. Wide social media network with a single (one) hidden layer with several LSTM neurons varying between 30 and 300.
2. Deep social media network with the number of hidden layers varying between 2 and 20.

In the case of a wide LSTM network, the number of neurons in the single layer is initially set to be 30. The social media network is trained on 75% of the data and validated on the remaining 25%. The weights of model were saved for testing using the test data set. The second iteration of the wide social media network sets the number of neurons in the hidden layer to be 40 and the rest of the parameters remain the same. A pattern is followed by increasing the number of neurons in increments of 10 until the social media network contains 300 neurons in one hidden layer. The saved models are tested and the accuracies are stored for comparison. Hence, the comparison plot for the wide model of the LSTM social media network is generated regarding the variation of test data accuracy concerning the increasing number of neurons in one hidden layer. The experiment is done for both binaries as well as multi-class classification.

A similar approach is followed for the implementation of a deep LSTM social media network as well. Here, the number of LSTM neurons in each layer is kept fixed, with variations done in the number of hidden layers. The number of LSTM neurons is set to be 55 for better results compared to others. Therefore, the first model of the deep social media network contains 2 hidden layers with 55 LSTM neurons. The second iteration contains 3 hidden layers with each having the same number of LSTM neurons following up to 20 hidden layers by keeping the same remaining parameters. The social media network is trained on 75% of the data and validated on the remaining 25%. The model weights are saved for testing the model using the test data set. Afterward, A comparison plot for the deep model of the LSTM social media network is generated in terms of variation of test data accuracy concerning the increasing number of hidden layers. In this way, the number of LSTM neurons in each hidden layer is fixed. This experiment is done for both binaries as well as multi-class classification.

Hence, there are four models of the LSTM social media network namely:

1. Wide social media network for binary classification
2. Deep social media network for binary classification
3. Wide social media network for multi—class classification
4. Deep social media network for multi—class classification

In the proposed method, a gradient descent algorithm is combined with the back-propagation through time (BPTT). This combined theme is used for computing the gradients during the training phase. The stochastic gradient descent (SGD) is referred to as the incremental gradient descent. It follows an iterative approach to optimize a differentiable objective function as a stochastic approximation of gradient descent optimization. In comparison with the batch gradient descent, it recomputes the gradients for a similar type of instance before each parameter update. In this way, it can be executed in redundant computations for huge data sets. SGD performs one update only at a time without redundancy. It is faster and also be used to perform online learning. The frequent updates performed by SGD with a high variance make the objective of a function leads heavily fluctuate. In the case of batch gradient descent, it converges towards the minima of the basin. Here, the parameters are initialized in the case of SGD. This fluctuation helps in jumping to a new and possibly better local minimum. However, it makes the convergence to the exact minimum complicated as the SGD is bound to overshoot. However, if the learning rate is decreased slowly, it has been shown that the SGD displays similar convergence. Batch gradient descent often converges to a global minimum for convex optimization and local minima for non-convex optimization. We used the softmax activation function shown by Eq. 2. Similarly, for a sigmoid function, the softmax function compresses the output of each neuron unit to be between 0 and 1. In addition, it ensures the division of values at each output such that the total sum of the output values should be equal to 1.

$$f(x_i) = e^{x_i} / \sum_{j=1}^{k} e^{x_j} \qquad (2)$$

The softmax function's output corresponds to a categorical probability distribution. In this way, it gives the probability value to be true for all of the classes. The range of the output lies between 0 and 1 in the usage of softmax. This function is used in the case of multi-class classification. In this case, the probability of each class is returned. The class that contains the highest values of probability is the target class. The loss is monitored by the cross—entropy method given by Eqs. 3 and 4. A function, that is used to measure the performance of the classification model must contain the probability of the output value between 0 & 1. The probability of the prediction diverges from the actual label, the cross-entropy loss keeps increasing. In the case of binary classification, the number of classes is denoted by $M = 2$.

$$- (y \log(p) + (1 - y) \log(1 - p)) \tag{3}$$

In the case of multi-class classification, where $M > 2$, a separate loss for each class label per observation is calculated and the result is summed.

$$- \sum_{c=1}^{M} y_{o,c} \log(p_{o,c}) \tag{4}$$

where M is the no. of classes
y—binary value (0 or 1), if class label c is the correct classification for observation o
p—predicted probability observation o is of class c.

If the probability of the predicted value diverges, then the cross-entropy loss will increase. The loss value will be 0 for a perfect model.

The social media network has been trained for convergence. The number of epochs parameter is set to a high value. In this way, prevention can be provided under the training of the network. The training occurs in batches that consist of 1000 samples per batch. The reason is, that an approximation of the distribution of input data by a batch is better than a single input/sample. The weights are updated by the backpropagation process. If the batch size is large, when compared the approximation is said to be better and it can take more time to process the data. Hence, the batch size must be appropriately large but should not exceed memory. The training of the social media network stops when convergence is attained after the monitoring of validation loss and is termed as early stopping. In this way, 25% of training data is used for validation to prevent overfitting. It can cause the social media network to perform poorly on test data. The social media network architecture, weights, and configuration of parameters (optimizer and loss) are stored on the disk for later use during testing.

5   Results

In this section, we discuss the obtained results and compare the performance of the RNN with the proposed LSTM. In addition, we compared the proposed method with the other standard machine learning algorithms. The following figures (Figs. 2 and 5) show the graphs of accuracy obtained on testing is done with the test data.

Figure 2 shows a plot of the accuracy concerning the number of neurons. The simulation starts with 30 neurons and goes up to 300 in increments of 10 in an RNN model for binary classification. This model of the neural social media network comes under the category of wide neural networks. The average accuracy is observed

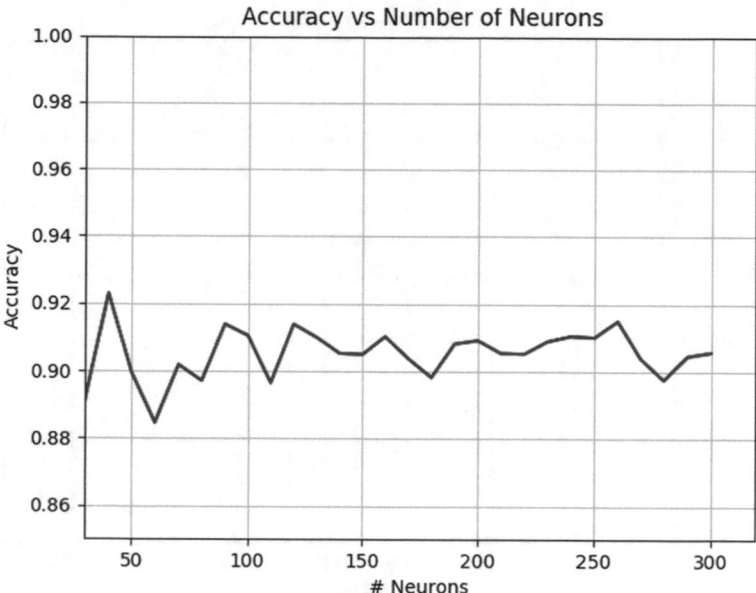

Fig. 2 Graph of accuracy versus increasing number of neurons for RNN binary classification

to be 90.53% while the minimum observed is 88.46% at 60 neurons. However, the maximum accuracy is 92.29% at 40 neurons.

Figure 3 shows a plot of the accuracy concerning the number of hidden layers. It begins from the 2nd layer in-depth and goes up to the 20th layer of depth. Each layer has a fixed number of neurons i.e. 40 in an RNN model for binary classification. The used class labels are normal and anomaly. This model of the neural social media network comes under the category of deep neural networks. The average accuracy is observed to be 87.98% while the minimum observed is 84.65% at depth of 4 layers. However, the maximum accuracy is 88.85% at depth of 13 layers.

Figure 4 shows the plot of accuracy concerning the number of neurons. It begins at 30 neurons and goes up to 300 in increments of 10 in an RNN model for multiclass classification. The class labels used by the model are named as normal and DoS, U2R, R2L, and Probe. The model of the neural social media network comes under the category of wide neural networks. The average accuracy is observed to be 83.98% while the minimum observed is 81.63% at 80 neurons. The maximum accuracy is 86.22% at 200 neurons.

Figure 5 shows the plot of accuracy concerning the number of hidden layers. It begins from the 2nd layer and goes up to 20 layers in depth. Each layer has 40 neurons in an RNN model for multi-class classification. The used class labels are normal and Dos, Probe, U2R, and R2L. This model of neural social media networks comes under the category of deep neural networks. The average accuracy is observed to be 85.36% while the minimum observed was 82.21% at depth of 2 layers. However, the maximum accuracy is 86.14% at the 6th layer depth.

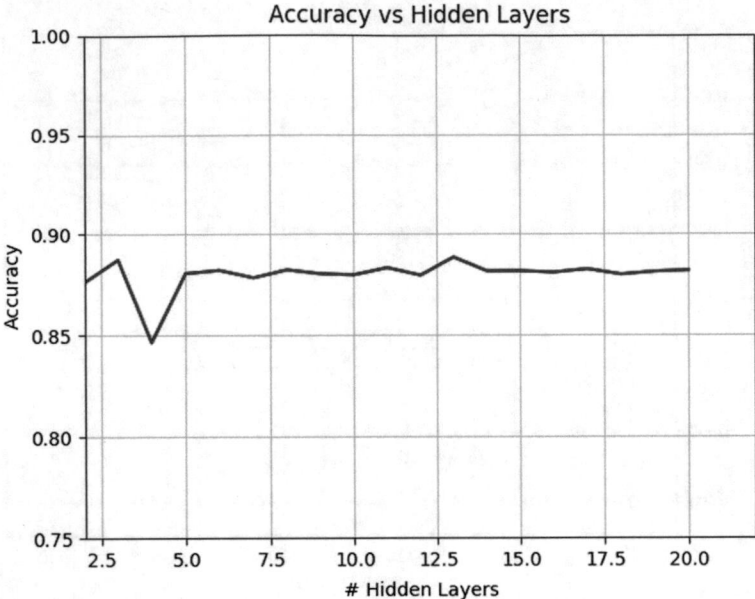

Fig. 3 Graph of accuracy versus increasing number of hidden layers for RNN binary classification

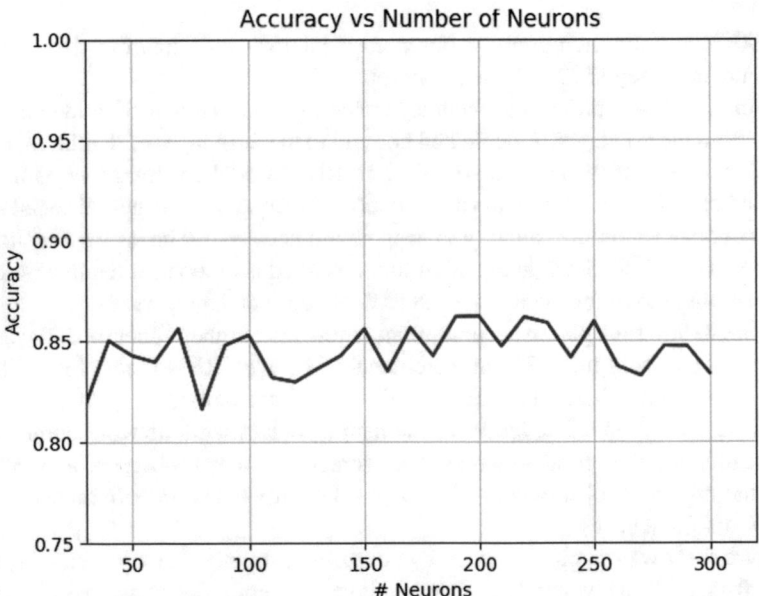

Fig. 4 Graph of accuracy versus increasing number of neurons for RNN multiclass classification

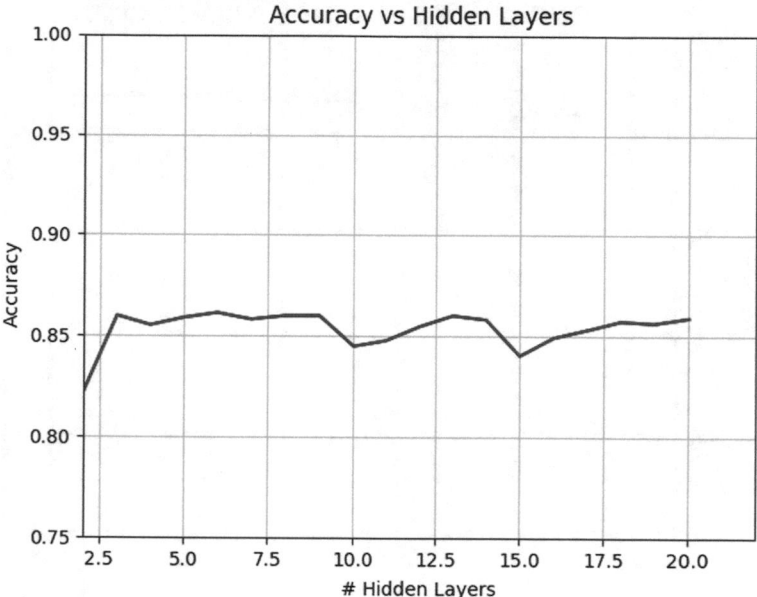

Fig. 5 Graph of accuracy versus increasing number of hidden layers for RNN multiclass classification

Figure 6 shows a plot of accuracy concerning the number of LSTM neurons. It begins at 30 neurons and goes up to 300 in increments of 10 in an LSTM neural social media network model for binary classification. The used class labels are normal and anomaly. This model of neural social media networks comes under the category of wide neural networks. The average accuracy is observed to be 96.48% while the minimum observed is 93.24% on LSTM neurons is 60. However, the maximum accuracy is 97.04% when the number of LSTM neurons is 270.

Figure 7 shows a plot of accuracy concerning the number of hidden layers. It begins from the 2nd layer in-depth and goes up to the 20th layer of depth. Each layer has a fixed number of LSTM neurons. The LSTM model for binary classification uses normal and anomaly class labels. This model of the neural social media network comes under the category of deep neural networks. The average accuracy is observed to be 97.66% while the minimum observed is 96.49% at a depth of the 4th layer. However, the maximum accuracy is 98.49% at a depth of the 18th layer.

Figure 8 shows a plot of accuracy concerning the number of LSTM neurons. It begins at 30 neurons and goes up to 300 in increments of 10 in LSTM. For multiclass classification, it uses class labels named normal and DoS, U2R, R2L, and Probe. This model of the neural social media network comes under the category of wide neural networks. The average accuracy is observed to be 95.14% while the minimum observed is 90.68% when the number of LSTM neurons is 40. However, the maximum accuracy is 96.09% when the number of LSTM neurons is 180.

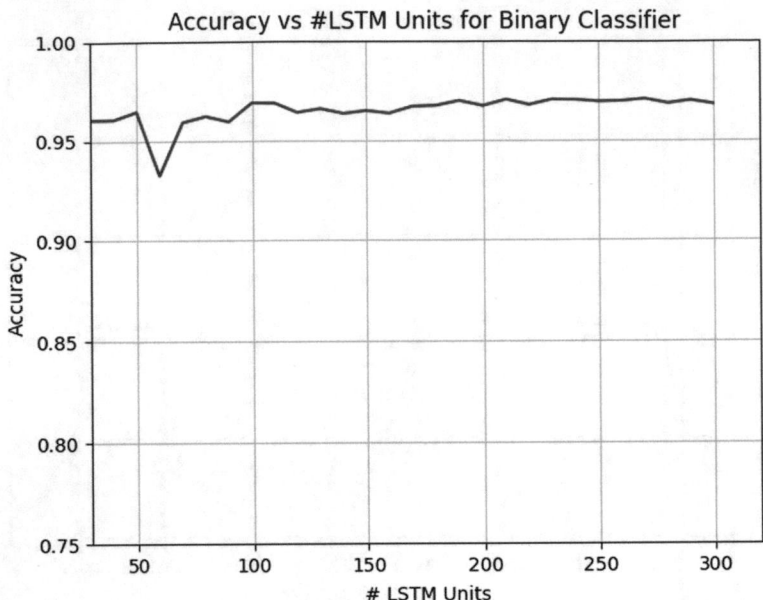

Fig. 6 Graph of accuracy versus increasing number of LSTM neurons for binary classification

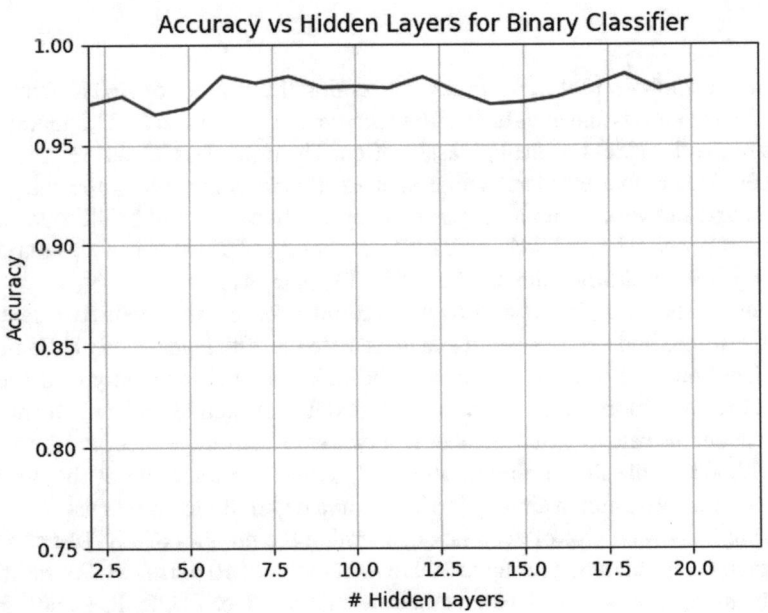

Fig. 7 Graph of accuracy versus increasing number of hidden layers for LSTM binary classification

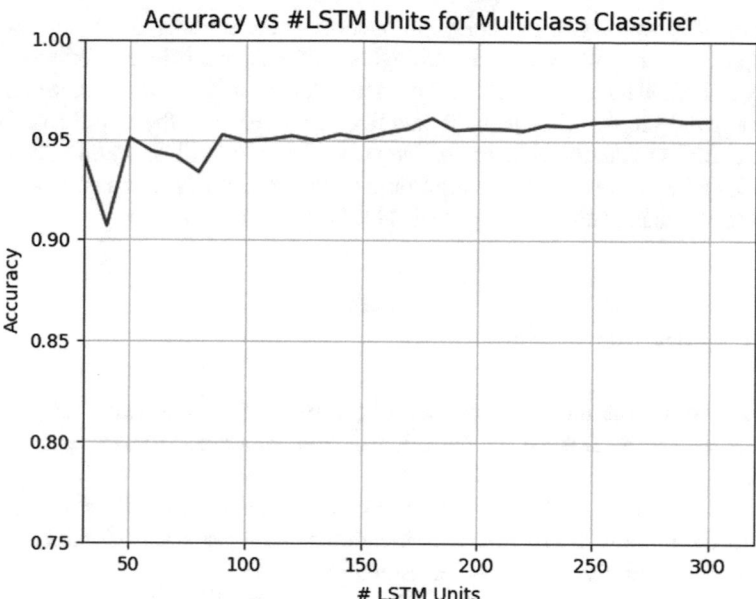

Fig. 8 Graph of accuracy versus increasing number of neurons for multiclass classification

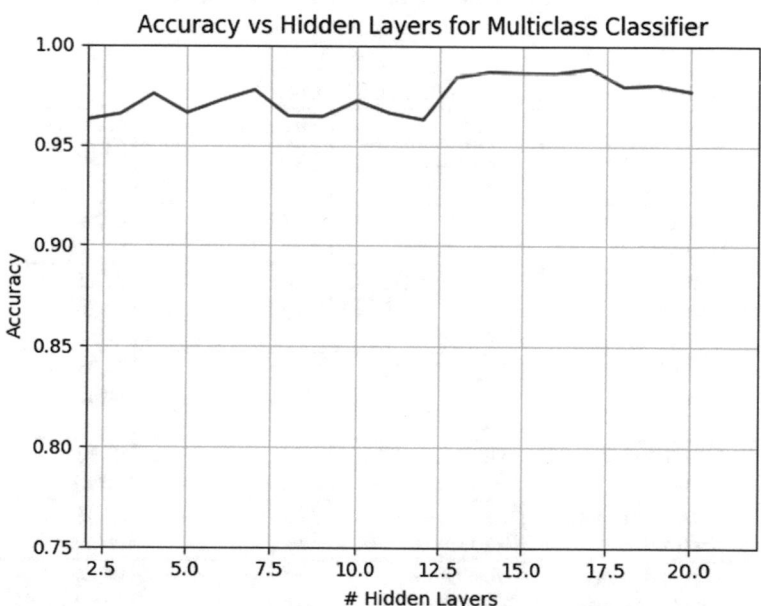

Fig. 9 Graph of accuracy versus increasing number of hidden layers for LSTM multiclass classification

Figure 9 shows a plot of accuracy concerning the number of hidden layers. It begins at 2 layers and goes up to 20 layers in depth. Each layer contains a fixed number of LSTM neurons in LSTM. For multi-class classification, it uses class labels named normal and DoS, Probe, U2R, and R2L. This model of the neural social media network comes under the category of deep neural networks. The average accuracy is observed to be 97.46% while the minimum observed is 96.27% at depth of 2 layers. However, the maximum accuracy is 98.84% at the 17 layer depth.

5.1   Discussion

Table 1 summarizes the accuracies of all the methods implemented in this paper on the NSL—KDD Dataset for detection of anomaly-based social media network intrusion.

From the table, we can easily observe that the Gaussian Naive Bayes classifier performs the poorest. It gives the classification accuracy of 85.3% for binary classification and 67.9% for multi-class classification.

The SVM performs better than the Naive Bayes classifiers. It gives accuracies of 86.25 and 82.75% for binary and multi-class classification respectively. However, the classification accuracies can be improved more. The value of parameter C is set

Table 1 Summarized accuracies of all methods implemented

S. No.	Method	Details	Binary accuracy	Multiclass accuracy
1	Naive Bayes	Gaussian Naive Bayes	85.30	67.90
2	Support vector machine	C = 1.0, Linear kernel	86.25	80.64
3	Multilayer perceptron	Wide network, 50, 150, 200 & 250 neurons	87.06 (avg.)	82.75 (avg.)
4	Multilayer perceptron	Deep network, 5, 10, 15 & 20 layers	87.96 (avg.)	84.35 (avg.)
5	Recurrent neural network	Wide network, 30–300 neurons	90.53 (avg.)	83.98 (avg.)
6	Recurrent neural network	Deep network, 2–20 layers	87.98 (avg.)	85.36 (avg.)
7	Long short term memory neural network	Wide network, 30–300 neurons	96.48 (avg.)	95.14 (avg.)
8	Long short term memory neural network	Deep network, 2–20 layers	97.66 (avg.)	97.46 (avg.)

to 1 and a linear kernel is used. SVMs do not perform well upon the dimensionality of the data is high.

The MLP of the artificial neural networks proves to show that it can provide better accuracy in this case of multi-class classification. The model of MLP is implemented as a wide model with one hidden layer. It takes the number of neurons at 50, 150, 200, and 250 for which the average accuracy has been recorded in the table for the cases of binary and multi-class classification. The deep model of the MLP contains a fixed number of 25 neurons and the number of hidden layers being 5, 10, 15, and 20. Afterward, an average accuracy has been recorded in the table for the cases of binary and multi-class classification. In this implementation, 75% of the training dataset is used for training and 25% for validation. The parameters are set and the hyper-parameters are fine-tuned to obtain the recorded results.

The deep and wide models of RNN perform better than the MLP models. The graph of accuracy versus increasing number of neurons for binary classification in RNN is shown by Fig. 2. It gives small fluctuation inaccuracies in the starting layers due to the initial phases of the network. It is not being able to learn well when the number of neurons is less. The graph of accuracy versus increasing number of hidden layers for binary classification in RNN is shown by Fig. 3. It indicates that the model does not show many fluctuation inaccuracies. The graph of accuracy versus increasing number of neurons for multi-class classification in RNN is shown by Fig. 4. It gives small fluctuation inaccuracies due to the inability of learning and converges. The graph of accuracy versus increasing number of hidden layers for multi-class classification in RNN is shown by Fig 5. It can be seen that not many fluctuation inaccuracies. From the graphs, we can conclude that the deep models of RNN for both binaries, as well as multi-class classification, perform better than the wider models of the RNN.

The deep and wide models of LSTM perform better than the RNN models. The graph of accuracy versus increasing number of LSTM neurons for binary classification in LSTM is shown by Fig. 6. The graph shows very small fluctuation inaccuracies in the beginning layers. The reason is at the initial phases, the social media network is not being able to learn well when the number of neurons is less. The accuracies seem to not saturate at the later layers. The graph of accuracy versus increasing number of hidden layers for LSTM binary classification is shown by Fig. 7. In this Figure, the model does not show many fluctuation inaccuracies. The graph of accuracy versus increasing number of LSTM neurons for multi-class classification in LSTM is shown by Fig. 8. In this Figure, the model shows small fluctuation inaccuracies. the reason is, the social media network is not being able to learn well and converge. The later layers do not show many variation inaccuracies. The graph of accuracy versus increasing number of hidden layers for multi-class classification in LSTM is shown by Fig. 9. In this figure, the model does not show many fluctuation inaccuracies. From these graphs, we can conclude that the deep models of LSTM for both binaries, as well as multi-class classification, perform better than the wider models of the LSTM. In addition, the LSTM neural social media network models perform better than the RNN models for anomaly-based intrusion detection.

6 Conclusion

Anomaly-based IDS is an effective method for the detection of previously known and also new/unknown attacks in the systems. It provides some procedures to detect the possible threats by continuous monitoring and modeling of behavior in networks. From the above plots, we can conclude that the LSTM with an accuracy of around 97% for both multi-class as well as binary classification. It can also be observed that the accuracy of binary classification is higher than that for multi-class classification. The deep networks generally perform better than the wide models of the networks in the case of RNN, LSTM, and MLP. It may occur because the deeper models of the networks can model the data representation better than the wider networks.

7 Future Scope

Feature selection algorithms can be applied and dimensionality reduction can be performed In our future work. In this way, we can reduce the load on the processors. It may also affect the performance of the models. The normalization methods can be varied to check the performance of the model. The model will work efficiently if a reduction will be applied on run time for the training and testing of the networks. In addition, experimentation can be done with the Bidirectional RNN, Neural social media network containing Gated Recurrent Units, and LSTM Autoencoders to observe any improvement in the accuracies.

References

1. Kuang, F., Xu, W., Zhang, S.: A novel hybrid KPCA and SVM with GA model for intrusion detection. Appl. Soft Comput. (2014)
2. Fiore, U., Palmieri, F., Castiglione, A., Santis, A.: Social media network anomaly detection with the RBM Neurocomputing. Neurocomputing 11 (2013). Elsevier B.V
3. Yin, C., Zhu, Y., Fei, J., He, X.: A deep learning approach for intrusion detection using recurrent neural networks. IEEE Access 5 (2017)
4. Thapa, N., Liu, Z., Kc, D. B., Gokaraju, B., Roy, K.: An anomaly-based social media network intrusion detection system using deep learning. In: 2017 International Conference on System Science and Engineering (ICSSE), Ho Chi Minh City (2017)
5. Reddy, R.R., Ramadevi, Y., Sunitha, K.V.N.: Effective discriminant function for intrusion detection using SVM. In: Proceeding of the International Conference on Advances in Computing, Communications and Informatics (ICACCI) (2016)
6. Alom, M.Z., Bontupalli, V., Taha, T.M.: Intrusion Detection using DBN. In: National Aerospace and Electronics Conference (NAECON) (2015)
7. Li, W., Yi, P., Wu, Y., Pan, L., Li, J.: A new intrusion detection system based on KNN classification algorithm in wireless sensor network. J. Elect. Comput. Eng. (2014)
8. Niyaz, Q., Sun, W., Javaid, A., Alam, M.: A deep learning approach for NIDS. In: Bio-inspired Information and Communications Technologies, (BIONETICS). Belgium, Brussels (2014)

9. Ingre, B., Yadav, A.: Performance analysis of NSL-KDD dataset using ANN. In: Proceedings of the International Conference on Signal Processing and Communication Systems (2015)

10. Buczak, A.L., Guven, E.: A survey of data mining and machine learning methods for cyber security intrusion detection. IEEE Commun. Surveys Tuts. **18** (2016)

11. Zhang, J., Zulkernine, M., Haque, A.: Random-forests-based social media network intrusion detection systems. IEEE Trans. Syst., Man, Cybern. C, Appl. Rev. (2008)

12. N. Farnaaz and M. A. Jabbar, "Random forest modeling for social media networkintrusion detection system," Procedia Comput. Sci., Jan. 2016

13. Khan, J.A., Jain, N.: A survey on intrusion detection systems and classification techniques. Int. J. Sci. Res. Sci., Eng. Technol. (2016)

14. Tang, T.A., Mhamdi, L., McLernon, D., Zaidi, S.A.R., Ghogho, M.: Deep learning approach for social media network intrusion detection in software defined networking. In: Proceedings of the International Conference on Wireless Networks and Mobile Communications (WINCOM) (2016)

15. Javaid, A., Niyaz, Q., Sun, W., Alam, M.: A deep learning approach for social media network intrusion detection system. In: 9th EAI International Conference on Bio-inspired Information and Communications Technology (BIONETICS) (2016)

16. Sheikhan, M., Jadidi, Z., Farrokhi, A.: Intrusion detection using reduced-size RNN based on feature grouping. Neural Comput. Appl. (2012)

17. Dhanabal, L., Shantharajah, S.P.: A study on NSL-KDD dataset for intrusion detection system based on classification algorithms. Int. J. Adv. Res. Comput. Commun. Eng. **4** (2015)

18. Tavallaee, M., Bagheri, E., Lu, W., Ghorbani, A.: A detailed analysis of the KDD CUP 99 data set. In: Submitted to Second IEEE Symposium on Computational Intelligence for Security and Defense Applications (CISDA) (2009)

19. Sahu, S.K., Mohapatra, D.P., Rout, J.K., Sahoo, K.S., Pham, Q.V. and Dao, N.N.: A LSTM-FCNN based multi-class intrusion detection using scalable framework. Comput. Electr. Eng. **99**, 107720 (2022)

20. Alqahtani, A.S.: FSO-LSTM IDS: hybrid optimized and ensembled deep learning network-based intrusion detection system for smart networks. J. Supercomput. 1–18 (2022)

21. Chen, A., Yang, F., Zheng, X., Guoming, L.: An efficient network behavior anomaly detection using a hybrid DBN-LSTM network. Comput. Secur. **114**, 102600 (2022)

22. Meliboev, A., Alikhanov, J., Kim, W.: Performance evaluation of deep learning based network intrusion detection system across multiple balanced and imbalanced datasets. Electronics **11**(4), 515 (2022)

Deep Digging of Anomalous Transactions in Financial Networks with Imbalanced Data

Vaishali Kansal and Pradumn Pandey

Abstract Anomaly (user or transaction) detection is a major concern and has intrigued a wide spectrum of disciplines, for example, finance, security, etc. Anomalous instances in a dataset are very rare in general, which leads to imbalanced distribution of classes in the dataset, and poses a great challenge to anomaly detection. In the case of credit card transactions, it is of utmost importance to identify fraudulent transactions in order to save financial organizations from giving credit to users that are not likely to be repaid in the future. Most of the prevailing techniques to address the class imbalance problem aim to balance the dataset by either producing synthetic data samples of a minority class or eliminating some samples of the majority class. In this chapter, we provide a brief overview of the challenges faced due to class imbalance while learning deep learning-based prediction models and discuss the sampling strategies that deal with the class imbalance data problem, for example, oversampling, SMOTE, and its variants, and data ensembles. Data ensembles divide the dataset into small subsets that are in itself balanced. Then, corresponding to each balanced sub-dataset, a classifier is trained, and voting is used for the first level of prediction. As deep learning approaches have the ability to map complex non-linear relations within high-dimensional data, they can be leveraged to anomaly detection problems. To improve the classification task's performance and avoid overfitting, Deep Multilayer Perceptron (DMLP) and Deep Convolutional Neural Network (DCNN) are trained for each balanced sub-dataset, and second level voting is performed. Also, a detailed numerical analysis of the proposed architecture is provided.

Keywords Anomaly detection · Deep learning · Ensemble learning · Class-imbalance data · Credit card fraud detection

V. Kansal (✉) · P. Pandey
Department of Computer Science and Engineering, Indian Institute of Technology, Roorkee, India
e-mail: vaishali_k@cs.iitr.ac.in

P. Pandey
e-mail: pradumn.pandey@cs.iitr.ac.in

© The Author(s), under exclusive license to Springer Nature Switzerland AG 2022
T.-P. Hong et al. (eds.), *Deep Learning for Social Media Data Analytics*,
Studies in Big Data 113, https://doi.org/10.1007/978-3-031-10869-3_15

1 Introduction

Anomalies are those patterns in the data whose behavior may or may not vary considerably from the majority of the data. Still, identification and avoidance of such instances are important to circumvent unnecessary loss or damage. Various types of anomalies and their categorization is done by the authors [10, 40, 65] based on various characteristics, for example, behavior of anomaly [14, 17], nature of anomaly [10, 25], static/dynamic nature [65], interaction pattern [3], available information [14, 65], etc. Anomaly detection defines the problem of finding the unexpected hidden behavior of users by exploring different patterns. Over the past decades [1], anomaly detection has received a lot of interest due to occurrences of such events in a diverse range of disciplines—for example, in finance [1], security [68], etc. Financial security is a major concern for any financial organization. For credit card companies, any transaction which is likely to be fraudulent must be identified to save the company from giving out credit that may not be repaid.

In order to discriminate between various approaches applied for anomaly detection in a structured way, authors in [40] classified them into three categories. Proximity based methods perform analysis based on its neighbors. There are different similarity measures to find the proximity like Distance [24, 42], Density [8, 24, 32], etc. Various clustering-based approaches have also been applied [35, 56] in order to detect an anomaly. These are unsupervised methods. Proximity methods are highly dependent on similarity measures, and clustering methods incur a high computational cost for large data sets.

The classification problem has been applied to various fields, including anomaly detection. The idea is to be able to classify different data points into different classes, and anomaly detection is a binary classification problem where a class is either normal or abnormal. Classification requires labeled data and is hence referred to as a supervised approach. Different data science and machine learning techniques, such as Support Vector Machines (SVM) [48, 57], Decision Trees (DT) [22], Random Forest (RF) [46, 74], have been applied for predicting if a transaction is fraudulent, using the information of previous transactions and the ones among them that turned out to be fraudulent. The general idea for each of these approaches is to teach a model the similarity or difference within and between different classes based on the labeled data. Also, bayesian classifiers [45, 51] like basic Naive Bayesian classifiers have been used. Numerous deep learning techniques have also been applied in the anomaly detection domain, as discussed by authors in [9, 47], due to its ability to learn complex dynamics and outperform the classical models. Neural networks have the ability to learn more complex underlying structures of different classes. Several types of neural networks [38, 53, 55] are commonly used, such as Multi-Layer Perceptron (MLP) [5, 58, 69], Convolutional Neural Network (CNN) [21, 41, 52], etc.

Anomalies are very rare instances in comparison to normal instances in the data, which implies that anomaly detection always suffers from class-imbalanced data [37]. The performance of deep learning models majorly depends on training data, and such class-imbalanced data poses great challenges to anomaly detection problem [38]. Imbalanced class distributions degrade the performance of classification techniques [4, 67], as most learners tend to get biased towards the majority class and in worst cases tends to neglect minority class completely [30, 44]. Moreover, learners fail to capture the decision boundary between anomalies and non-anomalies. Predictions obtained from such highly imbalanced data might even lead to over-fitting on the anomalous class [44]. So it is of utmost importance to consider this critical fact of imbalanced-class in anomaly detection problems so as to train the detection model.

To address the class imbalance problem, various re-sampling techniques have been studied [30, 43]. Re-sampling techniques aim to balance the available skewed data in order to ease the learning process. Re-sampling techniques can be broadly classified into three categories [6, 12, 18, 75]: under-sampling, over-sampling, and hybrid. The over-sampling technique balance the skewed class distribution by producing new instances of minority class [33]. Two widely used over-sampling methods are Synthetic Minority Oversampling Technique (SMOTE) [11] and Adaptive synthetic (ADASYN) [34] that generates synthetic instances of minority class using the nearest neighbor approach for balancing the class distribution. Under-sampling balances the skewed class distribution by eliminating some instances of majority class [50]. One of the basic techniques to discard samples is random, followed by the Random under-sampling technique. Hybrid technique is integration of over-sampling and under-sampling techniques [15, 39].

1.1 Motivation

Imbalanced data is an important problem in classification tasks where the distribution of class is highly skewed. Detecting fraud is a data mining problem that has two major challenges. The first challenge is that the behavior profiles of normal and fraudulent behaviors alter persistently, and the second is that datasets available for a credit card are highly imbalanced. Hence, the efficiency of fraud detection is extremely affected by the selection strategy of sampling used for balancing the data and the selection of classification techniques.

This chapter explores different sampling strategies, including dealing with the class imbalanced data that neither requires producing synthetic data nor eliminates the existing data. It divides the dataset into subsets in such a way that every subset contains a balanced distribution of both classes. Then, the classification model M is trained on each subset. Thus, corresponding to ith subset, we have a trained model M_i. Further, predictions made by individual learners (M_i) are aggregated by taking the majority vote. Here, we also aim to analyze the performance of classification tasks using Hybrid Ensemble-based learning classifiers, which combines several base learners (In this chapter, we consider two learners, DMLP and DCNN), to improve the performance by two-level of voting. Numerical analysis suggests that the voting ensemble produces better results than single classifier predictions. Also, out of two ways of voting employed at the second layer, Consensus For Acceptance

(CFA) method tends to identify true positive more accurately while Consensus For Rejection (CFR) method true negative without much compromising the counterparts. Different metrics and measures are considered for evaluating the performance, such as precision, recall, f1-score, and confusion matrix.

2 Literature Review

This section presents a brief review of the application of deep learning approaches in anomaly detection in the financial domain, such as credit card fraud detection and various challenges associated with classification for imbalanced class data.

2.1 Anomaly Detection Using Deep Learning

In the last few decades, computational intelligence in the finance industry has become a very popular topic [53]. Recently, deep learning, which is a sub-field of machine learning, has attracted many researchers due to its ability to outperform all the classical models [9, 47, 55]. Numerous deep learning models by various researchers have been proposed to address financial fraud detection [27, 53, 55].

Various deep learning approaches have been applied to address anomaly detection problems in various domains [9, 47]. Deep learning approaches offer several advantages, such as its ability to work in conjunction with high dimensional and multivariate data, which helps to integrate information from multiple sources and get rid of the challenges that are associated with an overhead of separate modeling of anomalies for each variable [9, 47]. The performance of deep learning models tend to scale with the availability of appropriate training data, which is its another advantage [38]. Also, deep learning methods offer the opportunity to model very complex non-linear relationships within data and leverage it for various anomaly detection problems [55].

In the literature [38, 54, 64], various deep learning models exist that can be broadly classified into several categories as shown in Fig. 1. Supervised methods, such as MLP [5, 58, 69], CNN [21, 52], RNN [71], (2) Unsupervised methods, such as DBN [36, 76], AE [77, 79], GAN [13, 19, 23, 72], RBM [59] and (3) Hybrid methods, such as ensemble learning [5, 20, 66, 68, 70, 73].

2.2 Imbalanced Data

As discussed before, class imbalanced problem [37] occurs when one class has an abundant number of instances compared to another class which has a very rare number of instances. A problem with imbalanced classification is that it is hard for a

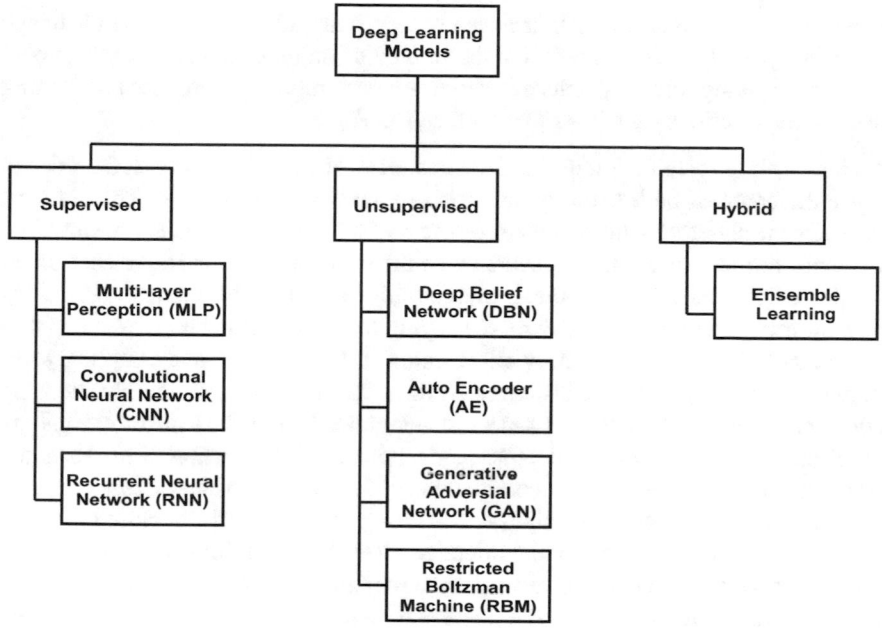

Fig. 1 Taxonomy of deep learning models applied to anomaly detection problem

model to learn the decision boundary effectively. This problem is usually observed in industries like medical [29], financial [12]. In all cases, the minority group which has a rare number of instances is actually the class of interest (positive class). In imbalanced class data, classifiers incorrectly over classify the majority group (negative class) as a consequence of its increased prior probability. Therefore, the classifier often misclassifies the positive class instances compared to negative class instances. Furthermore, basic evaluation metric, i.e., accuracy in such cases, which is common for performance evaluation of a classifier, may mislead an analyst by providing high scores that actually is an incorrect indication of good performance in case of highly imbalanced data. Therefore, using the right evaluation metrics is important in case of imbalanced data.

One of the critical challenges in countering credit card fraud detection is the highly skewed distribution of fraudulent and normal transactions, which degrades the potential of machine learning algorithms to discover patterns of fraudulent behavior. Sampling is a way to deal with imbalanced data. Major objective of the sampling strategy is to alter the distribution of the dataset in order to balance the majority and minority classes. There are several ways of sampling [75], such as oversampling the minority class [33], undersampling of majority class [50], a hybrid approach combining both undersampling and oversampling [15, 39].

Undersampling technique [50] under-represent the majority class by selecting instances from it uniformly. A significant advantage of undersampling is that it creates a model on the data which has been observed already. However, the challenge associated with this technique is that the amount of training data given to the model for learning is significantly reduced. Various undersampling techniques for learning skewed distributions have been well studied [6, 49].

Oversampling technique replicates the instances of the minority class in order to make the count of both majority and minority class instances equal. This, in turn, provides the classifier with adequate instances of both the classes to learn. However, there are various challenges associated with the oversampling approach, such as (a) it eventually may lead to over-fitting of a classifier, and (b) it doesn't incorporate any new information into the data by mere copying the same instances.

SMOTE [11] is one of the well-accepted state-of-the-art oversampling techniques. It produces synthetic instances for the minority class by randomly choosing some minority samples and their k-nearest neighbors. One of the k nearest neighbors is selected for each sample, and a line segment is drawn connecting them. Then, the instance is produced as a convex combination of the selected instance. This sampling technique is efficient as the newly produced instances are relatively close in feature space to prevailing instances of the minority class. A general limitation of this sampling method is that it does not consider the majority class for generating synthetic instances, resulting in ambiguous examples if there is a strong overlap between the classes.

Borderline SMOTE [31] The borderline instances and ones in close proximity to borderline are more prone to be misclassified than the instances far from borderline. These instances are generally ambiguous and lie around the border of decision boundaries where membership of classes tends to overlap. So, synthetic examples are only created along the decision boundary between the two classes in borderline SMOTE. So, based on the class of instances selected as nearest neighbors, there are many possible variants of SMOTE. Another main improvement made to SMOTE is the ADASYN [34] sampling technique in which the samples that are very difficult to learn are assigned more importance and are likely to be oversampled more often.

Numerous ways to handle imbalanced data have been proposed in the literature [30, 43, 44]. Note that no sampling method has a complete advantage over another. Most of the existing techniques that apply sampling strategy aim to balance the data either by generating synthetic data instances for the minority group or by eliminating some instances of the majority group. These techniques either suffer from overfitting or lose valuable information. In both cases, the classifier is likely to produce inaccurate predictions giving incorrect high accuracy. Therefore, it is of utmost importance to carefully choose sampling strategy as the classifier may suffer from overfitting or underfitting, which can lead to misclassification of fraudulent transactions as normal or vice-versa.

2.3 Ensemble Learning

Deep neural networks are non linear methods that offer increased flexibility as they can be modelled for high dimensional data and can be scaled in proportion to the amount of training data available. A major drawback of its flexibility is that it uses a stochastic gradient descent training algorithm for learning which signifies that they are highly dependent on the specifications of the training data. This, in turn, produces different predictions every time they are trained as they get different sets of weights. Therefore, neural network suffers from high variance problem. To deal with this problem, multiple models are trained instead of a single model, and then the predictions made by all the models are combined. This approach is called ensemble learning [78] which helps in reducing the variance as well as results in better predictions compared to any single model.

Ensemble methods combine various base classifiers to improve the performance of any individual classifier. Ensemble learning algorithms modify the training data distribution by dividing it into multiple subsets. Then, it deploys multiple base classifiers on every subset to make predictions and finally aggregate the result. Boosting and Bagging [7] algorithms are two common types of ensemble learning algorithms. The bagging technique works by taking random subsets with replacement. Many researchers have improved these existing algorithms to address the imbalanced class data problem [16]. Authors in [63, 76] employed ensemble learning techniques in the financial domain.

Empirically, ensemble methods produce better results when significant diversity is introduced among the models. Therefore, ensemble methods aim to promote diversity among the combined models. Although more random algorithms like random decision trees have proven to be stronger ensembles than any other decision tree algorithm. Thus, employing multiple robust learning algorithms tends to be more effective than applying techniques that attempt to downgrade the model performance to promote diversity.

This work employs an ensemble of DMLP and DCNN classifiers as base classifiers to carry out fitting on every subset of the actual dataset. Then, predictions obtained from each individual classifier are aggregated. Aggregation here is performed at two layers by employing a voting ensemble.

3 Methodology

Neural networks have been widely used for credit card fraud detection as they have the adaptive ability to learn from patterns of normal transactions which in turn helps in identifying patterns of fraud transactions [26, 60, 62].

MLP is a fully-connected feed-forward neural network that consists of a minimum of one hidden layer. Deep MLP (DMLP) is the first and basic deep learning model with respect to implementation. The only difference between MLP and DMLP is the number of layers which is more in DMLP compared to MLP. Each neuron in every layer has different input x, bias b, and weight w. Moreover, each neuron has a non-linear activation function that accumulates the weighted inputs from the neurons in the previous layer and generates the output. Output)(y) is calculated using Eq. 1. Some of the most preferred non-linear activation functions used in the literature [28, 61] are Rectified Linear Unit (ReLU), leakyReLU, sigmoid, softmax, hyperbolic tangent, and swish.

$$y_i = \sigma \left(\sum_i w_i x_i + b_i \right) \tag{1}$$

Multilayer deep ANNs help in achieving efficient classification performances. Its learning process is performed using the backpropagation process in which the output error from the output layer neurons is propagated back to the previous layer neurons. DMLP uses the Stochastic Gradient Descent (SGD) method to optimize learning for updating the weights of the connections between the layers.

Using deep learning, a predictive model can be designed for large complex data, which automatically extracts nonlinear features by deeply stacking layers. ANN has the ability to handle complex data and learn itself, but its learning process is very slow. While learning time is very less in the case of DCNN, it has the ability to avoid overfitting of the model [21, 52, 52]. DCNN is one of the multilayer deep learning models which removes local features through weighted filters in a hierarchical manner. It works by learning the deep nonlinear structure of the network, realizing complex function approximation, and representing the input-output mapping relationship. DCNN comprises convolutional, subsampling, and fully connected layers.

This chapter exploits an ensemble of the DCNN and DMLP approaches for classifying fraudulent transactions in credit card transaction data. The data for credit card transactions contains a specific pattern for the transactions made. The convolutional layer of DCNN tends to capture different features of the transaction data, and the pooling layer then merges features that are similar in nature into one. In order to extract basic features in the lower layer, several convolutional and pooling layers are stacked, and complex ones are derived in the upper layer, which is the fully connected layer. The DCNN model can classify the fraudulent user by removing discriminative features among them and learning patterns in the available transaction data. We have designed a one-dimensional DCNN to analyze its performance in credit card fraud detection and evaluated whether it can be generalized for other unseen transactions.

3.1 Data

We consider a data of size 284807 data points from Kaggle[1] which contains transactions made with credit cards by European cardholders. The data is highly skewed as

---

[1] https://www.kaggle.com/mlg-ulb/creditcardfraud.

it contains only 492 fraudulent (positive class) transactions out of 284807, which are only 0.17% of the considered data. The data is already pre-processed, and dimension reduction using Principal Component Analysis (PCA) is performed. Hence, it contains only numerical input values of 28 composite features. We eliminate the 'time' feature due to its insignificant importance. 'amount' feature is scaled using the Robust Scaler method [2] as this feature is not considered while applying PCA. Thus, we have a total of 29 features as input to a model to be trained. The data has two classes in which the value 0 represents a normal transaction, and the value 1 corresponds to a fraudulent transaction.

3.2 Subset Resampling

In this section, we discuss about a sampling method (subset resampling) to solve the problem of data imbalance without deleting a data point or adding synthetic minority class data points. Deleting a data point leads to information loss, and the addition of extra data points (synthetic data) causes an increase in training time. Here, we also do the performance analysis of the discussed sampling method and other state-of-the-art techniques for imbalanced data, such as oversampling, SMOTE, and its different variants.

Let we have r rare or minority and m majority class data points. In subset resampling, first, 20% of minority class data is picked for testing and same amount of data $(0.2 \times r)$ is considered from majority class. Remaining $(m - 0.2 \times r)$ majority class data points are divided into multiple subsets, let say S_1, S_2, ..., S_n, such that each subset has almost $|S_i| \approx 0.8 \times r$ data points, $\forall$ i. Now remaining minority class data points R_1 is combined with each S_i and $R_1 \cup S_i$ is used for training models M_i and M_i'. Figure 2 shows the proposed framework.

Further, we have used the hybrid ensemble method to enhance the performance of the classification task. We used two deep learning models, DMLP(M) and DCNN(M'), as the base classifiers. The resampled training subsets $(R \cup S_i, \forall i)$ are fed to base classifier models for learning. Each model carries outfitting on every subset and produces a set of predictions $P_i(M)$ and $P_i(M')$ for the same test set. In this way, n number of prediction sets are generated by each classifier (M and M') for the same given test set. Then, first-level voting V_1 is employed to aggregate each model's n set of prediction results separately. This, in turn, generates a set of predictions corresponding to each classifier. Then, second-level voting V_2 is applied to aggregate the prediction results of both the classifiers, see Fig. 2.

Two variants of SMOTE: SMOTE$_1$ and SMOTE$_2$, have been considered here for comparison. SMOTE is a sampling technique in which synthetic samples for minority classes are generated by taking nearest neighbor samples in minority classes only. This technique strengthens the minority class inside the decision boundary. In SMOTE$_1$, synthetic samples are generated by taking the nearest neighbor in the majority class only, which in turn helps to strengthen the border of the minority class in close proximity to the majority class, while SMOTE$_2$ is a hybrid approach of

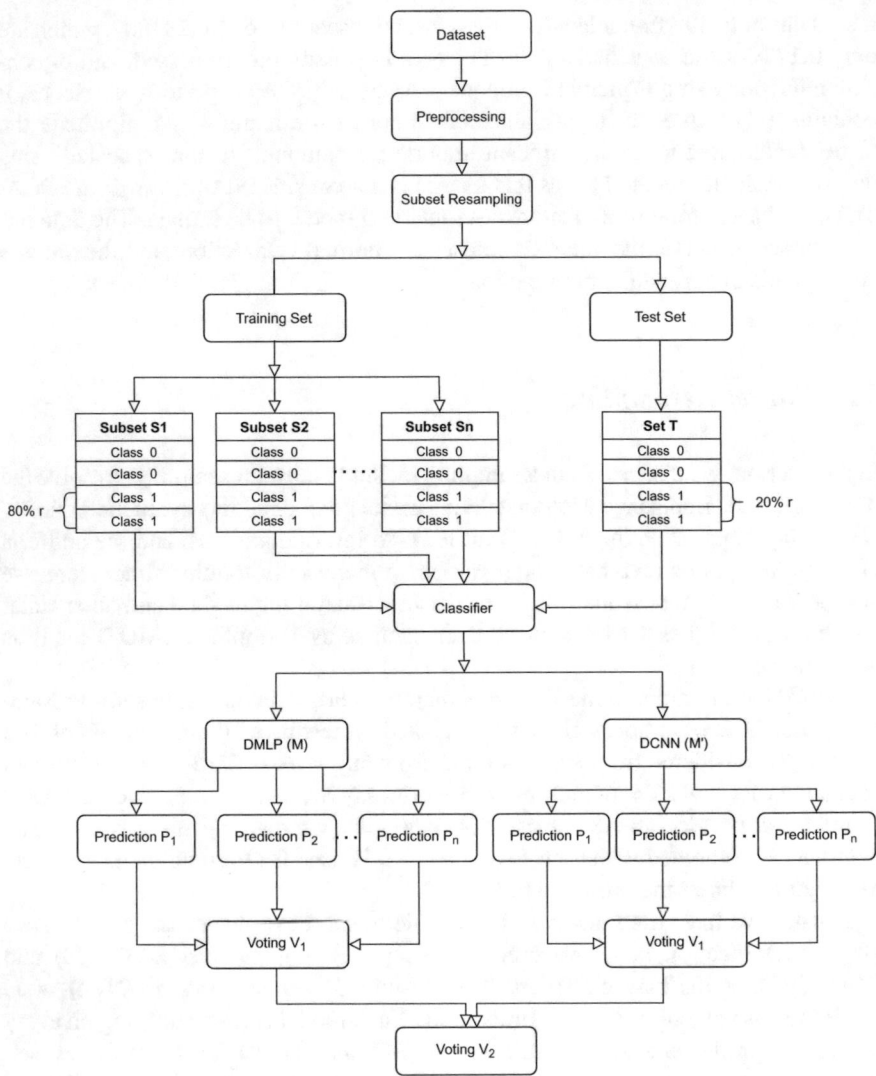

Fig. 2 Proposed framework to detect fraud in credit card transaction data which is sampled using subset resampling strategy to address class imbalanced data and is based on ensemble of two deep learning models DMLP(M) and CNN(M')

SMOTE and SMOTE_1. In SMOTE_2, synthetic samples are generated by considering nearest neighbor samples in both minority and majority classes. This strengthens the minority class within the boundary and at the boundary.

3.3 Performance Metrics

The metrics utilized for evaluation are Accuracy, Recall, Precision, Confusion Matrix, and f1-Score.

While evaluating the performance of various classification techniques, **Accuracy** (Ac) is the most conventionally used metric. It defines how many of the instances of the total instances are correctly classified as defined in Eq. (2).

$$Ac = \frac{TP + TN}{TP + TN + FP + FN},\tag{2}$$

where TP, TN, FP, and FN are true positive, true negative, false positive, and false negative, respectively. In the case of highly imbalanced data, Ac is not an adequate measure of performance, for example, let's assume in the considered data set only 0.17% are rare events, fraudulent transactions. In such a scenario, the model fails to classify fraudulent transactions and tends to produce biased predictions towards the majority class with more than 99.8% model accuracy if it labels all instances as negative (normal transactions). **Precision** (P) is another metric that measures the portion of positively identified class of interest (positive class or fraudulent transactions) from the total number of positively predicted instances, and defined as

$$P = \frac{TP}{TP + FP}.\tag{3}$$

Calculating precision alone is not sufficient for imbalanced class data as it does not consider the number of positive samples incorrectly labeled as negative. Similarly, **Recall** (R) measures the portion of accurately predicted positive instances from the total number of positive instances and defined as

$$R = \frac{TP}{TP + FN}\tag{4}$$

The recall metric is an important measure as it gives the ratio of anomalies detected (fraudulent transactions). It is not affected by imbalanced class data as it depends only on the positive class. It does not take into account the portion of negative samples that are incorrectly classified as positive, which actually creates a problem in cases where data is highly skewed with abundant negative samples and very few positive samples. For applications such as fraudulent transactions detection, rare disease identification, and rare earth event detection, we require a metric that has a high value if recall and precision are high; otherwise, not. **f1-Score** ($f1$) combines

Table 1 Confusion matrix

	Predicted negative (0)	Predicted positive (1)
Actual negative (0)	TN	FP
Actual positive (1)	FN	TP

recall and precision metric in order to define the relative importance of precision versus recall as stated in Eq. (5).

$$f1 = 2\frac{P \times R}{P + R}.$$ (5)

Confusion Matrix (C) A confusion matrix presents the performance of a classifier in the form of matrix in terms of predicted instances versus actual instances. The structure of a confusion matrix is presented in Table 1.

4 Experiments and Analysis

In this section, we do a performance analysis of the proposed architecture shown in Fig. 2 and various sampling techniques adopted to work with imbalanced data. We show the effectiveness of the two-layer voting system in the hybrid ensemble method. At the first level of voting (V_1), we consider majority voting to decide that the transaction is fraudulent or not, but at second level voting (V_2), we adopted two strategies: (1) Consensus For Acceptance (CFA), and (2) Consensus For Rejection (CFR). CFA defines that if after first level voting, all say that the transaction is fraudulent, then it is, otherwise not. CFR is opposite to CFA; it means that if, after first level voting, all say that the transaction is not fraudulent, then it is not; otherwise, it is.

In credit card fraud detection, detecting maximum possible fraudulent transactions is important, i.e., a number of predicted true positives (TP) should be close to actual positives. Also, for some finance industries, not losing any valuable customer at any cost is more important, so, in this case, TN should be close to actual negatives. Therefore, we have analyzed the performance on the basis of an average number of predicted TP and TN (as stated in confusion matrix C), P, and R for 10 consecutive executions. Moreover, in the case of imbalanced class distribution, f1-score is a better measure to evaluate the model performance, so we have evaluated f1-score also, $f1_0$ for class 0 and $f1_1$ for class 1.

First, we compare the performance of the subset resampling technique with over-sampling technique. It is observed from Table 2, the average number of TP predicted in the case of subset resampling is 86.7, which is higher than 79.1 obtained in over-sampling. Also, in the case of subset resampling, the value of recall R and f1-score for class 1, $f1_1$, that are 0.89 and 0.94, respectively, are significantly higher as com-

Table 2 Comparison of Oversampling technique with Subset Resampling technique for CFA voting ensemble of DCNN and DMLP classifier. Here, P is precision, R is recall, Ac is accuracy, $f1_0$ is f1-score for class 0, $f1_1$ is f1-score for class 1, C is the confusion matrix calculated in the same order as defined in Table 1

S.no.	Oversampling						Subset resampling					
	P	R	Ac	$f1_0$	$f1_1$	C	P	R	Ac	$f1_0$	$f1_1$	C
1	0.99	0.82	0.9	0.91	0.89	97 1 18 80	1.00	0.90	0.95	0.95	0.95	98 0 10 88
2	1.00	0.84	0.92	0.92	0.91	98 0 16 82	1.00	0.88	0.94	0.94	0.93	98 0 12 86
3	1.00	0.79	0.89	0.90	0.88	98 0 21 77	1.00	0.87	0.93	0.94	0.93	98 0 13 85
4	1.00	0.84	0.92	0.92	0.91	98 0 16 82	0.99	0.89	0.94	0.94	0.94	97 1 11 87
5	1.00	0.8	0.9	0.91	0.89	98 0 20 78	1.00	0.88	0.94	0.94	0.93	98 0 12 86
6	1.00	0.84	0.92	0.92	0.91	98 0 16 82	1.00	0.89	0.94	0.95	0.94	98 0 11 87
7	1.00	0.82	0.91	0.92	0.9	98 0 18 80	0.99	0.89	0.94	0.94	0.94	97 1 11 87
8	1.00	0.82	0.91	0.92	0.9	98 0 18 80	0.98	0.89	0.93	0.94	0.93	96 2 11 87
9	1.00	0.74	0.87	0.89	0.85	98 0 25 73	1.00	0.86	0.93	0.93	0.92	98 0 14 84
10	1.00	0.79	0.89	0.90	0.88	98 0 21 77	0.99	0.92	0.95	0.96	0.95	97 1 8 90
Avg.	1.00	0.81	0.90	0.91	0.89	97.9 0.1 18.9 79.1	1.00	0.89	0.94	0.94	0.94	97.5 0.5 11.3 **86.7**

pared to the oversampling, that are 0.81 and 0.89. Similarly, in Table 3, TP is 90.2 and $(R, f1_1)$ is $(0.92, 0.95)$ for subset resampling which is higher than 85.9 and $(0.88, 0.93)$ which are corresponding to oversampling. Therefore, subset resampling is performing better than oversampling. Also, it is observed that for subset resampling, the average number of TP predicted in CFR voting ensemble in Table 3 is 90.2, which is higher in comparison to the average number of TP 86.7 predicted in the CFA voting ensemble, see in Table 2. Similar observations are made for other considered sampling techniques also. After analyzing the contemplated performance of two-layer voting in oversampling and subset resampling method, we have then employed it on well known SMOTE sampling method and its variants. For SMOTE, it is observed from Table 4, that the average number of predicted TP in the case of CFR voting ensemble is 98, which is higher than 95.2 for CFA voting ensemble and $(97.8, 95.4)$ for individual classifiers DMLP and DCNN, respectively. Also, the average number of predicted TN in the case of CFA voting ensemble is 98, which is higher than 97 for CFR voting ensemble and $(97.7, 97.3)$ for individual classi-

Table 3 Comparison of Oversampling technique with Subset Resampling technique for CFR voting ensemble of DCNN and DMLP classifier. Here, P is precision, R is recall, Ac is accuracy, $f1_0$ is f1-score for class 0, $f1_1$ is f1-score for class 1, C is the confusion matrix calculated in the same order as defined in Table 1

	Oversampling						Subset resampling					
S.no.	P	R	Ac	$f1_0$	$f1_1$	C	P	R	Ac	$f1_0$	$f1_1$	C
1	0.99	0.92	0.95	0.96	0.95	97 1 8 90	0.99	0.93	0.96	0.96	0.96	97 1 7 91
2	0.98	0.94	0.96	0.96	0.96	96 2 6 92	1.00	0.91	0.95	0.96	0.95	98 0 9 89
3	0.99	0.87	0.93	0.93	0.92	97 1 13 85	1.00	0.9	0.95	0.95	0.95	98 0 10 88
4	1.00	0.92	0.96	0.96	0.96	98 0 8 90	0.98	0.93	0.95	0.96	0.95	96 2 7 91
5	0.99	0.85	0.92	0.92	0.91	97 1 15 83	1.00	0.91	0.95	0.96	0.95	98 0 9 89
6	1.00	0.9	0.95	0.95	0.95	98 0 10 88	0.98	0.92	0.95	0.95	0.95	96 2 8 90
7	0.99	0.85	0.92	0.92	0.91	97 1 15 83	0.96	0.92	0.94	0.94	0.94	94 4 8 90
8	0.98	0.86	0.92	0.92	0.91	96 2 14 84	0.95	0.94	0.94	0.94	0.94	93 5 6 92
9	0.98	0.82	0.90	0.91	0.89	96 2 18 80	0.95	0.91	0.93	0.93	0.93	93 5 9 89
10	0.99	0.86	0.92	0.93	0.92	97 1 14 84	0.97	0.95	0.96	0.96	0.96	95 3 5 93
Avg.	1.00	0.88	0.93	0.94	0.93	96.9 1.1 12.1 85.9	0.98	0.92	0.95	0.95	0.95	95.8 2.2 7.8 **90.2**

fiers DMLP and DCNN. Similarly, it is observed in other variants of SMOTE, i.e., $SMOTE_1$ in Table 5 and $SMOTE_2$ in Table 6, that the average number of predicted TP in the case of CFR voting ensemble is higher than CFA voting ensemble and individual classifiers.

In order to compare all the sampling techniques, we have summarized the results for all sampling techniques employed for CFA and CFR voting ensembles in Table 7. We have also calculated the variance of all the metrics considered. It is observed that the average number of predicted TN in all sampling techniques for the CFA voting ensemble is almost equal to actual negatives with very less variance. However, for the CFR ensemble, the average predicted TN is better in the case of SMOTE and Oversampling compared to other sampling techniques with very less variance. Moreover, the average number of predicted TN is better in the case of CFA than CFR as the variance is considerably low for CFA.

Next, considering the average number of predicted TP for the CFA voting ensemble, it is observed that SMOTE is significantly better than other sampling techniques.

Table 4 Performance analysis of SMOTE sampling technique for DMLP and DCNN classifier individually and their voting ensemble CFA and CFR. Here, P is precision, R is recall, Ac is accuracy, $f1_0$ is f1-score for class 0, $f1_1$ is f1-score for class 1, C is the confusion matrix calculated in the same order as defined in Table 1

S.no	DMLP P	R	Ac	$f1_0$	$f1_1$	C	DCNN P	R	Ac	$f1_0$	$f1_1$	C	CFA P	R	Ac	$f1_0$	$f1_1$	C	CFR P	R	Ac	$f1_0$	$f1_1$	C
1	1.00	1.00	1.00	1.00	1.00	98 0 / 0 98	1.00	0.98	0.99	0.99	0.99	98 0 / 2 96	1	0.98	0.99	0.99	0.99	98 0 / 2 96	1.00	1.00	1.00	1.00	1.00	98 0 / 0 98
2	1.00	0.99	0.99	0.99	0.99	98 0 / 1 97	0.99	0.99	0.99	0.99	0.99	97 1 / 1 97	1	0.98	0.99	0.99	0.99	98 0 / 2 96	0.99	1.00	0.99	0.99	0.99	97 1 / 0 98
3	0.99	1.00	0.99	0.99	0.99	97 1 / 0 98	1.00	0.97	0.98	0.98	0.98	98 0 / 3 95	1	0.97	0.98	0.98	0.98	98 0 / 3 95	0.99	1.00	0.99	0.99	0.99	97 1 / 0 98
4	1.00	1.00	1.00	1.00	1.00	98 0 / 0 98	1.00	0.97	0.98	0.98	0.98	98 0 / 3 95	1	0.97	0.98	0.98	0.98	98 0] / 3 95	1.00	1.00	1.00	1.00	1.00	98 0 / 0 98
5	0.99	1.00	0.99	0.99	0.99	97 1 / 0 98	0.99	1.00	0.99	0.99	0.99	97 1 / 0 98	1	1.00	1.00	1.00	1.00	98 0 / 0 98	0.98	1.00	0.99	0.99	0.99	96 2 / 0 98
6	0.99	1.00	0.99	0.99	0.99	97 1 / 0 98	0.99	0.97	0.98	0.98	0.98	97 1 / 3 95	1	0.97	0.98	0.98	0.98	98 0 / 3 95	0.98	1.00	0.99	0.99	0.99	96 2 / 0 98
7	1.00	1.00	1.00	1.00	1.00	98 0 / 0 98	0.98	0.96	0.97	0.97	0.97	96 2 / 4 94	1	0.96	0.98	0.98	0.98	98 0 / 4 94	0.98	1.00	0.99	0.99	0.99	96 2 / 0 98
8	1.00	1.00	1.00	1.00	1.00	98 0 / 0 98	0.99	1.00	0.99	0.99	0.99	97 1 / 0 98	1	1.00	1.00	1.00	1.00	98 0 / 0 98	0.99	1.00	0.99	0.99	0.99	97 1 / 0 98
9	1.00	0.99	0.99	0.99	0.99	98 0 / 1 97	0.99	0.94	0.96	0.97	0.96	97 1 / 6 92	1	0.93	0.96	0.97	0.96	98 0 / 7 91	0.99	1.00	0.99	0.99	0.99	97 1 / 0 98
10	1.00	1.00	1.00	1.00	1.00	98 0 / 0 98	1.00	0.96	0.98	0.98	0.98	98 0 / 4 94	1	0.96	0.98	0.98	0.98	98 0 / 4 94	1.00	1.00	1.00	1.00	1.00	98 0 / 0 98
Avg.	1.00	1.00	1.00	1.00	1.00	97.7 0.3 / 0.1 97.8	0.99	0.97	0.98	0.99	0.98	97.3 0.7 / 2.6 95.4	1	0.97	0.99	0.99	0.99	**98 0** / 2.8 95.2	0.99	1.00	0.99	0.99	0.99	97 1 / 0 **98**

Table 5 Performance analysis of another variant of SMOTE, i.e., SMOTE$_1$ for DMLP and DCNN classifier individually and their voting ensemble CFA and CFR. Here, P is precision, R is recall, Ac is accuracy, $f1_0$ is f1-score for class 0, $f1_1$ is f1-score for class 1, C is the confusion matrix calculated in the same order as defined in Table 1

S.no.	DMLP						DCNN						CFA						CFR					
	P	R	Ac	$f1_0$	$f1_1$	C	P	R	Ac	$f1_0$	$f1_1$	C	P	R	Ac	$f1_0$	$f1_1$	C	P	R	Ac	$f1_0$	$f1_1$	C
1	0.99	1.00	0.99	0.99	0.99	97 1 / 0 98	0.98	0.99	0.98	0.98	0.98	97 1 / 1 97	0.99	0.99	0.99	0.99	0.99	97 1 / 1 97	0.98	1.00	0.99	0.99	0.99	96 2 / 0 98
2	0.97	1.00	0.98	0.98	0.98	95 3 / 0 98	1.00	0.95	0.97	0.98	0.97	98 0 / 5 93	1.00	0.95	0.97	0.98	0.97	98 0 / 5 93	0.97	1.00	0.98	0.98	0.98	95 3 / 0 98
3	0.99	0.98	0.98	0.98	0.98	97 1 / 2 96	0.99	0.95	0.97	0.97	0.97	97 1 / 5 93	0.99	0.95	0.97	0.97	0.97	97 1 / 5 93	0.99	0.98	0.98	0.98	0.98	97 1 / 2 96
4	0.99	1.00	0.99	0.99	0.99	97 1 / 0 98	0.99	0.92	0.95	0.96	0.95	97 1 / 8 90	0.99	0.92	0.95	0.96	0.95	97 1 / 8 90	0.99	1.00	0.99	0.99	0.99	97 1 / 0 98
5	0.98	0.98	0.98	0.98	0.98	96 2 / 2 96	0.96	0.96	0.96	0.96	0.96	94 4 / 4 94	0.99	0.96	0.97	0.97	0.97	97 1 / 4 94	0.95	0.98	0.96	0.96	0.96	93 5 / 2 96
6	0.98	0.99	0.98	0.98	0.98	96 2 / 1 97	0.98	0.96	0.97	0.97	0.97	96 2 / 4 94	0.99	0.96	0.97	0.97	0.97	97 1 / 4 94	0.97	0.99	0.98	0.98	0.98	95 3 / 1 97
7	0.99	0.99	0.99	0.99	0.99	97 1 / 1 97	0.98	0.94	0.96	0.96	0.96	96 2 / 6 92	0.99	0.94	0.96	0.97	0.96	97 1 / 6 92	0.98	0.99	0.98	0.98	0.98	96 2 / 1 97
8	0.99	0.97	0.98	0.98	0.98	97 1 / 3 95	0.99	0.86	0.92	0.93	0.92	97 1 / 14 84	0.99	0.86	0.92	0.93	0.92	97 1 / 14 84	0.99	0.97	0.98	0.98	0.98	97 1 / 3 95
9	0.98	1.00	0.99	0.99	0.99	96 2 / 0 98	0.98	0.95	0.96	0.96	0.96	96 2 / 5 93	0.99	0.95	0.97	0.97	0.97	96 2 / 5 93	0.97	1.00	0.98	0.98	0.98	95 3 / 0 98
10	1.00	1.00	1.00	1.00	1.00	98 0 / 0 98	0.93	0.99	0.96	0.96	0.96	91 7 / 1 97	1.00	0.99	0.99	0.99	0.99	98 0 / 1 97	0.93	1.00	0.96	0.96	0.97	91 7 / 0 98
Avg.	0.99	0.99	0.99	0.99	0.99	96.6 1.4 / 0.9 97.1	0.98	0.95	0.96	0.96	0.96	95.8 2.2 / 5.3 92.7	0.99	0.95	0.97	0.97	0.97	97.2 0.8 / 5.3 92.7	0.97	0.99	0.98	0.98	0.98	95.2 2.8 / 0.9 **97.1**

Table 6 Performance analysis of hybrid variant of SMOTE, i.e., SMOTE$_2$ for DMLP and DCNN classifier individually and their voting ensemble CFA and CFR. Here, P is precision, R is recall, Ac is accuracy, $f1_0$ is f1-score for class 0, $f1_1$ is f1-score for class 1, C is the confusion matrix calculated in the same order as defined in Table 1

S.no.	DMLP						DCNN						CFA						CFR					
	P	R	Ac	$f1_0$	$f1_1$	C	P	R	Ac	$f1_0$	$f1_1$	C	P	R	Ac	$f1_0$	$f1_1$	C	P	R	Ac	$f1_0$	$f1_1$	C
1	1.00	1.00	1.00	1.00	1.00	98 0 / 0 98	0.99	1.00	0.99	0.99	0.99	97 1 / 0 98	1.00	1.00	1.00	1.00	1.00	98 0 / 0 98	0.99	1.00	0.99	0.99	0.99	97 1 / 0 98
2	0.99	0.99	0.99	0.99	0.99	97 1 / 1 97	1.00	0.99	0.99	0.99	0.99	98 0 / 1 97	1.00	0.99	0.99	0.99	0.99	98 0 / 1 97	0.99	0.99	0.99	0.99	0.99	97 1 / 1 97
3	0.99	1.00	0.99	0.99	0.99	97 1 / 0 98	0.99	0.95	0.97	0.97	0.97	97 1 / 5 93	0.99	0.95	0.97	0.97	0.97	97 1 / 5 93	0.99	1.00	0.99	0.99	0.99	97 1 / 0 98
4	0.99	0.98	0.98	0.98	0.98	97 1 / 2 96	0.99	0.94	0.96	0.97	0.96	95 1 / 6 92	1.00	0.94	0.97	0.97	0.97	98 0 / 6 92	0.98	0.98	0.98	0.98	0.98	96 2 / 2 96
5	1.00	0.99	0.99	0.99	0.99	98 0 / 1 97	1.00	0.93	0.96	0.97	0.96	98 0 / 7 91	1.00	0.92	0.96	0.96	0.96	98 0 / 8 90	1.00	1.00	1.00	1.00	1.00	98 0 / 0 98
6	0.99	0.99	0.99	0.99	0.99	97 1 / 1 97	0.96	0.92	0.94	0.94	0.94	94 4 / 8 90	0.99	0.92	0.95	0.96	0.95	97 1 / 8 90	0.96	0.99	0.97	0.97	0.97	94 4 / 1 97
7	0.97	1.00	0.98	0.98	0.98	95 3 / 0 98	0.98	0.97	0.97	0.97	0.97	96 2 / 5 95	0.99	0.97	0.98	0.98	0.98	97 1 / 3 95	0.96	1.00	0.98	0.98	0.98	94 4 / 0 98
8	0.99	1.00	0.99	0.99	0.99	97 1 / 0 98	1.00	0.95	0.97	0.98	0.97	98 0 / 5 93	1.00	0.95	0.97	0.98	0.97	98 0 / 5 93	0.99	1.00	0.99	0.99	0.99	97 1 / 0 98
9	0.98	1.00	0.99	0.99	0.99	96 2 / 0 98	1.00	0.95	0.97	0.98	0.97	98 0 / 5 93	1.00	0.95	0.97	0.98	0.97	98 0 / 5 93	0.98	0.99	0.98	0.98	0.98	96 2 / 0 98
10	0.99	0.99	0.99	0.99	0.99	97 1 / 1 97	0.98	0.93	0.95	0.96	0.95	96 2 / 7 91	1.00	0.93	0.96	0.97	0.96	98 0 / 7 91	0.97	0.99	0.98	0.98	0.98	95 3 / 1 97
Avg.	0.99	0.99	0.99	0.99	0.99	96.9 1.1 / 0.6 97.4	0.99	0.95	0.97	0.97	0.97	96.9 1.1 / 1.7 93.3	1.00	0.95	0.97	0.97	0.97	97.7 0.3 / 4.8 93.2	0.98	0.99	0.99	0.99	0.99	96.1 1.9 / 0.5 **97.5**

Table 7 Comparative analysis of different sampling techniques with respect to CFA and CFR voting ensemble methods. Here, TP is the number of predicted test positive, TN is the number of predicted test negative, P is precision, R is Recall, $f1_0$ is f1-score for class 0, $f1_1$ is f1-score for class 1. The average and variance of all the metrics considered for both voting ensemble CFA and CFR is noted in Table 7

Sampling technique		TP	TN	P	R	$f1_0$	$f1_1$
Oversampling	Avg. CFA	79.1	97.9	1	0.81	0.91	0.89
	Var CFA	8.32	0.1	1.00×10^{-5}	9.78×10^{-4}	1.21×10^{-4}	3.51×10^{-4}
	Avg. CFR	85.9	96.9	1	0.88	0.94	0.93
	Var CFR	14.99	0.54	5.44×10^{-5}	1.50×10^{-3}	3.82×10^{-4}	6.18×10^{-4}
Subset resampling	Avg. CFA	86.7	97.5	1	0.89	0.94	0.94
	Var CFA	2.68	0.50	5.00×10^{-5}	2.68×10^{-4}	6.78×10^{-5}	9.33×10^{-5}
	Avg. CFR	90.2	95.6	0.98	0.92	0.95	0.95
	Var CFR	2.40	3.96	3.96×10^{-4}	2.40×10^{-4}	1.21×10^{-4}	8.44×10^{-5}
SMOTE$_1$	Avg. CFA	92.7	97.2	0.99	0.95	0.97	0.97
	Var CFA	13.79	0.18	2.07×10^{-5}	1.44×10^{-3}	3.40×10^{-4}	4.52×10^{-4}
	Avg. CFR	97.1	95.2	0.97	0.99	0.98	0.98
	Var CFR	1.21	3.73	3.37×10^{-4}	1.26×10^{-4}	1.05×10^{-4}	9.28×10^{-5}
SMOTE$_2$	Avg. CFA	93.2	97.7	1	0.95	0.97	0.97
	Var CFA	7.51	0.23	2.66×10^{-5}	7.82×10^{-4}	1.95×10^{-4}	2.28×10^{-4}
	Avg. CFR	97.5	96.1	0.98	0.99	0.99	0.99
	Var CFR	0.50	1.88	1.81×10^{-4}	5.21×10^{-5}	7.85×10^{-5}	7.43×10^{-5}
SMOTE	Avg. CFA	95.2	98	1	0.97	0.99	0.99
	Var CFA	4.18	0	0	4.35×10^{-4}	1.02×10^{-4}	1.16×10^{-4}
	Avg. CFR	98	97	0.99	1	0.99	0.99
	Var CFR	0	0.67	6.67×10^{-5}	0	1.77×10^{-5}	1.7×10^{-5}

Similarly, for the CFR voting ensemble, SMOTE is performing better than other sampling techniques. Here, for both ensembles performance of oversampling is significantly less than other sampling. Also, the average number of predicted TP is better in the case of CFR than CFA as both average and variance are better in the case of CFR for all respective samplings.

Therefore, the CFR voting ensemble is performing better than the CFA voting ensemble in the case when the number of fraudulent transactions needs to be detected as maximum as possible. So, considering the predicted count of TP, the considered sampling techniques are shown in increasing order in Table 7. Also, Oversampling is the least performing, and SMOTE is the best performing sampling method among the considered ones for handling imbalanced data. SMOTE sampling technique is performing better in case of CFR voting ensemble predicting all 98 fraudulent transactions and 97 normal transactions with resulting R value 1, $f1_0$ score 0.99 and $f1_1$ score 0.99.

5 Conclusion

In this chapter, we analyze fraudulent transaction detection in the financial network using various sampling methods to address the class imbalance problem that has been considered in various studies. Also, hybrid ensemble learning has been shown to improve the efficiency of classification tasks by combining multiple classifiers. Considering this, we have analyzed the performance of the voting ensemble of two models, DCNN and DMLP, for different sampling techniques. The results obtained show that combining multiple classifiers tends to improve the prediction accuracy in terms of accurate identification of the class of interest. Although it also depends on the strategy used by the voting system CFR or CFA, in the case where it is hard to reach on consensus from prediction results. In this chapter, we have compared the performance of two voting ensemble strategies, CFR and CFA. We have concluded that in case of conflict in choosing a transaction as normal or fraudulent, marking it as fraudulent gives better predictions, i.e., CFR. For future work, several other deep learning classification models can be ensemble, such as LSTM, AE, to incorporate diversity to improve performance.

References

1. Ahmed, M., Mahmood, A.N., Islam, M.R.: A survey of anomaly detection techniques in financial domain. Futur. Gener. Comput. Syst. **55**, 278–288 (2016)
2. Ahsan, M.M., Mahmud, M., Saha, P.K., Gupta, K.D., Siddique, Z.: Effect of data scaling methods on machine learning algorithms and model performance. Technologies **9**(3), 52 (2021)
3. Akoglu, L., McGlohon, M., Faloutsos, C.: Oddball: spotting anomalies in weighted graphs. In: Pacific-Asia Conference on Knowledge Discovery and Data Mining, pp. 410–421. Springer, Berlin (2010)
4. Ali, A., Shamsuddin, S.M., Ralescu, A.: Classification with class imbalance problem: a review **7**, 176–204 (2015)
5. Bagga, S., Goyal, A., Gupta, N., Goyal, A.: Credit card fraud detection using pipeling and ensemble learning. Procedia Comput. Sci. **173**, 104–112 (2020)
6. Barandela, R., Valdovinos, R.M., Sánchez, J.S., Ferri, F.J.: The imbalanced training sample problem: Under or over sampling? In: Joint IAPR International Workshops on Statistical Techniques in Pattern Recognition (SPR) and Structural and Syntactic Pattern Recognition (SSPR), pp. 806–814. Springer (2004)
7. Bauer, E., Kohavi, R.: An empirical comparison of voting classification algorithms: bagging, boosting, and variants. Mach. Learn. **36**(1), 105–139 (1999)
8. Breunig, M.M., Kriegel, H.P., Ng, R.T., Sander, J.: Lof: identifying density-based local outliers. In: Proceedings of the 2000 ACM SIGMOD International Conference on Management of Data, pp. 93–104 (2000)
9. Chalapathy, R., Chawla, S.: Deep learning for anomaly detection: a survey (2019). arXiv:1901.03407
10. Chandola, V., Banerjee, A., Kumar, V.: Anomaly detection: a survey. ACM Comput. Surv. (CSUR) **41**(3), 1–58 (2009)
11. Chawla, N.V., Bowyer, K.W., Hall, L.O., Kegelmeyer, W.P.: Smote: synthetic minority oversampling technique. J. Artif. Intell. Res. **16**, 321–357 (2002)
12. Chawla, N.V., Japkowicz, N., Kotcz, A.: Special issue on learning from imbalanced data sets. ACM SIGKDD Explorations Newsl **6**(1), 1–6 (2004)
13. Chen, J., Shen, Y., Ali, R.: Credit card fraud detection using sparse autoencoder and generative adversarial network. In: 2018 IEEE 9th Annual Information Technology, Electronics and Mobile Communication Conference (IEMCON), pp. 1054–1059. IEEE (2018)
14. Chen, Z., Hendrix, W., Samatova, N.F.: Community-based anomaly detection in evolutionary networks. J. Intell. Inf. Syst. **39**(1), 59–85 (2012)
15. Choirunnisa, S., Buliali, J.: Hybrid method of undersampling and oversampling for handling imbalanced data (2018). https://doi.org/10.1109/ISRITI.2018.8864335
16. Du, H., Zhang, Y., Gang, K., Zhang, L., Chen, Y.C.: Online ensemble learning algorithm for imbalanced data stream. Appl. Soft Comput. **107**, 107378 (2021)
17. Eberle, W., Holder, L.: Anomaly detection in data represented as graphs. Intell. Data Anal. **11**(6), 663–689 (2007)
18. Estabrooks, A., Jo, T., Japkowicz, N.: A multiple resampling method for learning from imbalanced data sets. Comput. Intell. **20**(1), 18–36 (2004)
19. Fiore, U., De Santis, A., Perla, F., Zanetti, P., Palmieri, F.: Using generative adversarial networks for improving classification effectiveness in credit card fraud detection. Inf. Sci. **479**, 448–455 (2019)
20. Forough, J., Momtazi, S.: Ensemble of deep sequential models for credit card fraud detection. Appl. Soft Comput. **99**, 106883 (2021)
21. Fu, K., Cheng, D., Tu, Y., Zhang, L.: Credit card fraud detection using convolutional neural networks. In: International Conference on Neural Information Processing, pp. 483–490. Springer (2016)
22. Gaikwad, J.R., Deshmane, A.B., Somavanshi, H.V., Patil, S.V., Badgujar, R.A.: Credit card fraud detection using decision tree induction algorithm. Int. J. Innov. Technol. Explor. Eng. (IJITEE) **4**(6) (2014)

23. Gangwar, A.K., Ravi, V.: Wip: generative adversarial network for oversampling data in credit card fraud detection. In: International Conference on Information Systems Security, pp. 123–134. Springer (2019)
24. Ganji, V.R., Mannem, S.N.P.: Credit card fraud detection using anti-k nearest neighbor algorithm. Int. J. Comput. Sci. Eng. 4(6), 1035–1039 (2012)
25. Gao, J., Du, N., Fan, W., Turaga, D., Parthasarathy, S., Han, J.: A multi-graph spectral framework for mining multi-source anomalies. In: Graph Embedding for Pattern Analysis, pp. 205–227. Springer (2013)
26. Georgieva, S., Markova, M., Pavlov, V.: Using neural network for credit card fraud detection, 2159, 030013 (2019). https://doi.org/10.1063/1.5127478
27. Gómez, J.A., Arévalo, J., Paredes, R., Nin, J.: End-to-end neural network architecture for fraud scoring in card payments. Pattern Recogn. Lett. 105, 175–181 (2018)
28. Goodfellow, I., Bengio, Y., Courville, A.: Deep Learning. MIT Press, Cambridge (2016)
29. Grzymala-Busse, J.W., Goodwin, L.K., Grzymala-Busse, W.J., Zheng, X.: An approach to imbalanced data sets based on changing rule strength. In: Rough-Neural Computing, pp. 543–553. Springer, Berlin (2004)
30. Haixiang, G., Yijing, L., Shang, J., Mingyun, G., Yuanyue, H., Bing, G.: Learning from class-imbalanced data: Review of methods and applications. Expert Syst. Appl. 73, 220–239 (2017)
31. Han, H., Wang, W.Y., Mao, B.H.: Borderline-smote: a new over-sampling method in imbalanced data sets learning. In: International Conference on Intelligent Computing, pp. 878–887. Springer (2005)
32. Hautamaki, V., Karkkainen, I., Franti, P.: Outlier detection using k-nearest neighbour graph. In: Proceedings of the 17th International Conference on Pattern Recognition, 2004. ICPR 2004, vol. 3, pp. 430–433. IEEE (2004)
33. He, G., Han, H., Wang, W.: An over-sampling expert system for learing from imbalanced data sets. In: 2005 International Conference on Neural Networks and Brain, vol. 1, pp. 537–541. IEEE (2005)
34. He, H., Bai, Y., Garcia, E.A., Li, S.: Adasyn: adaptive synthetic sampling approach for imbalanced learning. In: 2008 IEEE International Joint Conference on Neural Networks (IEEE World Congress on Computational Intelligence), pp. 1322–1328. IEEE (2008)
35. He, Z., Xu, X., Deng, S.: Discovering cluster-based local outliers. Pattern Recogn. Lett. 24(9–10), 1641–1650 (2003)
36. Hinton, G.E., Osindero, S., Teh, Y.W.: A fast learning algorithm for deep belief nets. Neural Comput. 18(7), 1527–1554 (2006)
37. Japkowicz, N., Stephen, S.: The class imbalance problem: a systematic study. Intell. Data Anal. 6(5), 429–449 (2002)
38. Johnson, J.M., Khoshgoftaar, T.M.: Survey on deep learning with class imbalance. J. Big Data 6(1), 1–54 (2019)
39. Junsomboon, N., Phienthrakul, T.: Combining over-sampling and under-sampling techniques for imbalance dataset, pp. 243–247 (2017). https://doi.org/10.1145/3055635.3056643
40. Kaur, R., Singh, S.: A survey of data mining and social network analysis based anomaly detection techniques. Egyptian Inf. J. 17(2), 199–216 (2016)
41. Kim, J.Y., Cho, S.B.: Towards repayment prediction in peer-to-peer social lending using deep learning. Mathematics 7(11), 1041 (2019)
42. Knorr, E.M., Ng, R.T., Tucakov, V.: Distance-based outliers: algorithms and applications. VLDB J. 8(3), 237–253 (2000)
43. Kong, J., Kowalczyk, W., Menzel, S., Bäck, T.: Improving imbalanced classification by anomaly detection. In: International Conference on Parallel Problem Solving from Nature, pp. 512–523. Springer (2020)
44. Krawczyk, B.: Learning from imbalanced data: open challenges and future directions. Prog. Artif. Intell. 5(4), 221–232 (2016)
45. Kruegel, C., Mutz, D., Robertson, W., Valeur, F.: Bayesian event classification for intrusion detection. In: 19th Annual Computer Security Applications Conference, 2003. Proceedings, pp. 14–23. IEEE (2003)

46. Kumar, M.S., Soundarya, V., Kavitha, S., Keerthika, E., Aswini, E.: Credit card fraud detection using random forest algorithm. In: 2019 3rd International Conference on Computing and Communications Technologies (ICCCT), pp. 149–153 (2019). https://doi.org/10.1109/ICCCT2.2019.8824930

47. Kwon, D., Kim, H., Kim, J., Suh, S.C., Kim, I., Kim, K.J.: A survey of deep learning-based network anomaly detection. Clust. Comput. **22**(1), 949–961 (2019)

48. Li, K.L., Huang, H.K., Tian, S.F., Xu, W.: Improving one-class svm for anomaly detection. In: Proceedings of the 2003 International Conference on Machine Learning and Cybernetics (IEEE Cat. No. 03EX693), vol. 5, pp. 3077–3081. IEEE (2003)

49. Lin, W.C., Tsai, C.F., Hu, Y.H., Jhang, J.S.: Clustering-based undersampling in class-imbalanced data. Inf. Sci. **409**, 17–26 (2017)

50. Liu, X.Y., Wu, J., Zhou, Z.H.: Exploratory undersampling for class-imbalance learning. IEEE Trans. Syst. Man Cybern. Part B (Cybernetics) **39**(2), 539–550 (2008)

51. Maes, S., Tuyls, K., Vanschoenwinkel, B., Manderick, B.: Credit card fraud detection using bayesian and neural networks. In: Proceedings of the 1st International Naiso Congress on Neuro Fuzzy Technologies, pp. 261–270 (2002)

52. Modi, K., Dayma, R.: Review on fraud detection methods in credit card transactions. In: 2017 International Conference on Intelligent Computing and Control (I2C2), pp. 1–5. IEEE (2017)

53. Mubalaike, A.M., Adali, E.: Deep learning approach for intelligent financial fraud detection system. In: 2018 3rd International Conference on Computer Science and Engineering (UBMK), pp. 598–603. IEEE (2018)

54. Niu, X., Wang, L., Yang, X.: A comparison study of credit card fraud detection: supervised versus unsupervised (2019). arXiv:1904.10604

55. Ozbayoglu, A.M., Gudelek, M.U., Sezer, O.B.: Deep learning for financial applications: a survey. Appl. Soft Comput. **93**, 106384 (2020)

56. Peng, Y., Kou, G., Shi, Y., Chen, Z.: Improving clustering analysis for credit card accounts classification. In: International Conference on Computational Science, pp. 548–553. Springer (2005)

57. Piciarelli, C., Micheloni, C., Foresti, G.L.: Trajectory-based anomalous event detection. IEEE Trans. Circuits Syst. Video Technol. **18**(11), 1544–1554 (2008)

58. Pillai, T.R., Hashem, I.A.T., Brohi, S.N., Kaur, S., Marjani, M.: Credit card fraud detection using deep learning technique. In: 2018 Fourth International Conference on Advances in Computing, Communication & Automation (ICACCA), pp. 1–6. IEEE (2018)

59. Qiu, X., Zhang, L., Ren, Y., Suganthan, P.N., Amaratunga, G.: Ensemble deep learning for regression and time series forecasting. In: 2014 IEEE Symposium on Computational Intelligence in Ensemble Learning (CIEL), pp. 1–6 (2014). https://doi.org/10.1109/CIEL.2014.7015739

60. Raghavan, P., Gayar, N.E.: Fraud detection using machine learning and deep learning. In: 2019 International Conference on Computational Intelligence and Knowledge Economy (ICCIKE), pp. 334–339 (2019). https://doi.org/10.1109/ICCIKE47802.2019.9004231

61. Ramachandran, P., Zoph, B., Le, Q.V.: Searching for activation functions (2017). arXiv:1710.05941 (2017)

62. RB, A., K R, S.: Credit card fraud detection using artificial neural network. Global Trans. Proc. **2** (2021). https://doi.org/10.1016/j.gltp.2021.01.006

63. Sanabila, H.R., Jatmiko, W.: Ensemble learning on large scale financial imbalanced data. In: 2018 International Workshop on Big Data and Information Security (IWBIS), pp. 93–98. IEEE (2018)

64. Sarker, I.H.: Deep learning: a comprehensive overview on techniques, taxonomy, applications and research directions. SN Comput. Sci. **2**(6), 1–20 (2021)

65. Savage, D., Zhang, X., Yu, X., Chou, P., Wang, Q.: Anomaly detection in online social networks. Soc. Netw. **39**, 62–70 (2014)

66. Sohony, I., Pratap, R., Nambiar, U.: Ensemble learning for credit card fraud detection. In: Proceedings of the ACM India Joint International Conference on Data Science and Management of Data, pp. 289–294 (2018)

67. Sun, Y., Wong, A.K., Kamel, M.S.: Classification of imbalanced data: a review. Int. J. Pattern Recog. Artif. Intell. **23**(04), 687–719 (2009)
68. Vanerio, J., Casas, P.: Ensemble-learning approaches for network security and anomaly detection. In: Proceedings of the Workshop on Big Data Analytics and Machine Learning for Data Communication Networks, pp. 1–6 (2017)
69. Varmedja, D., Karanovic, M., Sladojevic, S., Arsenovic, M., Anderla, A.: Credit card fraud detection-machine learning methods. In: 2019 18th International Symposium INFOTEH-JAHORINA (INFOTEH), pp. 1–5. IEEE (2019)
70. Wang, H., Zhu, P., Zou, X., Qin, S.: An ensemble learning framework for credit card fraud detection based on training set partitioning and clustering. In: 2018 IEEE SmartWorld, Ubiquitous Intelligence & Computing, Advanced & Trusted Computing, Scalable Computing & Communications, Cloud & Big Data Computing, Internet of People and Smart City Innovation (SmartWorld/SCALCOM/UIC/ATC/CBDCom/IOP/SCI), pp. 94–98. IEEE (2018)
71. Wiese, B., Omlin, C.: Credit card transactions, fraud detection, and machine learning: modelling time with lstm recurrent neural networks. In: Innovations in Neural Information Paradigms and Applications, pp. 231–268. Springer (2009)
72. Xie, X., Xiong, J., Lu, L., Gui, G., Yang, J., Fan, S., Li, H.: Generative adversarial network-based credit card fraud detection. In: International Conference in Communications, Signal Processing, and Systems, pp. 1007–1014. Springer (2018)
73. Xie, Y., Li, A., Gao, L., Liu, Z.: A heterogeneous ensemble learning model based on data distribution for credit card fraud detection. Wireless Commun. Mobile Comput. **2021** (2021)
74. Xuan, S., Liu, G., Li, Z., Zheng, L., Wang, S., Jiang, C.: Random forest for credit card fraud detection. In: 2018 IEEE 15th International Conference on Networking, Sensing and Control (ICNSC), pp. 1–6. IEEE (2018)
75. Yap, B.W., Abd Rani, K., Abd Rahman, H.A., Fong, S., Khairudin, Z., Abdullah, N.N.: An application of oversampling, undersampling, bagging and boosting in handling imbalanced datasets. In: Proceedings of the First International Conference on Advanced Data and Information Engineering (DaEng-2013), pp. 13–22. Springer (2014)
76. Yu, L., Zhou, R., Tang, L., Chen, R.: A dbn based resampling svm ensemble learning paradigm for credit classification with imbalanced data. Appl. Soft Comput. **69**, 192–202 (2018)
77. Zamini, M., Montazer, G.: Credit card fraud detection using autoencoder based clustering. In: 2018 9th International Symposium on Telecommunications (IST), pp. 486–491 (2018). https://doi.org/10.1109/ISTEL.2018.8661129
78. Zhou, Z.H.: Ensemble learning. In: Machine Learning, pp. 181–210. Springer, Berlin (2021)
79. Zou, J., Zhang, J., Jiang, P.: Credit card fraud detection using autoencoder neural network (2019). arXiv:1908.11553